The
SOLDIER

The

SOLDIER

A HISTORY OF COURAGE, SACRIFICE AND BROTHERHOOD

DARREN MOORE

ICON BOOKS

Published in the UK in 2009 by
Icon Books Ltd, Omnibus Business Centre,
39–41 North Road, London N7 9DP
email: info@iconbooks.co.uk
www.iconbooks.co.uk

Sold in the UK, Europe, South Africa and Asia
by Faber & Faber Ltd, Bloomsbury House, 74–77 Great Russell Street,
London WC1B 3DA or their agents

Distributed in the UK, Europe, South Africa and Asia
by TBS Ltd, TBS Distribution Centre, Colchester Road
Frating Green, Colchester CO7 7DW

Published in Australia in 2009
by Allen & Unwin Pty Ltd, PO Box 8500,
83 Alexander Street, Crows Nest, NSW 2065

Distributed in Canada by
Penguin Books Canada, 90 Eglinton Avenue East,
Suite 700, Toronto, Canada M4P 2YE

ISBN: 978-184831-079-7

Typeset in 11.5 on 16pt Minion by Marie Doherty

Printed and bound in the UK by
CPI Mackays, Chatham ME5 8TD

CONTENTS

About the author

Darren Moore served in the Australian Army for seventeen years and held the rank of major when he left the service. He is a graduate of the Australian Defence Force Academy, the Royal Military College, Duntroon and the Australian Command and Staff College. Highlights of his military service include operational deployments to Papua New Guinea and East Timor and serving for two and a half years (2001–03) with the United States Army on an army-to-army exchange. Darren is currently a PhD candidate at the University of New South Wales. This is his second book.

For Theresa and Benjamin

Personal Acknowledgements

The basis of this book had a lengthy gestation, gradually informed by extensive reading of narrative accounts of war but also by periods of personal reflection. Although my period of active service in East Timor with the Australian Army was essentially risk-free and relatively comfortable, the long tropical nights on shift, watching the geckos scurry across the walls, produced a kind of melancholy, as I thought of my wife many miles away. My separation from home was measured in months rather than years, but I began to better appreciate the burden borne by those who had gone before me. Later, when I was serving on exchange with the United States Army, I experienced the seismic impact of the 11 September 2001 terrorist attacks and their aftermath. But in 2003, when some of my friends in the US Army deployed to an uncertain future in Iraq, and I had to tell some of the US soldiers under my command that they too needed to go, I began to reflect further on the burden that the soldier must bear. What particularly struck me was the apparently oblivious nature of much of American society – once the initial headlines arising from the invasion had receded – to the risks and hardships endured by their soldiers on a daily basis. I felt that society's knowledge of what the soldier must endure was becoming increasingly abstract. I think that for the rest of us, myself included, the least that we owe these men and women who are sent off to fight and possibly die in foreign fields on our behalf is to appreciate what it is that we ask of them. This book, in its own small way, seeks to address this factor.

A work of this nature is greatly dependent on the scholarship of others, in particular the books produced by a remarkable cluster of historians that congregated in the military history department of the Royal Military Academy Sandhurst from the 1960s to the mid-1980s, specifically Sir John Keegan and Sir Richard Holmes. I am indebted to them for marking out the path that I have followed. However, it would be remiss of me not to acknowledge the influence upon this work of those on the other side of the Atlantic, such as S.L.A. Marshall, Morris Janowitz, J. Glenn Gray, Gwynne Dyer and Dave Grossman, who have delved into the nature of the relationship between soldiers that enables armies to function. But my greatest debt is to those who have fought in war and who had the moral courage to bequeath their triumphs, their fears and their frustrations to print. Men such as Robert Graves, Eugene Sledge, William Manchester, Philip Caputo, Tim O'Brien, Hugh McManners, Anthony Swofford and Nathaniel Fick – and many others. Without their efforts, this work would be little more than conjecture, lacking even the vestiges of authenticity.

On a more personal level, a number of people gave generously of their time to read and discuss drafts of the manuscript. For the encouragement and advice they provided I would particularly like to thank Vice Admiral Sir Christopher Morgan of the Royal Navy, Professor Gary Sheffield, Dr Peter Caddick-Adams, Dr Alan Ryan, Andy Marino and Barry Holmes. In addition, several of my former colleagues from the Australian Army read the manuscript and offered frank but well-received advice on how it could be improved; specifically lieutenant colonels Rupert Hoskin and Murray Thompson and Major Patrick Lun. I would especially like to single out and thank Lieutenant Colonel Robert Worswick for the assistance he provided me at a crucial stage of this project. Although I am grateful for all the help I have received on this project, for errors of fact, and for the opinions expressed, I alone take full responsibility.

Andrew Lownie, my agent, was prepared to take a chance on me, and I thank him for the advice and encouragement he has provided me over the years that this book was being developed. I would also like to thank the team at Icon, particularly Simon Flynn, Duncan Heath and Sarah Higgins. Simon was an enthusiastic supporter of this book from the start, Duncan was instrumental in knocking off some of the rough edges and shepherding the book through to publication, while Sarah provided invaluable assistance in relation to tracking down permissions and the finalisation of the text.

Lastly, I would like to thank my wife Theresa, who has been steadfast in her belief that I would succeed.

LIST OF ILLUSTRATIONS

INTRODUCTION

No visitor to the Australian War Memorial at Canberra could fail to be moved by the national bereavement represented by the names inscribed on the long series of bronze panels leading the eye towards the Hall of Memory and the final resting place of the Unknown Soldier. Nine names from the Sudan, 589 from the Boer War, six from the almost forgotten Boxer Rebellion, 61,508 names from the First World War, 39,767 from the Second World War, and so on until we reach the pair of names from the conflict in Iraq. 102,804 names in all; collectively the greatest single catalogue of loss due to war anywhere in the world.

That soldiers may be killed in war is self-evident, especially to the soldiers themselves. That soldiers may be wounded on the battlefield is also common knowledge, although the image of crippled veterans begging on the streets is one that has largely faded from our collective memory. Anyone who has ever read a soldier's memoir or watched a war film would instinctively understand that the soldier's life is characterised by discipline, hardship and sacrifice. Death, disfigurement, discipline and duty: is this the sum total of the burden that soldiers bear for society?

This book will not concern itself with strategy, nor battles, nor generals, nor tactics, nor weapons, nor logistics. Rather, this book will examine in detail topics that, in almost all comparable books, have been ignored or accorded cursory treatment. It examines the nature of the relationship between the soldier and the state, and it explores which segments of society serve in the armed forces. It follows the soldier from basic training to discharge or death and reveals how soldiers confront the possibility of being mutilated or killed; the mental and social conditioning that enables them to kill in battle; and the anguish of killing your comrades, whether this is deliberately through the application of the death penalty or accidentally as a result of friendly fire. It examines the relationship between love, sex and war and reveals the 'trial by media' faced by modern soldiers in relation to their decisions and actions on the battlefield.

1

War has been a constant scourge of humanity but we have only literary fragments of the experiences of common soldiers before the Napoleonic Wars. This lack of written accounts is partly because most soldiers were illiterate, due to the absence of state-sponsored education. It was not until 1870, for example, with the passage of the Education Act, that basic primary education was made compulsory in Britain. Furthermore, apart from a few noted generals, their respective societies did not place much value on the stories their soldiers had to tell. Therefore, this book is generally confined to examining the wars of the 19th to 21st centuries, drawing heavily on published personal accounts of war.

Wars are fought on and below the ground, in the air, and on the surface of and under the sea. The soldiers, sailors and airmen who actually confront the enemy are backed up by support infrastructure that extends from logistical and headquarter elements located in the area of operations, back to the industries at home that supply the fighting elements with the required matériel. All these disparate elements contribute to the waging of war. Although accounts of combat from sailors and airmen will be woven into the narrative, the preponderance of material will be drawn from the accounts of soldiers, specifically ground combat soldiers. This narrow focus is not intended to devalue or slight the contributions made to the conduct of war by the sailor, the airman or the civilian, but rather is a reflection of two factors. First, the experiences of a soldier, particularly an infantry soldier, are better documented than those of other combatants and provide a rich trove of source material and analysis from which to draw out the essence of this book. Second, the nature of ground combat, whereby the soldier must seek out and close with his enemy and either kill or capture him, is the essence of the brutality of war and it is such soldiers who must bear the greatest burden when a nation is committed by its leaders to military action. For example, riflemen comprised 68.5 per cent of the men of an American infantry division of the Second World War. Statistics from 57 such divisions showed that riflemen suffered 94.5 per cent of the division's casualties.[1]

The nuclear capability of the world's major powers reversed the trend towards total war that reached its apotheosis during the Second World

War, whereby whole nations were directly exposed to war's destructiveness. Subsequently, at least among developed nations, wars have largely been fought by small professional armies. Furthermore, the death in battle of the members of these professional armies is regarded by the public as an occupational hazard rather than a tragedy. The common rationalisation is that, after all, they did volunteer to join the military and were aware that fulfilling their duty might entail them being wounded or killed.

The military is increasingly unrepresentative of the society it serves, and the burden borne by soldiers is becoming more distant and inconsequential to society in general. For the vast majority of citizens, war will remain an abstraction, a topic of dinner conversation, an item of momentary interest when watching the evening news on television. But soldiers will be required to risk their lives, and it is their families that will have to shoulder a disproportionate amount of the cost so that their fellow citizens may feel secure.

The Duke of Wellington, reflecting on his victory at Waterloo, lamented that, 'I always say that, next to a battle lost, the greatest misery is a battle gained.'[2] This comment encapsulates the divergence between how civilians and how soldiers view military endeavours, and is the crux of this book. For the civilian, wars are a series of defined battles that are retrospectively labelled either a defeat or a victory. But the soldier is denied the benefit of this strategic simplification. In *The American Soldier: Combat and its Aftermath*, the sociologist M. Brewster Smith stated that what the media may proclaim as a victory is actually the amalgamation of numerous little victories and many local defeats. Furthermore, a victory is not achieved without casualties. Therefore, the elation on the home front when a victory is proclaimed may appear somewhat obscene to the soldiers who took part in the fighting and who measure the cost of the victory in dead and wounded comrades.[3] An American soldier of the Second World War bitterly observed that:

When they [civilians] hear about an airstrip being taken or a piece of land taken, they are happy, and they should be, but I often wonder if they stop to think there have been a lot of boys blown all apart and killed and lost arms and legs and eyesight and a lot more I'm not even going to mention. Let them stop and think about just how a group of guys takes a piece of land. Let them think of the machine gun bullets they are ducking, of the 88s and mortars.[4]

Your opinion as to the cost of a military action depends very much on how you tally up the ledger. I recall a conversation I once had with a senior military officer concerning the validity of the American participation in the Vietnam War. He stated that studies have shown an overwhelming number of Americans who fought in Vietnam found the war a positive experience, to which I retorted: 'Except for the 58,000 whose names are carved on The Wall in Washington, DC.' 'If I should be the only casualty in a victorious army of a million men, the victory would be without interest for me,' wrote Guy Sajer in *The Forgotten Soldier*, his memoir of his service with the Wehrmacht on the Eastern Front during the Second World War. He continued: 'The percentage of corpses, in which the generals sometimes take pride, doesn't alter the fate of the men who've been killed. The only leader I know of who finally made a sensible remark on this point, Adolf Hitler, once said to his troops: "Even a victorious army must count its victims".'[5]

Soldiers must face the possibility that their occupation may bring about their death or disfigurement. They suffer enforced separation from their loved ones. They are required to seek out and kill men whom they have most likely never met. At times they will be ordered to kill their comrades. At other times the death of their comrades will arise from a tragic mistake. And through it all, soldiers will have to bear these tribulations under the harsh and unforgiving gaze of the media. These are the burdens of war.

Chapter 1

THE SOLDIER AND THE STATE

The physical demands of soldiering, particularly among the combat arms, require that militaries are largely composed of young men. Many of these young men, if not the majority, will join the military straight out of school. They will be lacking in life experience and many will have a poor understanding of what they are signing up for. When soldiers accept the 'king's shilling' they enter into a social contract of unlimited personal liability. Tobias Wolff captured the naivety of young soldiers when he commented in his memoir of the Vietnam War, *In Pharaoh's Army*, that 'it never occurred to me … that we would be used stupidly or carelessly or for unworthy ends. Our trust was simple, immaculate, heartbreaking.'[1]

There is a range of expectations about what tasks a soldier may be called upon to undertake. At one end of this range are the expectations of politicians. They view the military as an element of national power and expect servicemen to carry out any task they assign, noting that defence of the realm is the foremost duty of any government. Their viewpoint is that the military is an apolitical instrument of the state. The politicians expect professional soldiers to be dedicated servants of the state and to cultivate a professional indifference as to whom they are called upon to fight and why. Timothy Gowing, on his way to the Crimea to protect the Ottoman Empire from Russian aggression, felt 'we were going out to defend a rotten cause, a race that almost every Christian despises. However, as soldiers we had nothing to do with politics.'[2]

'I am a soldier. I obey orders'

The role of the soldier is to win the battle, not to justify the war. A case in point is a comment made by Petty Officer Robert, a twenty-year-old pilot in the French Navy, who was shot down by the Viet Minh over Dien Bien Phu in April 1954. Robert was captured and was accused by a Viet Minh officer of having committed 'barbarous acts', for which he deserved to be executed. Robert replied that he had not wanted the war. The Viet Minh officer asked, 'Well, then, why did you fight in it?' Robert stated: 'I am a soldier. I obey orders.'[3] An almost identical response was given by a German general captured by Patton's 3rd Army when asked why he had not surrendered earlier, so as to spare Germany from further destruction. The German impassively replied: 'I am a professional, and I obey my orders.'[4]

This attitude of professional indifference can generate a moral ambivalence among soldiers in relation to the missions they are required to carry out. 'I wasn't supposed to think about whether these were criminal orders or not criminal orders. The commander was the one and only commander, subordinates were supposed to fulfil his orders,' maintained the Soviet soldier Nikonor Perevalov when questioned about the acts he carried out during the Second World War.[5] J. Glenn Gray noted that he frequently heard his fellow American soldiers of the Second World War remark: 'When I raised my right hand and took the oath, I freed myself from the consequences for what I do. I'll do what they tell me and nobody can blame me.'[6] D.M. Mantell, who studied the combat motivation of American Green Berets (Special Forces) during the Vietnam War, detected that:

I was not talking to fervent and ideologically engaged persons but rather to socially and politically disinterested professional soldiers who were uninformed on the social and political issues of the day ... They made little attempt to disguise the fact that they saw themselves as hired guns, paid killers who were not particularly concerned with their employers or their victims.[7]

David Hackworth put it more brusquely when he wrote of his comrades in the Korean War: 'The politics or purpose of the war was not our concern. We didn't understand or care about the big picture ...'[8]

The hesitance of soldiers to take a stand on the righteousness or otherwise of a conflict is a reflection of the constraints on their freedom of action. To refuse to deploy on an operation is to invite all manner of social and legal rebuke; at the very least, their courage and professionalism will be questioned by their peers and superiors. If social pressure does not sway their decision then they are likely to face court martial and professional disgrace. Rather, most soldiers will just accept the hand dealt by fate. Nathaniel Fick, who served with the US Marine Corps during the 2003 invasion of Iraq, wrote: 'Pro-war. Anti-war. War for freedom. War for oil. Philosophical disputes were a luxury I could not enjoy. War was what I had. We didn't vote for it, authorise it, or declare it. We just had to fight it.'[9]

An effective coping mechanism for the soldier is to compartmentalise his involvement in the conflict. The soldier therefore concentrates on events that he can control, such as his own actions in combat, and leaves the moral questioning to others. Fick commented that his men 'had to know that, whether or not they supported the larger war, they had fought their little piece of it with honor and had retained their humanity'.[10]

War presents professional soldiers with an opportunity to put into practice what they have trained for; to slake their professional curiosity about how well their weapons and tactics will work against an enemy. Gary McKay, a young Australian Army officer, recalled his frustration at not being deployed on active service: 'By now I was determined to go to Vietnam. I was beginning to feel like an athlete who was spending all his time training and never being allowed to compete ... I wanted to see if I could do the job I had been trained to do.'[11] James Newton, a pilot with the Royal Navy, expressed similar sentiments as he prepared to sail to the Persian Gulf in 2003.

[T]here was no escaping the excitement of it. After all, we hadn't joined up to spend a few decades flying around Salisbury Plain … before ending our career filing papers … That would be fine in years to come, but first I wanted a slice of the real action, to put myself on the line. Imagine training every day for ten years as a Premiership footballer but never getting to play in a competitive match. That's how it comes to you after a while.[12]

Moreover, David Bradley, an officer with the British Army, wrote of his relief when told his unit was deploying to Iraq after months of speculation. 'As a soldier and in particular an infantryman it is an unspoken desire to be in contact, to fire your weapon in anger and test yourself in combat.'[13]

For many soldiers, particularly those in elite units, war presents them with an opportunity to prove that they are the best at what they do. On the eve of the 2003 American-led invasion of Iraq, Major General David Petraeus, Commanding General of the 101st Airborne (Air Assault) Division, was asked by an embedded journalist how he would feel if Saddam backed down and the confrontation with Iraq was resolved without gunfire. Petraeus answered:

There would be relief at not putting these wonderful soldiers in harm's way. There might also be a bit of a letdown, and it would only be that: a *bit* of a letdown. This is the biggest prizefight in our careers, and every soldier at every level has been training for this for months, if not years.[14]

The American citizen soldiers who fought the Second World War expressed a similar reaction. In *The Men of Company K*, Harold Leinbaugh and John Campbell wrote that in August 1944 it appeared that the German retreat from France had become a rout.

[I]t had seemed the war could end any day, possibly before we got overseas. Battle was the last challenge, the payoff of two years of training, the test we had to pass before we could call ourselves soldiers. We had questions about ourselves that could be answered only in combat. After journeying this far and working so long and hard to become soldiers, we would have felt cheated to miss out on the fighting.[15]

Taking the oath

Some soldiers will enlist in the military in a time of war, in which case they will hold few illusions about the nature of the task required of them. But many soldiers enlist in a time of peace.

Soldiers swear an oath (or affirmation) as part of their ritualised entry into the military. These oaths are somewhat vague about the exact nature of the tasks a soldier may be ordered to undertake, but most soldiers will have a broad understanding of what they may be called upon to do, based on the historical usage of the military in that country. A soldier who enlists in the army of a nation with a foreign policy of enforcing its will upon other nations, and which regularly uses military force to fulfil that policy, can hardly claim to be surprised when he is sent off to invade another country. Conversely, a soldier who enlists in the military of a country with a long history of neutrality can rightly claim to be taken aback if his country's military is ordered by its politicians to act in an expeditionary manner.

At the other end of the range of expectations about what tasks a soldier may be called upon to carry out are those who believe that the military is a 'defence' force and not an 'offence' force. That is, they feel the military should be used only to defend the territorial integrity of the nation and not in an expeditionary manner to invade other nations. Soldiers who feel this way are placed in a moral quandary when ordered to carry out a mission that falls outside what they believe to be the legitimate role of their country's military.

On 6 June 1982, elements of the Israel Defense Forces (IDF) invaded Lebanon. The official objective of the campaign was to overrun the Lebanese-based units of the Palestine Liberation Organisation (PLO), who had been firing rockets and mortar shells into northern Israeli settlements with impunity. The IDF's stated intention was to create a 40-kilometre buffer zone to protect these settlements. The strategic aim, however, was to link up with the Christian militias in the Lebanese capital of Beirut and eliminate the PLO presence in Lebanon.

The IDF's invasion of Lebanon – ironically named, in view of later events, 'Operation Peace for Galilee' – would prove to be extremely divisive in Israeli society, with large public protests staged in major Israeli cities. Previously, the IDF's wars had been ones of necessity and were in accordance with the theme of *en brera* (no choice), which meant there was no alternative to a military strike if the Israeli nation was to survive. Writing of the 1967 Six Day War, Yael Dayan stated: 'What for other countries would have been defeat, for us would mean extermination. It was not possible for us to lose the war and survive and each man carried this knowledge in his heart when we moved west.'[16] Reuven Gal, a former chief psychologist of the IDF, points out in *A Portrait of the Israeli Soldier* that:

> From the 1948 War of Independence through the 1973 Yom Kippur War, the Israeli soldier who was fighting on the front literally knew that he was fighting for the defense of his own home and family ... The 1982 War of Lebanon was a deviation from this concept. For the first time in its history, the IDF launched an offensive attack from a position of numerical superiority and not in response to an actual threat to the very existence of Israel.[17]

The 1982 invasion was denounced by elements of Israeli society as a *milchemt brera* (war of choice). During the three-year occupation of Beirut, 170 Israeli reservists refused to serve out their month-long active duty call-up if it involved service in Lebanon. All were convicted by court martial and sentenced to one to three months' confinement in a military prison. A significant number of reservists openly refusing to serve was previously unheard of in the IDF. But a much more substantial demonstrative protest, which directly questioned the legitimacy of the invasion of Lebanon, was that of Colonel Eli Geva in July 1982.

Some ten days into the invasion, it had become evident that significant elements of the PLO had withdrawn to Beirut and secreted themselves among the densely populated Muslim section of the city. Geva was the commander of an armoured brigade which had led the invasion into Lebanon and was now tasked to be one of the first combat formations to enter West Beirut if the order was given. Geva was torn between his commitment to the military, which entailed following orders, and his moral compass. He felt he could not accept responsibility for the inevitable civilian casualties that would arise from the IDF entering Beirut and targeting the PLO infrastructure. Nor did he feel that he could justify the legitimacy of the mission to the families of the men in his brigade who would undoubtedly be killed.[18] Geva asked to be relieved of his command, though he offered to stay and fight with his brigade as a tank commander. When this request was refused, he asked to be attached to the brigade's medical staff. This request went unanswered. Geva was then one of the youngest colonels in the IDF with the potential to serve on the General Staff. Because of his stand, he was called before the IDF's Chief of Staff and then the Defense Minister to explain why he had requested to be relieved. Two days after meeting with the Israeli Prime Minister to discuss his position, Geva was informed that the IDF's Chief of Staff had personally ordered his immediate discharge from the military. Geva was not permitted to return to Beirut to bid farewell to his men and was denied a position in the reserves.[19]

At some point soldiers will come to believe that their society asks too much of them. When an individual reaches this point it leads to insubordination and desertion. When an army reaches this point it leads to mutiny. Fifty-four divisions of the French Army mutinied in April–May 1917 after the failure of the Aisne offensive, which resulted in 187,000 French casualties. This mutiny was largely because of the soldiers' loss of confidence in the French commander, General Robert Nivelle, who was replaced in mid-May by General Henri Philippe Pétain. The soldiers made it clear that they would hold the line but would not undertake any offensive action. Their complaints were largely addressed. For the remainder of the year the French Army was spared taking part in any major offensives and its morale was nursed back to health. This rebuilding of morale was helped by the fact that the German Army was undefeated and still in possession of large swathes of French territory. But when a clear sense of purpose, which serves to justify the continued sacrifices demanded of soldiers, is missing, a widespread and often insurmountable breakdown in discipline may occur. A germane example is the mutinies among the Allied forces sent to fight the Bolsheviks in Russia following the end of the First World War.

Following the signing of the Treaty of Brest-Litovsk, Russian forces withdrew from the war. Under the terms of the treaty, German forces were free to occupy large areas of European Russia. German advances were perceived by the Allied powers to constitute a threat to the northern Russian ports of Murmansk and Archangel, where there were large stockpiles of Allied war matériel that it was feared would now fall into the hands of the Germans. In mid-1918, a small British–French–American expeditionary force seized the two ports. Yet the Allied Expeditionary Force in Russia was not withdrawn following the signing of an armistice with Germany in November 1918; rather, they were left to endure the Russian winter. The British soldiers, sailors and airmen, many of whom had enlisted for the duration of the First World War, did not understand why they were fighting in Russia – specifically why they were fighting the Bolsheviks – and why they had not been demobilised like their comrades on the Western Front. To make things even worse, the Russian campaign did not have popular support, and this sentiment would have been evident in the letters received by the men from their families asking when they were coming home. In February 1919, a detachment of Royal Marines protested about their continued presence in Russia, and widespread disobedience spread to the French and American contingents, with elements of both refusing to engage the Red Army forces. A British sailor commented: 'We are not at war with Russia and had nothing to fight them for. It was generally said that the wealthy people of England would benefit by our fighting and we had had enough of that during the last five years.'[20] In March 1919 the decision was made to withdraw the force and a volunteer North Russian Relief Force was raised to safeguard the evacuation. The Allied forces were withdrawn by October 1919.

The contract between soldier and state

The socialisation process of the military is such that a soldier's self-esteem and self-identity are, ironically, very much dependent on his reputation among his fellow soldiers. If a soldier refuses to carry out a mission he risks being branded a coward by his companions, which in itself often provides the necessary impetus to overcome any moral scruples. Soldiers may also rationalise that they owe a debt to their society to carry out the assigned mission. This was certainly the view of Carl Von Clausewitz, who wrote: 'The end for which a soldier is recruited, clothed, armed and trained, the whole object of his sleeping, eating, drinking, and marching is *simply that he should fight at the right place and the right time*.'[21] A sailor who served in the Falklands War commented:

> I said to the wife when I took her to the station that I'd been in the navy for fifteen years and I couldn't just say: 'Look I don't want to go now'. You don't join a club and, when the going gets tough, say: 'Well, I'm not a member any more'. I said: 'I don't want to go, but I've got to go. The taxpayers have been paying my wages all these years and okay they've called my number, they've called all our numbers. It's our duty.'[22]

A similar sentiment was expressed by John Nichol, a Tornado navigator with the Royal Air Force during the 1991 Gulf War, when asked how he felt about killing or being killed:

> The answer was very simple: we had taken the Queen's shilling. We had been trained, at enormous expense to the British taxpayer ... The taxpayers had to have some return for all that money they had pumped into us – something more than our shining presence at airshows. Yes, it was a shock, and, no, we had never expected to fight, but that fact, the simple fact of being a professional, with a job there to be done, that was for us the overriding factor in the whole business.[23]

The key clause in the metaphorical contract between the soldier and the state is this: the state will pay you and allow you, to an extent, to have a self-regulating society. In return, if the state requires your service for any matter, you are to provide it willingly. Soldiers do not have the freedom to choose which war they will fight. When the bugle sounds they are required to fall in. It is not a clause of this contract that the state will attach the same value to soldiers' lives as soldiers do. As his platoon of US Marines raced across central Iraq to attack an airfield, Nathaniel Fick pondered this factor.

> I was a Marine. I would salute and follow orders. Without knowing the big picture, we had to trust that they made sense. My Marines and I were willing to give our lives, but we preferred not to do so cheaply. The fear was a realisation that my exchange rate wasn't the only one being consulted.[24]

The salient point is whether the achievement of a military objective is worth the lives it costs. But there is a wide discrepancy between the front-line soldier's definition of what is worth dying for and what is not, and the opinion of those further back. As Addison Terry surveyed the burning hulks of North Korean tanks and the 'lumps of blackish brown' scattered around the rice paddies that were formerly North Korean soldiers, he allowed himself a few moments' reflection as to the purpose of the Korean War.

> No one knew why we were here and, although the policy makers in Washington had published statements of policy that might be accept-able to the party supporters at home, I knew that the same statements would sound mighty hollow in this valley where the smell of death was so heavy. I wondered how many GIs would accept Washington's expla-nation of the war as adequate justification for the price they individu-ally were paying over here in the filth.[25]

Gwynne Dyer tell us that soldiers willingly accept that combat may result in their death or maiming: 'But what they require in return is the assurance (or the illusion) that their death will not be wasted or unnoticed or meaningless. Yet most soldiers' deaths in civilised warfare are all of those things.'[26]

To an extent, the disciplinary apparatus of the military removes a soldier's ability to refuse a mission. So it is not surprising that soldiers usually accept the hand that fate has dealt them. As General Colin Powell remarked about the Vietnam War: 'Questioning the war would not have made fighting it any easier.'[27]

Soldiers are required to risk life and limb in the fulfilment of their government's objectives. It is therefore not surprising that many of them are unable to consider the merits of the government's position from a detached viewpoint. An American National Guard soldier who was severely wounded by an explosion in Iraq emphasised that: 'I support the war completely. I'm prouder than hell. At least I got hurt doing something I believe in.' He made these comments after enduring eighteen operations and the breakdown of his marriage due to the strain of his recuperation.[28] Another US soldier, who lost a leg following an ambush in Iraq, unequivocally rejected a suggestion that the basis for the war was false. 'They're telling me I went out there and I got my leg blown off for a liar, and I know that's just not true.'[29] To feel otherwise about the validity of the war would serve only to increase the psychological torment of these soldiers, nor would it make their recovery any easier.

A poignant example of the psychological torment that can arise when soldiers begin to question their government's rationale for sending them off to war are the experiences of Lewis B. Puller Jr. – the only son of the US Marine Corps' most decorated officer, Lieutenant General Lewis 'Chesty' Puller. In October 1968 Puller Jr. was severely wounded by a booby-trapped howitzer round while serving with the Marines in Vietnam. During his long period of hospitalisation he became disillusioned with the conduct of the war and felt that the sacrifices demanded by the war were being borne by only a small portion of society. Following his discharge from the military, Puller re-entered university. He became aware that many of his fellow students had a 'prevailing attitude that American involvement in the Vietnam War was, if not downright immoral, certainly a mistake of epic proportions'. Puller's bitterness increased as he

> began to feel that my own sacrifice and that of all of us who had fought the war were meaningless. Unable then to discover any higher purpose for the wasted lives of the dozen men whom I counted as friends who had not come home, I begin to despise the government and the Marine Corps, which had asked of many of us everything we had and given back almost nothing.[30]

He became increasingly withdrawn and 'tried to dispel with alcohol the magnitude of the obscene fraud of which I had been a willing victim'.[31] Puller eventually became an alcoholic and contemplated suicide. Although he was later to recover from his alcoholism and kept the disease at bay for thirteen years, in early 1994 he relapsed and also struggled with an addiction to painkillers. He committed suicide in May 1994. His wife, Toddy Puller, said in a statement: 'To the list of names of victims of the Vietnam War, add the name of Lewis Puller … He suffered terrible wounds that never really healed.'

Consider also the furore that arose among veterans following the publication in 1995 of Robert McNamara's mea culpa, *In Retrospect: The Tragedy and Lessons of Vietnam*. McNamara, the US Secretary of Defense from 1961 to 1968, wrote that, in his opinion, the Vietnam War was a 'terrible mistake'. Although many veterans claimed they were angered that McNamara had not publicly voiced his misgivings about the war during his period in office, McNamara clearly explained that his stated opposition to the war was heavily influenced by hindsight. Rather, the source of the veterans' anger was probably more the fact that one of the key architects of American involvement in Vietnam now admitted this involvement was a mistake, thereby further diminishing the validity of the cause for which their country had sent them to war. It would seem that some groups would rather believe a comforting lie than an unpalatable truth. Once the basis for going to war was undermined, the reasons the veterans had clung to in order to justify the sacrifices they were asked to make and the hardships they endured were also undermined. This led to anger and resentment.

Alternatively, a circular argument may be embraced asserting that the sacrifices made by soldiers are inherently noble; therefore the death of a soldier in any conflict, regardless of the morality of the cause, is also noble. Even McNamara succumbed to the moral simplicity of this argument.

> In the end, we must confront the fate of those Americans who served in Vietnam and never returned. Does the unwisdom of our intervention nullify their effort and their loss? I think not. They did not make the decisions. They answered their nation's call to service. They went in harm's way on its behalf. And they gave their lives for their country and its ideals. That our effort in Vietnam proved unwise does not make their sacrifice less noble.[32]

Acceptance of this argument, while reassuring to veterans, disguises the harsh reality that wars may be fought for a variety of reasons, not all of them noble, and that often a soldier's death in combat accomplishes nothing and means nothing – except to the bereaved.

Emotional investments: families and governments

The families of deployed soldiers are often numbered among the staunchest supporters of their government's decision to commit military forces. They have a deep-seated emotional need to believe (or at least convince themselves) that the 'cause' justifies the risk of death or disfigurement faced by their loved ones. They are unable to change the fact that their family member has been sent overseas to fight and so are left with two options: embrace the 'cause' or become embittered. It is not surprising that most, perhaps even subconsciously, will choose the former. The father of a US Army officer killed in Iraq in November 2004 proclaimed in his son's obituary that the United States was 'a moral nation, steadfast in our principles … we do not shirk our duties to commit our blood to just and necessary causes. Because that is what keeps us free.'[33] Likewise, the father of another American serviceman killed in Iraq stated in March 2007, 'My brother died in vain in Vietnam; that won't happen to my son', and voiced his steadfast support for continuing the war in Iraq to honour the dead and complete the mission.[34] Kevin Mervin, a British soldier who took part in the 2003 invasion of Iraq, when reflecting upon whether the invasion was 'worth it', wrote:

> [I]t had to be worth it because many coalition soldiers paid the ultimate price for that simple word we take for granted – freedom. If you disagree with that and think they died for nothing, then argue with the families who lost their husbands, fathers, sons and brothers in the conflict. Anyone who says they died for nothing should think again.[35]

This deep-seated need of the families of deployed soldiers to believe in the 'cause' can manifest itself in hostility towards those who have the temerity to question the conduct of the war. In *Warrior Race: A History of the British at War*, Lawrence James notes that during the First World War, anti-war gatherings usually degenerated into violence.

> Pacifist sentiment provoked rage and violence in a population that found the peacemongers' message intolerable. If what they said was true, then the public had been duped and hundreds of thousands were being killed and maimed for no reason … According to the anti-war propagandists, all this suffering was futile, a possibility that was as unthinkable as it was grotesque.[36]

Militaries are generally among the most conservative of society's institutions and it is rare for senior military officers to make a public statement expressing their opposition to a mission. Indeed, Article 88 of the United States' *Uniform Code of Military Justice* states:

> Any commissioned officer who uses contemptuous words against the President, the Vice President, Congress, the Secretary of Defense, the Secretary of a military department, the Secretary of Transportation, or

the Governor or legislature of any State, Territory, Commonwealth, or possession in which he is on duty or present shall be punished as a court martial may direct.[37]

The US military's *Manual for Courts-Martial* states that the maximum punishment for this offence is 'dismissal, forfeiture of all pay and allowances, and confinement for one year', and without a hint of irony goes on to elaborate that 'the truth or falsity of the statements is immaterial'.[38]

The preferred option for many soldiers who disagree with their government's use of the military is to resign quietly. In 1854, Major General Ethan Allen Hitchcock confided in his diary his reservations about the threatened war between Spain and the United States:

> As to leaving the Army: I may do so if I choose at this time and no one notice me, for I am unknown except to a few friends. If I wait and a war with Spain be forced on us by the headlong ambition of false policy of the Cabinet at Washington it might be hazardous to retire, even though in principle opposed to the war, not only as unjustifiable towards Spain, but as impolitic and injurious as respects ourselves.[39]

Moreover, the Algerian-born French soldier Jules Roy confessed in *Dienbienphu* that:

> In 1952, I had, for the first time, declined the honor of bearing arms when I realised, after a few weeks' travel across Indochina, that we were serving an unjust cause ... I could fight for the West only if I was not at the same time supporting the sordid combination of interests which soiled its cause.[40]

Roy resigned from the French Army in 1953.

The abstract forces that propel nations towards war, such as nationalism, pride, hatred and fear, extort an immense emotional investment from the public; even more so from politicians and leaders. It is with good reason that accounts of conflict speak of nations risking their 'treasure' (generally measured in the lives of young men) to go to war. This emotional investment, whether through pride or obstinacy, may result in an escalating commitment to a failing course of action. Thus wartime leaders will find it difficult to acknowledge that, in many situations, the number of men previously wounded and killed will have no effect on the subsequent cost of continuing an offensive, and will continue to commit more men and matériel until the conflict reaches its bloody conclusion. In *Achilles in Vietnam*, the psychologist Jonathan Shay summarises this paradox as: 'So many men have died that we can't give up now; we've got to go on "so they will not have died in vain". Surprisingly, this argument has no appeal to soldiers actually fighting, even though it is utterly compelling to rear-echelon noncombat soldiers and civilians.'[41] A contemporary example of this sentiment is provided by an address made by President George W. Bush to a campaign rally in Missouri shortly before the American death toll in Iraq passed 1,000. 'My promise to them [family members who had lost someone to the conflict in Iraq] is that we will complete the mission so that their child or their husband or their wife has not died in vain. We will be there until the mission is finished, and then we're coming home.'[42]

Such comments may help alleviate the grief of those mourning soldiers who will never come home. But it is also worth reflecting upon some comments made by Bush's challenger for the presidency in 2004, John Kerry. In April 1971 Kerry, who had been awarded the Silver Star for his service in Vietnam, testified before the Senate's Committee on Foreign Relations on behalf of a group called the Vietnam Veterans Against the War.

> [Each day] someone has to give up his life so that the United States doesn't have to admit something that the entire world already knows, so that we can't say that we have made a mistake. Someone has to die so that President Nixon won't be, and these are his words, 'the first President to lose a war.' We are asking Americans to think about that because how do you ask a man to be the last man to die in Vietnam? How do you ask a man to be the last man to die for a mistake?[43]

The American political and military leadership largely realised that the war in Vietnam was unwinnable after the strategic shock of the enemy's Tet offensive of February 1968. Yet it was another five years before the US completed the withdrawal of its ground forces. In that five-year period another 37,000 US servicemen died in the paddy fields and jungles of Vietnam. Tellingly, a generation later, a United States Marine who took part in the 2003 invasion of Iraq predicted: 'If these people [the Iraqis] don't want for themselves what we want for them, then this *will* be Vietnam. We'll get our pride and our credibility involved, and then we'll keep throwing men and money down the pit long after everybody else knows we're fucked.'[44]

Coming home to an ungrateful public

Another clause not present in the contract between the soldier and the state is that society will be appreciative of the privations endured and the sacrifices made by the soldier. 'The only trouble is that people don't appreciate you after the fighting is over. I know, I've been through it before', bitterly observed an American Army officer to Frederick Downs during the Vietnam War. 'When the war gets over the people don't give a shit for you anymore. Just be there to die for them when the next one starts, that's all they care about. In between, a soldier is just an embarrassment to them.'[45]

An aspect of a soldier's homecoming that many soldiers find particularly difficult to accept is the realisation that most civilians are largely indifferent to the hardships endured by soldiers on their behalf. Society is generally supportive of its fighting men and women during a war, particularly if the nation has been fully mobilised to meet the war effort, as occurred in Britain during the First and Second World Wars. During the First World War, soldiers returning to Britain from the front greatly appreciated being met with free chocolates, cigarettes and food at Victoria railway station, being shouted drinks in pubs, and the fact that the mansions of the aristocracy were being used to house convalescing soldiers. At least initially, civilians may be temporarily caught up in the national euphoria following the successful prosecution of a war. Jean-Baptiste Barres, a member of Napoleon's Imperial Guard, provided a detailed account in his memoirs

21

of the triumphal return of the French Army to Paris following the defeat of the Russians at the battle of Friedland in June 1807. They entered the city through a specially erected triumphal arch to the cheers of multitudes of Parisians, then proceeded to a luncheon at tables set for 10,000 along the Champs Elysées, the day concluding with fireworks and public dances. 'All the theatres of the capital were open to the Guard', enthused Barres. 'The parterre, orchestra, and the first tier of the boxes were reserved for us, and the first row in the other boxes.'[46] A decade or so later, it would be the English rather than the French who were staging victory celebrations. James Anton fondly recalled the cheering crowds and the generosity of citizens as his regiment made a triumphal march from Ramsgate in Kent to their base at Edinburgh upon their return from Waterloo. At Edinburgh, Anton noted:

> We entered the castle, proud of the most distinguished reception that ever a regiment had met with from a grateful country. Two nights we were admitted free to the Theatre, two to the Olympic Circus, two days to the panoramic view of Waterloo; and to conclude our triumphal rejoicings … an entertainment was provided for us in the Assembly Rooms, George's Street, to which the noblemen and gentlemen contributors came, and witnessed the glee with which we enjoyed their hospitality.[47]

The victory parade over, it may come as a shock to many soldiers that society's long-term concern for their welfare is less forthcoming. An Australian soldier protested in a letter home in 1916 that 'There is nothing in the world more short lived & fleeting than a nations [sic] remembrance of her fighting men after peace is declared, the Public has no gratitude'.[48]

The difficulties faced by returned soldiers as they attempt to reintegrate into society are well documented, particularly the impediments to employment experienced by maimed veterans. Frederick the Great's veterans were given a licence to beg, while many of Wellington's Peninsular veterans likewise ended up on the streets. The historian Bill Gammage notes in *The Broken Years*, his sympathetic account of the Australian soldiers of the First World War, that the sight of invalid soldiers begging on the street after the war was distressing for those who did not serve. He surmises that 'stay-at-home Australians, weary of war, recoiling from its horror, and sickened by the number of its victims tried to forget those tragic years as quickly as possible ... they wanted a return to normalcy'.[49] Rather than honouring these men for the sacrifices they made, the good citizens wished they would just go away. In England, George Coppard, who had noted the extraordinary support given to soldiers during the First World War by the home front, bitterly condemned the abandonment of his comrades once victory was secured.

> Lloyd George and company had been full of big talk about making the country fit for heroes to live in, but it was just so much hot air. No practical steps were taken to rehabilitate the broad mass of demobbed men, and I joined the queues for jobs as messengers, window cleaners and scullions. It was a complete let-down for thousands like me, and for some young officers too. It was a common sight in London to see ex-officers with barrel organs, endeavouring to earn a living as beggars.[50]

The implementation of the Beveridge Plan by Clement Attlee's Labour government, which laid the foundations for the so-called Welfare State, was instrumental in sparing the demobbed British veterans of the Second World War many of the hardships experienced by their predecessors from the First World War. Even so, some felt that society had largely forgotten them, particularly in relation to pensions and medical care. One Second World War veteran commented in a letter to the historian Sean Longden that:

> I feel quite cynical about what's happened since the war. I believe that it is totally wrong that they should be dependent on charities like the Poppy Day appeal. Leaving wreaths in memory of those who died is sentimental and does nothing for the men who survived ... I think that soldiers are used as tools – bureaucrats who have no experience of war make decisions and gamble with other people's lives. After they're

gambled, those remaining are forgotten and some have to depend on charity to survive. I think that young people today looking back at what's happened to ex-servicemen wouldn't come out and fight for their country.[51]

The British soldiers of the Second World War, however, were treated with much greater compassion than their Russian counterparts. In mid-1946, Soviet authorities estimated that there were some 2.75 million surviving invalids from the war. These men (and women) were all provided with a pension and many were also entitled to supplementary food packages. This subsistence support kept the veterans alive, but for those unable to work, future prospects were limited. Although in theory entitled to the best medical support the Soviet Union could offer, the reality was a shortage of hospitals, doctors and prosthetic limbs. Maimed veterans begging on the streets became a common sight in Soviet towns and cities, particularly Moscow. In 1947 Stalin ordered the streets of Soviet cities to be cleared of beggars. The invalid former soldiers were forced onto trains heading north, many ending up on the island of Valaam on Lake Ladoga near the border with Finland.[52] Valaam was the site of a former monastery and fitted the requirement of keeping the crippled veterans out of sight and out of mind in a country that was desperately trying to put the destruction of the war behind it.

But it is not only the veterans of the world wars who have had cause to bemoan their treatment on returning from combat. Robert Lawrence, a British officer who was awarded the Military Cross for his actions during the battle to capture Mount Tumbledown during the Falklands War, was shot in the head in the final stages of the battle. Lawrence became embittered because of his treatment by the British Army and the British government's bureaucracy after he was wounded:

I knew one could argue that the victims of war had never been well looked after. But the difference between the last two world wars, it seemed to me, and the Falklands conflict was that the former had been fought fairly close to home, had gone on for a number of years and affected the whole of the nation. Most people had either lost someone close or knew somebody who had, and everyone was involved. There just wasn't the same degree of national understanding for those who had been wounded in a brief war thousands of miles away. People either wouldn't, or couldn't, see the Falklands victims of the eighties in the same light. As a result, we were battling on our own.[53]

What I didn't realise, until, like so many others, I came back crippled after doing my bit for my country, was the extent to which we had been conned. Conned into believing in a set of priorities and principles that the rest of the world and British society in general no longer gave two hoots about. We had been 'their boys' fighting in the Falklands, and when the fighting was over, nobody wanted to know.[54]

Kayla Williams, who served with the US Army in Iraq, also became embittered by the bureaucratic inertia and seeming lack of concern shown by the army towards one of her comrades who was severely wounded when his vehicle convoy was ambushed near Mosul:

[T]he bureaucracy he has to negotiate to get therapy programs has been horrible. This is a man who almost made the ultimate sacrifice for his country. Now he has to fight for everything ... Does the Army expect a man with a traumatic brain injury to advocate on his own behalf for the care and treatment he deserves? There are days he can barely get out of bed in the morning, the pain is so intense. Watching how shabbily the Army treats Shane – not to mention so many other seriously wounded veterans of this war – has been the deepest disillusionment for me.[55]

The indifference and hostility encountered by soldiers returning to the United States from Vietnam is well documented. In this instance, a segment of the population actively opposed to the Vietnam War, and who were frustrated with their inability to alter government policy, took out their anger on those citizens most affected by the war: the soldiers who had to fight it. Robert Mason, who served as a helicopter pilot with the

US Army, recalled that his return flight from Vietnam had a brief stopover in Honolulu. 'I also grabbed a *Newsweek* and went to the counter to pay. The clerk, a young woman, took my money and asked if I was return-ing from Vietnam. I said yes, proudly. She suddenly glared at me and said, Murderer.'[56] Frank McCarthy, who served as a rifleman with the 1st Infantry Division in Vietnam, returned to America in March 1967. Not appreciating the strength of the opposition to the war, he decided to visit his brother at Berkeley while still wearing his army uniform. 'I was spit on. This gang of guys walking behind me threw peanuts at me', McCarthy ruefully commented in a later interview. 'I went into a bar and phoned my brother. I almost didn't make it out of there. The real shock of coming back was in that bar. These guys weren't going to let me out. They wanted to kick my ass. Calling "You kill any women? You kill any kids?"'[57]

The absence of an organised welcome home was particularly galling for the returning Vietnam veterans. David Donovan tells us that when he stepped off the plane after arriving back in America he was ready for anything. 'Anything that is, but what we got. We stepped off the plane to an empty tarmac.'[58] He bitterly commented: 'I felt my country didn't give a damn about me or the sacrifice I and thousands of others were mak-ing in their name. I found I didn't like being shot at for low pay and less thanks.'[59]

✶ ✶ ✶

Soldiers are required to risk their lives and limbs for causes they frequently do not understand and may not support. They are required to seek out and kill men with whom they have no personal conflict and are required to go on killing and being killed until men, far removed from the realities of conflict, tell them to stop. They are required to bear these burdens on behalf of their societies. This is the nature of the contract between the soldier and the state.

Chapter 2

WHO SERVES

In medieval Europe the feudal tenure system imposed an obligation on various segments of the population for military service. Under this system the king possessed all land, and all other landowners were considered the king's tenants. The king granted fiefs to the peers of the kingdom (princes, dukes, earls and barons) and to the heads of the Church (bishops and abbots), with the stipulation that they were to provide a specified number of knights on demand. Evolving out of the feudal system was the concept of *noblesse oblige*: that privilege entails responsibility. The application of this concept to martial matters meant that a number of the aristocracy served in the military, not for financial gain, but because of a sense of class responsibility. In *A War of Nerves*, Ben Shephard comments that the landed aristocracy was 'a class which had always regarded war as its ultimate justification, and which kept in training when there was not a war to be fought by duelling, hunting and dangerous activities of various kinds'.[1]

The aristocracy generally did not serve in the military to make their fortunes; indeed, many of them had to pay for the privilege of serving. In December 1653, following the English Civil War and the execution of Charles I, Oliver Cromwell installed himself as England's first Lord Protector. A parliament was formed the following year, but when it failed to ratify the Protector's powers Cromwell dissolved it. In 1655 Cromwell, in response to a series of minor royalist uprisings, set up eleven military governors – England was effectively under a military dictatorship. Following Cromwell's death in 1658, a power struggle among the generals resulted in a period of instability leading up to the restoration of the monarchy in 1660. This period of military rule left a profound aversion among the English to standing armies. So when a standing army was reformed in 1683, Parliament implemented a system to ensure that the army would never again be in the hands of men who were likely to reimpose military rule. The system put in place was the purchasing of commissions.

The British purchase system

By requiring a substantial amount of money to be outlaid for a commission, the purchase system ensured that officers were generally men of independent means who, because of their stake in the maintenance of the status quo, were unlikely to foment military unrest. Furthermore, officers' salaries were deliberately kept low and fees for the officers' mess kept high to discourage soldiers of fortune from seeking entry into the army as officers. In addition to the fee paid to purchase the initial rank of ensign or cornet, an applicant was required to provide evidence that he had received the 'education of a gentleman'. Each successive rank would require a further payment until the officer reached the rank of lieutenant colonel. Promotion to colonel and general officer was based on seniority – and social connections. In addition to the regulation price, which was set by and paid to the government, a premium was often paid to the previous incumbent of the position. The actual cost of the commission varied in accordance with the exclusivity and thus the desirability of the regiment. Despite appearances to the contrary, officers were generally required to pass written examinations and have a prescribed number of years' service before advancing in rank. Officers could sell their commission on retirement but their commission could also be forfeited for dishonourable behaviour. The purchase system underwent numerous modifications, some of which were to curtail obvious abuses such as the purchasing of commissions for children, but it was maintained in Britain for over two centuries.

One abuse of the purchase system that was never rectified was the manipulation of half-pay. Half-pay had been initiated as a retaining fee so that during a time of peace, officers who were not required for full-time duty would effectively be released from the service, though they could be called up at any time. Transfer to half-pay was at the discretion of the Secretary of War, though in the 19th century it was used by members of the aristocracy as a means of avoiding distasteful service and securing promotion. Although commissions were purchased, it was still necessary for an aspiring officer to progress through each subsequent rank. Cecil Woodham-Smith outlines in *The Reason Why* how ambitious young officers manipulated the system:

A young man would buy a vacant captaincy in a regiment in which he had no intention of serving, and next day he would go on half pay; though no service was done with the regiment, he had become a captain, which qualified him to buy his next step as a major in a more desirable regiment ... By going on half pay, or by exchanging, at a price, into another regiment, wealthy officers avoided uncomfortable service abroad. When a fashionable regiment had to do a turn of duty in India, it was notorious that a different set of officers went out from those who had been on duty at St James's Palace or the Brighton Pavilion. When the regiment returned, the Indian duty officers dropped out and a smarter set took their place.[2]

The purchase system was abolished in 1870. It had largely ensured that army officers were drawn from the wealthy, landed classes, as well as providing the country with a relatively inexpensive army. Enthusiasm for military service, however, does not necessarily equate with effectiveness. The British practice of drawing its military officers almost exclusively from the upper class had the inherent risk of determining rank by social standing and personal wealth rather than ability, though the bravery and gallantry of British Army officers was seldom questioned. Indicative of the danger of this approach was the performance of the British Army in the Crimean War. In *On the Psychology of Military Incompetence*, Norman Dixon, reflecting on the effectiveness of the British Army in this war, concludes: 'The poor quality of the officers, most of whom had bought their commissions … stood in marked contrast to the excellence of the men.'[3]

A notable example of such an officer was James Thomas Brudenell, the 7th Earl of Cardigan, who led the ill-fated charge of the Light Brigade at the Battle of Balaclava in October 1854. Brudenell joined the army in 1824. Taking full advantage of his wealth and the opportunities provided by the purchase system, he had risen to the rank of lieutenant colonel by 1830.

Brudenell's primary interest was in the pomp and ceremony of military service, and he made no secret of his desire for his regiment to be the most glamorous in the British Army. A firm proponent of the aristocracy's birthright to command and be obeyed, he would not tolerate the slightest deviation from his perfectionist desires for his regiment. He relentlessly persecuted, publicly ridiculed and taunted officers who did not have the desired social standing, or for imagined slights. In February 1834, Brudenell was dismissed from the command of the 15th Hussars, largely due to the outrageous manner in which he treated one of his officers. Remarkably, two years after being removed from command, because of his influence at court, he was gazetted to be the commanding officer of the 11th Light Dragoons. The purchase price was reported to have exceeded £40,000.

Brudenell, now the Earl of Cardigan, had seemingly learned nothing from his previous dismissal and was soon subjecting the men of his regiment to what was later determined by the War Office to be an unnecessarily harsh standard of discipline. Despite receiving a series of reprimands from the commander-in-chief and from other senior army officers concerning the inappropriate manner in which he treated his officers, Cardigan remained in command. Following the outbreak of the Crimean War in March 1854, Cardigan was promoted to brigadier general and placed in command of the Light Brigade of cavalry. He was promoted to major general in June 1854. Not one to share the discomforts of campaigning with his soldiers, Brudenell dined (he brought to the Crimea his personal French cook) and slept on board his luxury yacht moored in Balaclava harbour while many of his men succumbed to the bitter cold and shortage of food ashore.

A tradition of service

But of greater significance here is the fact that, despite some noted failures, the British aristocracy continued to view military service as a duty to their country. This tradition of service extended to the British royal family. George II was the last British monarch to lead his troops into battle, in 1743 at the age of 60, when the British Army defeated the French at the Battle of Dettingen in Germany (in the War of Austrian Succession). Later monarchs (Edward VIII and George VI) saw active service in the First World War while waiting to ascend the throne. The future Queen Elizabeth II served as a second subaltern in the Auxiliary Territorial Service for a few months at the end of the Second World War and her husband, Prince Philip, served on various battleships and destroyers throughout the war. All of Queen Elizabeth's children, with the exception of Princess Anne, have served in the military in some capacity. Notably, Prince Andrew, the Duke of York, trained as a helicopter pilot with the Royal Navy and served with the taskforce sent to retake the Falkland Islands from Argentina in 1982. Princes Harry and William entered the British Army in May 2005 and January 2006 respectively; Prince Harry deployed to Afghanistan in December 2007, where he served as a Forward Air Controller directing jet attacks against Taliban insurgents.

The aristocratic nature of the officer class of the British Army was commented upon by Major General J.F.C. Fuller, who wrote that when he joined the British Army in 1898:

> It was an aristocratic army, feudal in the sense that it was grounded on leadership and fellowship, in which, with few exceptions, the leaders were the sons of gentlemen, and more frequently than not eldest sons – the privileged son. When I went to Sandhurst we were not taught to behave like gentlemen, because it never occurred to anyone that we could behave otherwise.[4]

In the initial months of the First World War, six peers (a member of one of the five degrees of British nobility: duke, marquis, earl, viscount or baron), sixteen baronets, six knights, 95 sons of peers, 82 sons of baronets and 84 knights' sons were listed among the dead.[5] In the majority of cases these men were serving as officers in the regular army prior to the outbreak of the war. Thirty-two peerages and 35 baronetcies were extinguished in the period 1914–20, not all as a consequence of the war, though some 300 peers or their eldest sons (of the 1,500 who served) were killed in action or died as a result of their war service.[6] In addition to the aristocracy, other prominent British families also lost members during the First World War. For example, the grandson of Charles Darwin, the only male child of Rudyard Kipling (whose body was never recovered), and the eldest son of the British Prime Minister, H.H. Asquith, were all killed in France during the war.

The carnage of the First World War exhausted the traditional avenues of supply for army officers. In August 1914 there were 15,000 officers in the British Army. During the course of the war a further 235,000 men were commissioned.[7] By necessity, an increasing number of officers were commissioned from the ranks. By the end of the war, it was estimated that 40 per cent of officers came from working- or lower-middle-class backgrounds.[8] Many of those commissioned during the war, particularly former 'rankers', were given temporary commissions – that is, they held their commission only for the duration of the war – and were disparagingly known as 'Temporary Gentlemen' by their social superiors who formed the old officer caste of the British Army.

There was little change in the background of officers in the regular army between the two world wars. For example, over 84 per cent of entrants to Sandhurst in 1939 had attended a public school. In *The Sharp End: The Fighting Man in World War II*, John Ellis concludes that initially in the Second World War, officer selection 'was purely a matter of a brief interview in which family connections, sporting accomplishments and possession of those nebulous characteristics that constitute a "decent chap" counted for far more than brains and application'.[9]

But the growing number of casualties and the expansion of the army meant that in 1942 a new system of officer selection based on meritocratic criteria was introduced. Interviews were conducted over three days and each selection board included a non-regular officer as well as a psychiatrist. By 1943–44 only about a quarter of applicants commissioned in the army had a public school background.

Recruiting the lower ranks

While a sense of *noblesse oblige* may have helped fill the officer ranks of the British services, it certainly did not produce recruits for the lower ranks. Prior to the mid-19th century, the British government refused to fund the maintenance of a large peacetime navy, therefore it needed to rapidly build up the force in times of war. For example, the Royal Navy increased its strength from 45,000 men in 1793 to 140,000 men in 1815. Additionally, over this same period, in excess of 100,000 sailors died (the majority from disease) and tens of thousands deserted. Conditions on board British warships during the Napoleonic period were so bad, and the pay so poor, that there were never enough volunteers to satisfy the navy's manpower requirements. The navy therefore turned to 'press gangs'. Since the 16th century the navy had had the legal right to press men for service for the defence of the realm. Press gangs operated in ports and generally targeted merchant seamen because, as trained mariners, they could be employed on board straight away. Merchant ships could even be stopped at sea and their sailors taken off for service in the Royal Navy. However, officers and apprentices aboard merchant ships were exempted from being pressed and the navy was obliged to leave enough men on board the merchant vessel to 'navigate the ship'.

The Impress Service covered every port in Britain and was under the command of Royal Navy officers not attached to seagoing ships. The senior naval officer of the Impress Service at each port was known as the Regulating Officer, and he would hire some of the local thugs as 'gangers' to form a press gang of eight to twelve men. The press gang was accompanied by a commissioned naval officer when it ventured out to seize suitable recruits. A bonus was paid for every man pressed, but corruption was widespread, many men securing their liberty by paying a bribe to the press gang once seized. The press gang used a number of techniques to obtain 'recruits', including knocking them unconscious, threatening them with weapons and plying them with alcohol. Running battles were often fought between locals and the press gangs, with the locals trying to release the seized men. Having been 'pressed', most men were then taken to a worn-out hulk moored offshore. These ships were used because moving the pressed men away from the shore reduced the chances of a rescue attempt or the commencement of legal action. Once enough men were pressed, a receiving ship, generally an old man-of-war that was no longer capable of putting to sea, would take the pressed men on board and transfer them to a ship of the line.

In theory, those aged under eighteen and over 55 were exempted from being pressed, as were seamen employed by the government, seamen with less than two years' experience at sea, apprentices, fishermen, tradesmen and men who had enlisted in the militia. In practice, a so-called 'exempt' man would often disappear to sea before legal proceedings could begin. In addition to legal action, pressed men could also obtain their release if they furnished a substitute. A pressed man remained in the service until the war ended or he managed to desert. To prevent desertions, sailors were usually denied shore leave. The efforts of the press gangs left families destitute and careers disrupted or destroyed.

Even with the press gangs, the Royal Navy was unable to satisfy its manpower demands. In March 1795, Parliament passed the two Quota Acts, whereby each borough and county had to provide an annual quota of men for service with the navy. The quotas were based on the population of the borough/county and, where relevant, its number of seaports and hence number of sailors. In the absence of volunteers, local magistrates often gave men convicted of petty crimes the choice between going to sea or going to jail.

The Royal Navy had priority for manning, with the British Army forced to take what was left. The army could not resort to press-gangs and therefore the majority of its recruits came from debtors' prisons, work-houses and the fields. Well might the Duke of Wellington write in one of his dispatches: 'We have in the service the scum of the earth as common soldiers.'[10] Shelford Bidwell writes in *Modern Warfare* that these soldiers were military surrogates for the ordinary citizens ... this duty was delegated to the poor and unfortunate who could not contrive to earn a living in any other way. They did as they were told in their dumb, obedient way, leading degraded and squalid lives, marching across the chess-board of the theatre of operations and killing or being killed as the situation demanded.[11]

A survey in 1846 revealed that two-thirds of the army's recruits enlisted because they were destitute and that between half and two-thirds of recruits were unskilled labourers.[12] Lieutenant Henry Clifford described the soldiers he served alongside in the Crimea as having come from 'the plough, the slums of London, or the bogs of Paddyland'.[13]

The recruitment pool for the British Army had barely improved by the turn of the century. John Baynes estimates that 5 per cent of the soldiers in the 2nd Scottish Rifles pre-1914 were from the lower middle class, 25 per cent from the working class and 70 per cent from the 'real lower class': generally unskilled labourers from the slums of Glasgow and Lanarkshire. For such men the choice was often join the army or face starvation.[14] John Keegan concludes that the motivation for most men to enlist was simply to escape poverty, and that:

Almost any other sort of employment was thought preferable, for soldiering meant exile, low company, drunkenness or its danger, the surrender of all chance of marriage – the removal in short, of every gentle or improving influence upon which the Victorian poor had been taught to set such store.[15]

According to Gerald Oram: 'Far from being an honourable profession, soldiering was considered worthless by most classes, but especially among the working class who regarded the army as a refuge for drunkards and criminals rather than a respectable trade.'[16] This sentiment echoes the comments of Field Marshal Lord Wavell, who affirmed of the pre-1914 British Army: 'There was in the minds of the ordinary God-fearing citizen no such thing as a good soldier; to have a member who had gone for a soldier was for many families a crowning disgrace.'[17] The view that the English soldier was drawn from the dregs of society was also evident among the continental armies. A German officer of the First World War smugly wrote in his memoirs that: 'All British common soldiers were professionals, consisting mostly of men unable to make a living in any other profession.'[18]

From volunteers to conscripts

In Britain, the outbreak of the First World War in August 1914 produced, in the words of Lawrence James, an 'emotional intoxication that infected the country during the first weeks of the war. Thousands of men succumbed to or were swept along by a mood that verged on hysteria.'[19] The sense of crisis engendered by the initial retreat from Mons by the outflanked British Expeditionary Force (BEF) of four divisions (the so-called 'Old Contemptibles') produced a surge of volunteers for the army. 750,000 men enlisted in the war's first two months, with almost 1.2 million men 'flocking to the colours' in 1914. This influx of recruits was sorely needed, as the battles of 1914, such as the Marne and First Ypres, had all but decimated the BEF.

In theory, volunteers had to be at least nineteen years old to enlist in the British Army and be sent overseas, but the matter of under-age soldiers was usually overcome by a simple expedient. George Coppard was sixteen years and seven months old when he enlisted in the army in August 1914. He recalled that his initial attempt at enlisting was not successful. 'The sergeant asked me my age, and when told replied, "Clear off son. Come back tomorrow and see if you're nineteen, eh?" So I turned up again the next day and gave my age as nineteen.'[20] Coppard was duly enlisted. However, his family later petitioned the War Office for Coppard's discharge, as he had already served in France for eight months and had only just turned eighteen. The War Office replied that, 'as his age on attestation is 19 years and 7 months, that is therefore his official age'. Coppard drily noted: 'Apparently the production of my birth certificate cut no ice with them.'[21]

Another 1.3 million volunteers were accepted into the army in 1915. England's public schools, the traditional source of officers for the British services, rose to the challenge. All but eight of the 539 boys who graduated from Winchester College between 1909 and 1915 joined up prior to the introduction of conscription. Many of these wealthy young men joined to seek adventure and out of a sense of class duty – they were to provide the junior officers. For the less well-off, who would form the bulk of the soldiery, some joined out of a sense of obligation and duty, while many joined up simply because their friends and workmates had enlisted. For such men the army provided a diversion from the drudgery of their current occupation, a chance to escape family concerns and/or trouble with the law, an opportunity for adventure, plus regular pay and regular meals.[22]

Britain was the only major European power without mandatory military service when the war started in August 1914. The French had introduced universal military service in 1793 via the *levée en masse*, inaugurating the era of mass continental land armies. The other European powers had to follow suit (Sweden introduced conscription in 1812, Prussia and Norway in 1814, Spain in 1831 and Russia in 1874) or risk their armies being vastly outnumbered. But for Britain the Royal Navy, rather than the British Army, was the principal guarantor of national security. Consequently, the vast majority of defence spending was directed towards the navy, while the army's primary role was to maintain long-service overseas garrisons, a role unsuited for short-service conscripts.

The British Army remained a volunteer force until January 1916, when conscription was introduced via the Military Service Act. The introduction of conscription was brought about by the realisation, following the failed Allied offensives of 1915 – such as the British defeat at Loos – that the war would be a drawn-out struggle of attrition. The German Army would need to be bled dry by constant offensive action. An attritional struggle demanded more manpower than could be provided by voluntary recruitment. In the memorable simile of Siegfried Sassoon: 'What in earlier days had been drafts of volunteers were now droves of victims.'[23]

The Military Service Act made single men and childless widowers aged between eighteen and 41 liable for compulsory military service. In addition to married men (or those widowed with children), several protected professions (including the clergy) were excluded from being called up. An amendment to the Act in May 1916 extended its provisions to married men, as well as enabling the retention of time-expired servicemen (that is, those men who had already fulfilled their service requirement) for the duration of the war, and the right to re-examine men previously declared unfit for service. In April 1917, further modifications to the Act permitted Home Service Territorials to be examined to determine if they were fit for service abroad (technically the Territorials were required to serve only in England and previously had to volunteer for service abroad). The April 1917 modifications also permitted the re-examination of men who had been invalided out of the service on account of wounds or ill-health to determine whether they were now fit enough to resume active service. Furthermore, a revised list of protected occupations was published.

The fourth version of the Act was passed in January 1918, enabling the government to quash all occupational exemptions at its discretion, as well as abolishing the two-month grace period that previously applied when an occupation lost its exemption. As the number of casualties incurred by the British Army was still greater than the number of replacements, a fifth and final version of the Act was passed in April 1918. This amendment dropped the minimum age for conscription to seventeen and extended the maximum age to 50, as well as extending the provisions of the Act to Ireland (though for political reasons this was never enforced). Conscription ended in 1919. Some 2.3 million British men were conscripted into the British armed forces during the First World War.

Provision was made in the Military Service Act for Conscientious Objectors (COs) – those who were opposed to participating in warfare because of their ethical, moral or religious beliefs – to undertake 'work of national importance', generally manual labour or agricultural work, instead of military service. Approximately 16,000 COs were formally recognised by the conscription tribunals; 3,300 served in the army as non-combatants, mostly as medical orderlies and stretcher-bearers, while others served in essential industries. Some 1,500 COs refused to undertake any service whatsoever in conjunction with the war effort and were sent to prison. The release of the COs from prison was deliberately held back until mid-1919 to give returning soldiers a head start at securing peacetime employment.[24]

In the interwar period, the British Army was not the career of first choice for the working class – unsurprisingly, given the widespread aversion to military service after the slaughter of the First World War – and so the army returned to the recruiting grounds whence it had drawn its pre-1914 members, in particular the unemployment line. Lord Carrington, who joined the Grenadier Guards in January 1939, wrote of the regular soldiers he encountered in his regiment: 'I think a large proportion had joined because they were hungry in civilian life; the army meant food, a roof and a job; and, once they got used to it, a job in which they could take pride as compared to the dole in which there could be no pride at all.'[25]

The Military Training Act, passed in April 1939, required all British men aged 20 and 21 who were physically capable of doing so to undertake six months' military training. When war broke out, however, the British Army could only muster less than a million men: the French raised 5 million. The National Service (Armed Forces Act) was passed in September 1939 to increase the number of men in uniform. The terms of this Act were almost identical to the Military Service Act of January 1918, making all able-bodied men between the ages of eighteen and 41 liable for conscription, with a provision that single men would be called up before married men.

Following the outbreak of the Second World War in September 1939, British women voluntarily entered military service as auxiliaries (the army had its Auxiliary Territorial Service, the navy the Women's Royal Naval Service and the air force the Women's Auxiliary Air Force). Many women also served as nurses with the services. Although prohibited from a direct combat role (the Soviet Union was the only nation during the Second World War to employ women in combat roles), the women auxiliaries served as drivers, switchboard operators, signallers, searchlight operators, photo-reconnaissance interpreters, plotters for fighter units, and with anti-aircraft batteries (women served as range-finders and fire coordinators but were prohibited from firing the guns). But not enough women volunteered to satisfy the pressing need for personnel.

The National Service Act of December 1941 resulted in Britain becoming the first country to order a general female conscription (the Soviet Union later followed suit). All single women between the ages of 20 and 30 (with the exception of mothers of children under fourteen) who were not already in essential work were affected by this act. These women were given the choice of enlisting in the auxiliary services, the civil defence organisation or the Women's Land Army (as agricultural workers), or being placed in government-approved jobs in one of the war industries, thereby releasing men for active duty. Conscription was later extended to include married women, though exemptions for pregnancy and for women raising young children remained in place. In August 1942 conscription was extended to include women aged up to 45, and fire warden duty was made compulsory for all women aged 20 to 45 (less those with children under the age of fourteen), unless employed for a minimum of 55 hours a week. Some 7.5 million British women were mobilised during the Second World War, of whom 450,000 were conscripted into the armed forces. By 1943, 375,000 women were serving with the Civil Defence Service (a quarter of its strength), 36,000 were ambulance drivers and 73,000 were in the fire services. By 1944, 80,000 women were serving with the Women's Land Army. The remainder of the mobilised women served in the war industries or as fire wardens.[26]

Although conscripted women successfully replaced many of the men who had left jobs in vital war industries to serve with the armed forces, women were physically unsuitable for work in certain industries, such as mining. At the start of the war, many coal miners were conscripted into the armed forces, and vacancies in cleaner and less dangerous occupations meant there were few volunteers to go down the mines. By mid-1943, the coal mines had lost 36,000 workers and with the coming of winter, shortages of coal were beginning to affect the war effort. In December 1943, the Minister for Labour and National Service, Ernest Bevin, announced a scheme whereby 10 per cent of all new conscripts aged eighteen to 25, chosen at random, would be sent to work in the coal mines. Colloquially known as 'Bevin Boys', some 48,000 young men served in the mines under this scheme, which continued until 1948.

Conscription continued after the end of the Second World War to provide the manpower needed to fulfil occupation duties in Germany and Japan, as well as maintain garrisons throughout the British Empire. The National Service Act of July 1947 legislated a period of one year's full-time service in the armed forces, followed by five years in the reserves, to come into effect from 1949. But the emergence of the Cold War, coupled with the Malayan Emergency, resulted in the passage of the National Service Amended Act in December 1948, under which from 1 January 1949 all eighteen-year-olds – less those living in Northern Ireland, who were excluded from the provisions of National Service to avoid exacerbating the existing civil unrest – were expected to serve full-time in the armed forces for eighteen months (the increased period of service enabled more effective use of conscripts in overseas theatres). Following their full-time service, the National Servicemen would remain in the reserves, such as the Territorial Army, for another four years, where they would be subject to an immediate recall in the event of a national emergency.

In response to the manpower demands of the Korean War, in 1950 the period of full-time duty for National Servicemen was increased to two years, though the period to be spent in the reserves was reduced by six months as an offset. The period of full-time service remained at two years until the end of National Service. The majority of National Servicemen went into the army, and by 1951 they constituted half the force. Unlike the military draft in the United States, few exemptions were granted for National Service, with the men being traced through their National Health records to ensure that no one had skipped registering with the Ministry of Labour and National Service. By the end of the 1950s, Britain's withdrawal from her colonies and the end of occupation duties in Germany and Japan made a large conscript army unnecessary. The Defence Review of 1957 identified the need for a transition to a rapid deployment force staffed with professional soldiers rather than conscripts. National Service officially ended on 31 December 1960, though the last National Servicemen were not discharged from the army until May 1963. Since the end of the Second World War, some 1.3 million men had undertaken National Service, 395 of whom were killed on active service (the Malaya, Korea, Suez and Aden conflicts). The British military has been an all-volunteer force since the end of National Service.

Commonwealth recruits

In 1878, the British politician W.E. Forster argued that the British government was relying on 'Gurkhas, Sikhs and Mussulmen to fight for us' rather than upon 'the patriotism and spirit of our own people'.[27] Adding credence to Forster's claim is the fact that in the half-century before 1914, Indian troops had served in more than a dozen imperial campaigns, ranging from China to Uganda.

Although the glory days of the British Empire had long since passed, the British Army at the end of the 20th century had an ever-increasing proportion of foreigners serving in its ranks. A vibrant British economy had reduced the number of Britons seeking a career in the armed forces, their place being taken by young citizens of the Commonwealth attracted by the pay and conditions, as well as by the chance of gaining British citizenship. One such soldier was Private Johnson Beharry, a citizen of the Caribbean nation of Grenada, who served with the 1st Battalion, Princess of Wales's Royal Regiment. On 18 March 2005 Beharry was awarded the Victoria Cross for saving members of his unit from ambushes on 1 May and 11 June 2004 in Al Amarah, Iraq. In his autobiography, *Barefoot Soldier*, Beharry wrote of his reasons for enlisting in the British Army: 'If I joined the army it would solve my problems at a stroke. I can remain in the UK. I might even get a British passport. I'll also get a reasonable wage.'[28]

The difficulty in attracting British nationals to serve in the armed forces in the mid-1990s, combined with falling retention levels, resulted in the raising of Gurkha reinforcement companies to be attached to under-strength infantry battalions. Unlike members of the British Army in general, the vast majority of soldiers in the Brigade of Gurkhas serve for a minimum of fifteen years and so the Gurkhas were able to help fill the depleted mainstream regiments.[29] A maximum of three Gurkha Reinforcement Companies served at any one time and temporarily supplemented the strength of some of the British Army's most distinguished infantry regiments.

In 1998, recruitment shortfalls resulted in the removal of the restriction requiring citizens of Commonwealth nations to have lived in the United Kingdom for at least five years before applying to join the British Army. The number of soldiers drawn from Commonwealth nations consequently soared from just over 200 to 6,600 within a decade. Indeed, so great was the influx of foreign soldiers that the Army General Staff put forward a proposal that their numbers be capped at 10 per cent of the total strength to protect the 'Britishness' of the army.[30] The largest foreign contingents were drawn from Fiji, Jamaica, South Africa and Ghana. In January 2001, Tim Collins noted that when he took over command of the 1st Battalion of the Royal Irish Regiment a total of nineteen nationalities were represented in the unit.[31] In 2004, when the 1st Battalion of the Princess of Wales's Royal Regiment deployed to Iraq, it had 116 foreign soldiers serving in its ranks, 20 per cent of its total strength.[32]

Recruitment in the American Civil War

In Great Britain, with its tradition of *noblesse oblige*, there is a long-standing expectation that the sons of wealthy and powerful families will serve in the armed forces. This is not generally the case for the equivalent segment of society in the United States.

During the American Civil War, the one-year enlistment of nearly half the troops in the Confederate forces was scheduled to end in mid-1862. To forestall this potentially disastrous weakening of the Confederate Army, on 16 April 1862 the Confederate Congress enacted the first conscription law in American history. Yet not all eligible men would be required to serve. One option for avoiding service was to hire a substitute from the pool of persons not liable for duty, such as men outside the specified age group or immigrant aliens. Additionally, within a week the congress passed a supplementary law exempting from the draft men employed in certain occupations, such as miners, teachers and the clergy, who were considered vital to maintaining the war effort.

The option of substitution was much more socially divisive than that of occupational exemption, as the wealthy could buy their way out of the army regardless of whether or not they were assisting the war effort. This issue gave rise to the bitter slogan: 'A rich man's war but a poor man's fight.' Because of a sustained public outcry, the Confederate Congress repealed substitution as an option for avoiding the draft in December 1863, making those who had previously engaged substitutes liable for conscription. However, the congress had earlier added a provision that exempted from the draft one white man (euphemistically termed an overseer) for every plantation with twenty or more slaves. This exemption was a blatant concession in favour of the plantation class that dominated southern society.

By early 1863 the Union Army had reached the same impasse that the Confederate forces had faced the previous year. 'The men likely to enlist for patriotic reasons or adventure or peer-group pressure were already in the army', writes James McPherson in *The Battle Cry of Freedom: The Civil War Era*. McPherson concludes:

> War weariness and the grim realities of army life discouraged further volunteering. The booming war economy had shrunk the number of unemployed men to the vanishing point. The still tentative enlistment of black soldiers could scarcely begin to replace losses from disease and combat and desertion during the previous six months.[33]

Congress responded by passing the Enrollment Act in March 1863 and establishing the Provost Marshal Bureau in the War Department to enforce conscription. Provost marshals were sent to each congressional district to enrol men eligible for the draft. These men formed the manpower pool from which each district would be required to fulfil its quota (four calls for new troops were made after the passage of the Enrollment Act). Initially a call for volunteers was made in the district. If not enough volunteers came forward, the remaining positions in the district's quota were filled by draft-eligible men chosen via a lottery. Daniel Crotty recounted the emotional scenes arising from the holding of the draft lottery in his memoirs of the Civil War.

With anxious faces the wheel commences to revolve, and those who are drafted have their names announced. Once in a while a poor fellow, when he hears his name, staggers to the door and makes his way to his humble home, that is soon to be left fatherless, to inform his loving wife and darling children of his bad luck in the wheel. They fall on his neck and weep as though their hearts would break at the loss of their only mainstay in this life. Oh, what misery this cruel war has spread all over the land.[34]

More than one fifth of the men chosen in the draft lotteries 'failed to report', many fleeing to the western territories or Canada. Many others were exempted from service because of a physical or mental disability, or because they were the sole means of support for a family member (such as a motherless child or an indigent parent). The remainder of the men selected for the draft had two further options if they wished to avoid service. As in the south, they could hire a substitute, which exempted them from any future draft calls, even if their substitute died or deserted the next day; or they could pay a commutation fee of $300, which exempted them from the current but not future draft calls. Substitutes were drawn from eighteen- and nineteen-year-olds and from immigrants who had not filed for citizenship; that is, men who were not liable for conscription. Andrew Carnegie, whose domination of the steel industry would later make him one of the richest Americans of the 19th century, received his draft notice in the summer of 1864. Not wishing to leave his comfortable existence as an official of the Pennsylvania Railroad, Carnegie hired a substitute through a Pittsburgh draft agent. Instead of joining the Union Army, Carnegie paid the draft agent $850 and an Irish immigrant served in his place.[35]

Of the 207,000 men who were drafted for service with the Union Army, 74,000 furnished substitutes and 87,000 paid the commutation fee. This left only 46,000 who actually joined the army, though the threat of being drafted undoubtedly encouraged others to enlist voluntarily (volunteers received a bonus for enlisting). In the Union states commutation proved even more divisive than substitution did in the Confederacy, particularly for those who could afford neither, as $300 amounted to almost a year's wages for an unskilled labourer. Widespread discontent with the draft system helped fuel the riot that broke out in New York City on 13 July 1863. The riot raged for four days and left at least 105 people dead. Draft offices were attacked, as were well-dressed and by implication wealthy men, who were hounded down the streets by cries of 'there goes a $300 man'. The police were unable to regain control of the streets, so the authorities brought in some army regiments from Pennsylvania. These regiments poured volleys of fire into the rioters, breaking up the mobs. By 17 July the riots were over. Congress eventually repealed commutation as a draft option in July 1864.[36]

US conscripts in the world wars

Following America's entry into the First World War in April 1917, many men volunteered to serve in the armed forces. But the manpower demands of the American Expeditionary Force (AEF) necessitated the return of the draft. Under the provisions of the Selective Service Act of 1917, all men between the ages of 21 and 30 were required to register for the draft by 5 June 1917. Twenty-four million men did so, of whom some 2.8 million were drafted, the majority (80 per cent) being drafted in 1918. Two thirds of the men who served in the AEF were draftees. Exemptions were permitted, usually on the basis of occupation (for example, 100,000 shipbuilders were given a blanket exemption) or for physical disabilities, while students could defer their service for up to three years while completing their studies. Exemptions and deferments were determined by local draft boards operating under federal guidelines. No legal alternative to serving in the military was provided for conscientious objectors and many were imprisoned.[37]

As had occurred during the Civil War, the rich and powerful who were inclined to do so avoided the draft. Joseph P. Kennedy (father of John F. Kennedy) secured an occupational exemption by leaving his job as the president of a bank for a lower-paid position as an administrative manager at a shipyard. But not all men in such a position took this route. For example, all four sons of Theodore Roosevelt, President of the United States from 1901 to 1909, served with the AEF during the war. The youngest son, Quentin, served as an Army Air Corps pilot and was shot down and killed on 14 July 1918 during the Second Battle of the Marne. Moreover, of the two sons of Roosevelt's successor as President, William Howard Taft, one tried to join the US Army but was rejected because of his poor eyesight, while the other was accepted into the artillery and served in France.

The draft lapsed after the 1918 Armistice and was not reinstated until September 1940, when President Franklin D. Roosevelt signed into law the Selective Training and Service Act. This Act resulted in America's first peacetime draft. The Act formally established the Selective Service System as a federal agency and required all eligible males to register for the draft. Registration day was 16 October 1940, upon which 16 million American men signed up. The first draft notices were issued two weeks later. In January 1945, Roosevelt proposed in his annual State of the Union Address that nurses be drafted to overcome the critical shortage of nursing staff in Western Europe. A bill to draft nurses for overseas war service was passed in the House of Representatives and came within one vote of passing in the Senate.

Between 1940 and 1945, approximately 10 million men were drafted into the American armed forces (draftees comprised two thirds of the US military in the Second World War), the majority in 1942 and 1943 – almost twice as many men were drafted in 1943 as would be drafted during the ten-year Vietnam period. Of the 10 million men drafted, roughly a quarter were trained for ground combat.

A provision in the Selective Training and Service Act allowed conscientious objectors (COs) to undertake alternative service with the Civilian Public Service (CPS). This initiative sought to provide a legal alternative to military service for COs, thereby sparing them the persecution and imprisonment suffered by COs during the First World War. The first lot of COs arrived at the CPS camps in 1941. Some would remain until released in 1947 – two years after the war ended.

As in the First World War, draft exemptions were allowed for workers in essential war industries and for men involved in agricultural production. College students were initially granted deferments, but as the manpower needs of the army increased, student deferments were restricted to men studying for careers in engineering, science and medicine. By 1944, the male college population was less than one third of its pre-war level. The manpower needed to fulfil occupation duties in Japan and Germany resulted in the draft continuing into 1946; though the length of service was set at eighteen months following the end of the war (previously men had been drafted for the duration of the war plus six months).

Many of America's rich and famous men served in the armed forces during the Second World War, though this service did not necessarily entail them being exposed to combat. Joe DiMaggio, the nation's pre-eminent baseball player, enlisted in the army in February 1942. He spent the war playing exhibition games for the troops and the public. Most other major league players of note had a similar wartime service career: only two major leaguers were killed in combat. Likewise, famous boxers, such as Joe Louis and Sugar Ray Robinson, spent their time in the army giving exhibition bouts for army camps and hospitals. But not all professional sportsmen spent their time in the military staging exhibition matches. Ted Williams, one of baseball's foremost hitters, volunteered to become a naval aviator, though he spent the war undergoing training or as a stateside instructor.[38]

Many prominent actors, such as John Wayne, were granted a draft deferment because the output of Hollywood was deemed essential for maintaining wartime morale. Wayne was initially granted a 3-A deferment (dependency or hardship) because of the need to support his family, but this was later changed to a 2-A: deferred because his civilian occupation as an actor was considered to be in the national interest. Wayne spent the war making movies and undertaking United Service Organisations (USO) tours. Other actors made films for the War Department, such as Ronald Reagan, who narrated training films for bomber pilots, and Clark Gable, who made films about aerial gunnery (though he volunteered to fly on at least one, and possibly as many as six, combat missions over Europe and was awarded the Distinguished Flying Cross). But again there were exceptions.

James 'Jimmy' Stewart, who in 1941 had received the Academy Award for Best Actor for *The Philadelphia Story*, volunteered to join the Army Air

Corps (he had to take the physical twice after he was initially rejected for being ten pounds underweight). Stewart enlisted in the army in March 1941. As a bomber pilot, he flew twenty combat missions with the Eighth Air Force over Europe. He was awarded the Distinguished Flying Cross with Oak Leaf Cluster (designating a second award) and the French Croix de Guerre. Tyrone Power, who in 1939 was the number two box office star in America, enlisted in the US Marine Corps in August 1942. He became a pilot and flew in the Iwo Jima campaign. Henry Fonda, who had been nominated for the Best Actor Academy Award in 1940 for *The Grapes of Wrath*, served as an officer in the US Navy in the Pacific theatre and was awarded the Bronze Star.

Eight sitting members of Congress served in the armed forces during the Second World War. In July 1942 Roosevelt ordered all members of Congress then serving with the military to return to their legislative duties. Four did so, including Congressman Lyndon B. Johnson (later to become the 36th President of the United States), who had been awarded the Silver Star for gallantry in action. Four chose to quit Congress to continue fighting. Additionally, in 1944, Senator Henry Cabot Lodge, who had left the armed forces following Roosevelt's 1942 directive, resigned from the Senate to resume active military service.

President Roosevelt's four sons all served in the armed forces during the Second World War. Furthermore, Marine Sergeant Peter G. Saltonstall, son of the senator for Massachusetts; Lieutenant Peter G. Lehman, son of the governor of New York; and Joseph P. Kennedy Jr., son of the US ambassador to the United Kingdom, were all killed in combat during the war.

Korea, Vietnam and dodging the draft

Congress, acting on the wishes of President Truman, allowed the Selective Service Act to expire on 31 March 1947 and the military reverted to voluntary enlistment. Within a year the military had fallen below the 2 million men level considered essential by the Department of Defense, so in 1948 Congress passed a new Selective Service Act. This Act was due to expire on 30 June 1950 but shortly before its expiration date was extended until July

1951, following the North Korean invasion of South Korea and the United States being committed to the Korean War. In June 1951 Congress again extended the provisions of the Selective Service Act, which now became known as the Universal Military Training and Service Act (renamed the Military Selective Service Act in 1967). Over 1.5 million men were drafted during the Korean War period (1950–53).

Educators, concerned about the national 'brain drain' caused by the depletion of college populations during the Second World War, lobbied for college students to be granted a deferment from the draft based on their academic performance. By 1950, college students could avoid the draft by ranking in the top half of their class or by scoring well on an aptitude test. Within a year more than three-quarters of the nation's male undergraduate students had deferred or were exempted from the draft.

Between 1954 and 1964 draft quotas were low. The government's policy was to draft all registrants who were classified as available immediately for military service (classification 1-A) so as to be able to justify their claim of a universal obligation of military service. But the size of the 1-A pool was controlled by liberalising the rules for exemptions and deferments. Deferments for postgraduate study were introduced, hardship and dependency deferments were extended to include fatherhood and, by raising the required medical entry standards, pre-induction physical rejection rates soared. But the maturing baby-boomer generation threatened to overwhelm the system and jeopardise the notion of the draft as a universal obligation. The solution adopted by the Selective Service was to place all married men in a low-priority 1-A category.

The escalation of the American military commitment to the Vietnam War resulted in draft calls rising from fewer than 10,000 per month in 1964 to more than 30,000 per month in 1966. Even with the baby-boomers now entering draft-eligible age, not enough men were being drafted under the existing guidelines. Therefore the system of exemptions and deferments that had evolved over the past decade began to be repealed. Occupational exemptions were minimised, most draft boards no longer gave deferments to married men, and fatherhood deferments were limited to those who had not previously been granted a student deferment. Additionally, the

Selective Service limited student deferments to those maintaining respectable grades or who had scored highly on the designated aptitude test. Three quarters of a million students took this test in 1966 and 1967, and many who scored poorly were reclassified and drafted. Eventually the aptitude test was abandoned and educational deferments were made available only to college students who maintained respectable grades. During this period, college grades were half-jokingly referred to as 'A, B, C, D, and Nam'.[39]

In the period from 4 August 1964, when the Gulf of Tonkin Resolution marked America's formal entry into the Vietnam War, to 29 March 1973, when the last American troops left Vietnam, 26.8 million American men came of draft age (women were excluded from the draft). 8.7 million men voluntarily enlisted (though many of these did so because of the threat of the draft, as volunteers were usually able to pick the branch in which they served) and some 1.8 million men were drafted into the army and the Marine Corps, the overwhelming majority of draftees entering the army. Of these men, 648,500 (35 per cent) served in Vietnam, where 17,725 draftees were killed (less than 1 per cent of those drafted). Draftees comprised a quarter of the total American military forces that served within the borders of South Vietnam (2.6 million) and accounted for 30 per cent of the total number of American deaths in this conflict.[40]

In contrast to the pattern of service during the First and Second World Wars, few members of the upper echelons of American society served in the military during the Vietnam War. In *Chance and Circumstance: The Draft, the War and the Vietnam Generation*, Lawrence Baskir and William Strauss write:

> The draft was not, however, an arbitrary and omnipotent force, imposing itself like blind fate upon men who were powerless to resist. Instead, it worked as an instrument of Darwinian social policy. The 'fittest' – those with background, wit, or money – managed to escape. Through an elaborate structure of deferments, exemptions, legal technicalities, and noncombat military alternatives, the draft rewarded those who manipulated the system to their advantage.

Among this generation, fighting for one's country was not a source of pride; it was misfortune. Going to Vietnam was the penalty for those who lacked the wherewithal to avoid it. A 1971 Harris survey found that most Americans believed that those who went to Vietnam were 'suckers, having to risk their lives in the wrong war, in the wrong place, at the wrong time'.[41]

The draft caused bitter divisions in American society, though these were largely along socio-economic rather than racial lines. By 1965, African-Americans comprised 31 per cent of all combat troops in the army and accounted for approximately a quarter of the men killed in action. The Defense Department, wary of being accused of racially targeted recruiting and mindful of the burgeoning civil rights movement, embarked on a campaign to reduce the percentage of combat troops belonging to minorities. By 1966, only 16 per cent of combat troops were drawn from minorities (a reduction of almost 50 per cent in a year). By 1970, the figure was 9 per cent.

While racial disparity among combat troops was somewhat addressed, social and economic inequalities never were. Poorly educated, low-income whites and poorly educated low-income African-Americans constituted a disproportionate share of the troops who served in Vietnam. Survey after survey found that men from disadvantaged backgrounds were much more likely to serve in the military, go to Vietnam and see combat than their better-off peers. Paul Fussell, who was a professor at an American university throughout the Vietnam conflict, tells us that:

> The class system was doing its dirty little work quite openly and nobody seemed to care. Nor were the rest of us morally clean. Early in the war, I had written disingenuous letters testifying to the deep religious convictions of sons of middle-class friends of mine to keep them out of the army, and later I was perfectly happy to see many of my students flee to Canada, leaving the less-fortunate boys ... the job of pursuing America's misbegotten course in Southeast Asia. I never knew anyone

whose son had been killed, wounded, or even badly inconvenienced by the war.[42]

A viewpoint affirmed by James McDonough, who served with the 173rd Airborne Brigade during the Vietnam War. McDonough wrote that:

The military draft was never concerned with equity. Its sole purpose was to obtain the required numbers of men with sufficient mental and physical qualities to do the job. Since the pool of men from which to draw was much greater than the numbers needed, a natural selection process allowed the educated and the privileged to avoid the draft altogether.[43]

Among the draft-eligible men of the Vietnam era, approximately 9 million avoided the draft through deferrals or exemptions. Undergraduate students were granted deferments for most of the war, though President Nixon abolished these in December 1971, by which time American involvement in the war was almost over. Initially postgraduate students were also granted deferments, but these were abolished in February 1968, with the exception of students attending divinity schools or studying in medical fields. Students at divinity schools remained exempt from the draft throughout the war, which had the effect of producing a notable increase in applications for these schools post-February 1968. Student deferments did not provide a permanent escape from the draft. It was almost always necessary for former students to obtain some other form of deferment or exemption to cover the period between graduation and when they reached the age of 26 and their vulnerability to the draft ended.

Altering one's family circumstances provided another way to avoid the draft. Military pay was low throughout the war and most draft boards gave hardship deferments to registrants who were the sole source of income for widowed mothers, younger brothers and sisters, or other dependants. Registrants could therefore qualify for a hardship deferment by inviting a disabled or elderly relative to live with them. Those applying for a hardship deferment were not subjected to a means test, and these deferments were made available to the poor and wealthy alike. For example, the actor George Hamilton received a hardship deferment because his mother lived in his Hollywood mansion and relied on his $200,000 annual income for support.[44]

Most draft exemptions were based on a physical, mental, psychiatric or moral defect of the registrant. Every prospective draftee had to be examined twice: once before the draft board could classify him 1-A, and again when he reported for induction. Pre-induction physicals provided various opportunities for those wishing to avoid being drafted. A common tactic was to gain or lose enough weight to fall outside the military standards (weight based on height), while others aggravated existing injuries or subjected themselves to a permanent though minor injury, such as slicing off half a thumb or being punched in the nose hard enough to cause a deviated septum. Engaging in petty criminal behaviour, being, or appearing to be, homosexual, or belonging to a subversive organisation would usually result in an exemption on the grounds of having a 'questionable moral character'. Family doctors could help by giving extra tests to find heart murmurs and other minor conditions that qualified registrants for a medical exemption. Many professional athletes qualified for a medical exemption because of their history of bone or joint injuries, such as Joe Namath (one of the National Football League's greatest-ever quarterbacks) who was exempted because of the condition of his knees, along with thousands of other former high school footballers. The law also provided an exemption for anyone under orthodontic care; thus getting braces on their teeth was a common last-minute tactic for registrants facing an immediate call-up.[45]

A million men avoided service in the active military by enlisting in the National Guard or the Army Reserve. At the end of 1968, with the draft still in full force, the Army National Guard had a waiting list of 100,000. After two years of shrinking draft calls the waiting list had vanished. Six months later the Guard was 45,000 men under strength. Many draft-vulnerable men who became guardsmen or reservists were able to leapfrog other men on the waiting list by having influential persons nominate them to fill the rare vacancies. Many professional athletes had billets in the National

Guard arranged for them – the Dallas Cowboys had ten players assigned to the same National Guard division at one time. Because of the inherent difficulty of getting a position with the National Guard during much of the Vietnam era, joining the Guard was not a viable option for many men from disadvantaged backgrounds. A disproportionate number of reservists were college graduates and only about 1 per cent of reservists were black.[46]

Guardsmen and reservists incurred an initial four- to six-month full-time duty obligation, plus yearly summer camps and monthly unit meetings extending over a six-year period. If they failed to show up for the camps and meetings they were technically absent without leave (AWOL) and subject to immediate call-up for active duty, but this was seldom enforced. For the first three years of the Vietnam War President Johnson, conscious of the political repercussions of calling up the often well-connected members of the reserve component, refused to activate guardsmen or reservists. Instead the government relied on increased draft calls to fill the demands for troops. In 1968, in the wake of the Tet offensive, 37,000 guardsmen and reservists were called to active duty, principally as a symbolic gesture. About 15,000 National Guard members were sent to Vietnam, 97 of whom were killed. As the United States began to decrease its military commitment to the war, the public pressure to activate reservists eased. No additional guard or reserve units were sent to Vietnam after 1968.

The draft exemption category that usually entailed the greatest personal hardship was that of being a conscientious objector (CO). In the Vietnam era approximately 172,000 men applied for and were granted an exemption from the draft based on their moral or religious beliefs. In 1965 the US Supreme Court declared that CO status had to be granted to all confirmed pacifists, regardless of their religious background. It was up to the applicant, with the assistance of supporting statements from his local priest, family friends etc., to convince the local draft board of his sincerity. To qualify for a CO classification the applicant had to oppose all wars – not just the Vietnam War. Most successful CO applicants were required to undertake two years' alternative service in the civil community, many working in low-paying jobs such as hospital orderlies or as construction workers in wilderness camps, though some COs served in the military in non-combatant roles.

By 1969 the military's manpower demands for service in Vietnam had begun to ease, draft calls had peaked, and the number of men eligible for the draft was increasing. Congress and the public were unwilling to let the Selective Service Agency keep the draft system in balance by liberalising deferments and exemptions, as had occurred in the period between the Korean and Vietnam wars. The system was now so discredited that the public demanded a fundamental restructuring. This resulted in the reintroduction of a draft lottery (the last draft lottery had been in 1942) to determine the order of call-up for induction.

The vulnerability of men to being drafted via the draft lottery was limited to one drawing. The potential pool of draftees was restricted to those turning nineteen and those aged under 26 who were coming off deferments. The draft lottery was a major change from the previous call-up system, whereby the oldest eligible man was drafted first, the theory behind this being that the younger men would remain eligible for subsequent call-ups. In 1973, following the American withdrawal from Vietnam, the draft was ended and the military converted to an all-volunteer force. No new draft orders were issued after 1972 and the last draftee to be inducted (from the 1971 draft lottery) entered the army on 30 June 1973.

Sixteen million young American men avoided serving in the active military during the Vietnam era through deferment, exemption or resistance. For each man the draft posed a deeply personal decision: what would they do when they came of age? Should they enlist, should they wait to be drafted, should they try for a medical deferment, should they seek a position in a Reserve Officer's Training Corps (ROTC) unit or in the reserves, or should they go into exile? Approximately 50,000 Americans (30,000 draft dodgers and 20,000 deserters) chose exile to escape military service, some living abroad for a decade or more. Canada was the most popular destination (as well as being the closest), chosen by approximately 30,000 of the draft evaders and deserters, with another 1,000 fleeing to Sweden. Despite official overtones of conciliation and forgiveness (in September 1974 President Gerald Ford established a clemency programme for those who fled abroad), many draft evaders were left with a lasting feeling of guilt about the men who went to Vietnam in their place.

Project 100,000

As noted, the escalation of the US military presence in Vietnam in 1966 required an increase in the number of men inducted into the military via the monthly draft calls. As the better-educated and wealthier elements largely avoided the draft, it fell upon some of the most disadvantaged segments of US society to provide the required manpower. In October 1966, the Secretary of Defense, Robert S. McNamara, launched Project 100,000. Under this scheme the Department of Defense would be forced to accept men who would have previously been rejected as being mentally or physically substandard. Men inducted through this programme would euphemistically be known as 'New Standards Men'. McNamara intended to admit 40,000 men during the first year to trial the concept, followed by 100,000 each year thereafter – hence the project's name – if the trial proved successful. Project 100,000 was terminated in December 1971 because of decreasing manpower requirements as the American military presence in Vietnam wound down. By that time 354,000 men had entered the military through this programme; over half of them were volunteers, the remainder being draftees. The volunteers largely entered the navy, air force (who did not accept draftees) and Marine Corps. Those who were drafted went into the army, with some entering the Marine Corps.[47]

Approximately one tenth of all enlisted recruits each year were Project 100,000 men. About half of these men came from the generally poorer southern states of the US, and about 40 per cent belonged to a minority (mostly African Americans). Almost all of these men were admitted under lowered aptitude or educational standards. The remainder (less than 10 per cent) were enlisted with temporary physical defects, such as being grossly overweight, seriously underweight or requiring minor surgery (for example, to treat a hernia or an undescended testicle). Only volunteers were accepted under the lowered medical standards. Over half of the Project 100,000 men had not graduated from high school and their median reading ability was at a sixth-grade level, with one in ten reading below the fourth-grade level. The well-publicised lowering of the military's entry standards resulted in the Project 100,000 men receiving the epithet of 'McNamara's Moron Corps'.[48]

Project 100,000 was an element of President Johnson's 'War on Poverty'. The stated primary goal of the project was to provide a means of upward mobility for economically and educationally disadvantaged men through the inherent training, discipline and socialisation of military service (roughly half the men who entered the services through this programme were unemployed when inducted). Other stated aims were to assist in meeting the military's manpower needs and to achieve greater equity by spreading throughout society the opportunities and obligations of military service, though this increased equity seemed to be directed towards the lower end of the socio-economic spectrum.[49]

Although the military's entrance standards were lowered, its performance standards were not revised. The Project 100,000 men had to meet the same criteria as regular enlistees to graduate from training courses and for advancement and retention, though remedial training, particularly in literacy skills, was made available. Approximately half of the Project 100,000 men assigned to the army and the Marine Corps were allocated to combat specialisations (mainly the infantry and the artillery). Lewis B. Puller Jr. had a Project 100,000 entrant serve in his platoon of Marines in Vietnam. Puller noted that the man was 40 years old, completely bald and 'seemed to me somewhat mentally deficient'. Puller would later remark: 'I had a hard time figuring out how his skills with a machine gun were going to help him earn a living after the Marine Corps'.[50]

The rest of the Project 100,000 men were assigned to areas that did not require extensive technical training, such as cooks, drivers, supply clerks etc. (though these military specialisations did have civilian equivalents and so provided transferable skills). A report to Congress in February 1990 noted that, when compared to their civilian counterparts from similar educational and socio-economic backgrounds, being in the military provided little, if any, advantage to the men enlisted under Project 100,000.[51] In *Long Time Passing: Vietnam and the Haunted Generation*, Myra MacPherson comments: '[T]he program provided political advantages for President Johnson. By the systematic, deliberate drafting of those with marginal minds and lives, the President was able to avoid the politically incendiary action of ending student deferments or calling up the reserves.'[52]

In his autobiography, *A Soldier's Way*, General Colin Powell wrote bitterly of how the offspring of the privileged had avoided combat in Vietnam, describing the American draft policies as an 'antidemocratic disgrace'.

> I can never forgive a leadership that said, in effect: These young men – poorer, less educated, less privileged – are expendable (someone described them as 'economic cannon fodder'), but the rest are too good to risk. I am angry that so many of the sons of the powerful and well placed and so many professional athletes (who were probably healthier than any of us) managed to wrangle slots in Reserve and National Guard units. Of the many tragedies of Vietnam, this raw class discrimination strikes me as the most damaging to the ideal that all Americans are created equal and owe equal allegiance to their country.[53]

The modern US military: a narrowing recruitment base

The under-representation of the American economic, cultural and political elite in the US military that became apparent during the Vietnam era was exacerbated by the change to an all-volunteer force in 1973. Nowhere is this more evident than in the decline in the number of veterans serving in Congress. The 110th Congress (which first sat in January 2007) had 130 veterans (including those with service in the Reserves) among its 540 members, nine fewer than the 109th Congress and 22 fewer veterans than the 108th Congress. By way of contrast, in the 1970s approximately 70 per cent of congressional members were veterans.

The current pitch of US military recruiters is that the military will provide applicants an opportunity to better themselves, either through educational opportunities that would have been too costly if they had remained civilians, or through personal development. This pitch is not often directed at the graduates of the Ivy League colleges or the sons and daughters of business and political leaders. Rather, it is aimed at the children of the lower-middle and working classes. Men like Joshua Key, who was raised in a two-bedroom trailer in rural Oklahoma and who went on to fight in Iraq. 'Before I had even graduated from high school, a string of army recruiters started showing up at our trailer, banging on the flimsy door that blew open on windy nights, promising health insurance and higher education in exchange for military service,' noted Key. 'They were smart men, those recruiters. They didn't waste time at the doors of doctors and lawyers, but came straight for me.'[54] Further evidence of the reluctance of America's elite to serve is provided by the demographics of the armed forces that fought the combat phase of Operation Iraqi Freedom in mid-2003.

Over 600 professional football players served in the US armed forces during the Second World War, and nineteen were killed while on active service. Only one US professional athlete was killed while serving with the military during the Vietnam War: James Robert Kalsu. First Lieutenant Kalsu, a former member of the Buffalo Bills (Buffalo, New York), enlisted in the army after the end of the 1968 football season to fulfill his ROTC obligation. Kalsu was trained as an artillery officer and deployed to Vietnam in November 1969. He was killed by North Vietnamese mortar fire at Firebase Ripcord in the Ashau Valley on 21 July 1970 while serving with the US Army's 101st Airborne Division.[55]

Only one former professional athlete, Pat Tillman, served with the US armed forces in Afghanistan and Iraq in 2003–04. Tillman was killed in Afghanistan on 22 April 2004 (by 'friendly fire') while serving with the US Army's 75th Ranger Regiment. Senator John McCain commented in his eulogy for Tillman that 'it was his uncommon choice of duty to his country over the profession he loved and the riches and other comforts of celebrity, and his humility that make Pat Tillman's life such a welcome lesson in the true meaning of courage and honor'.[56] Tillman chose to leave a multi-million-dollar career in the National Football League for a relatively low-paying job as an army ranger. It was so unexpected that a member of the American socio-economic elite would choose to give up a financially rewarding career to serve in the military that Tillman became an unwilling national celebrity.

Of the 535 members of Congress, only one, Senator Tim Johnson of South Dakota, had a child among the 150,000 servicemen and women deployed to the Iraqi theatre during the combat phase. His son, Brooks Johnson, was a staff sergeant with the 101st Airborne (Air Assault) Division.[57] When interviewed about his unique position, Brooks Johnson commented:

> I think it would be better for society if we had more people from a certain socioeconomic strata that chose military service – if not for a career, for a couple of years. But I think there's a lot of people that view things like military service – while they might think that it's essential to the nation – beneath their own abilities and talents.[58]

Coincidentally, only one senator's son served with the military in Vietnam: former Vice President Al Gore, son of Senator Al Gore Sr. of Tennessee. Gore served in a non-combatant role as a military journalist, writing articles for submission to military and hometown papers. Twenty-seven sons of House members also served in Vietnam – only one, the son of Maryland Congressman Clarence Long, was wounded. None was killed.

In *The American Soldier: A Social and Political Portrait*, the sociologist and political scientist Morris Janowitz noted that as the standard of living rises, tolerance for the discomforts of military life decreases.[59] So it is not surprising that the majority of current American servicemen and women are drawn from the lower-income demographic. In 1973, 23 per cent of the US military were drawn from minorities. In 2000 this figure had risen to 37 per cent. Currently the two greatest demographic growth areas for the US military are Hispanics and black women. In the decade to 2003, even though the total number of military personnel dropped by 23 per cent, the number of Hispanics in the military grew by 30 per cent. Furthermore, black women (who make up only 16 per cent of the American female population) now outnumber white women in the US Army.[60] A March 2003 article in the *New York Times* commented: 'A survey of the American military's endlessly compiled and analysed demographics paints a picture of a fighting force that is anything but a cross section of America. With minorities overrepresented and the wealthy and underclass essentially absent ...'[61] This sentiment is shared by Charles C. Moskos, Professor of Sociology at Northwestern University, Chicago, who concludes that:

> In World Wars I and II, the British nobility had a higher killed-in-action rate than the working class. Our enlisted ranks resemble the British: they're lower- to middle-class, working-class, intelligent people, who are joining for both the adventure and the economic opportunity. But the officer corps does not represent American nobility. These are not people who are going to be future congressmen or senators. The number of veterans in the Senate and the House is dropping every year. It shows you that our upper class no longer serves.[62]

During the First World War, France put 20 per cent of its entire population into uniform, while Germany had 18 per cent serving. More than 16 million men and women served in the US military during the Second World War, representing approximately 12 per cent of the American population of 131 million. Millions more were employed in vital defence industries. In 1968, when the Vietnam War was at its height, the active-duty US military numbered over 3.5 million, comprising 1.7 per cent of the US population (approximately 200 million). Note that this figure does not include the Reserve component, few of whom served in Vietnam. In mid-2008, the active-duty US military numbered 1.4 million, approximately 0.5 per cent of the US population of 304 million. Even when a Reserve component of 1.1 million is added, still less than 1 per cent of the US population had any possibility of being deployed to Iraq or Afghanistan. Additionally, during the Vietnam era there was the draft, which was probably more influential than any other factor in forcing mainstream America to take a personal interest in the conflict, and which was directly responsible for the growth of the anti-war movement. There is no draft in place to supply personnel to fight in Iraq and Afghanistan; so for the vast majority of Americans, the fighting and dying in Iraq and Afghanistan will remain an item of momentary interest on the news and nothing more.

Civilian and military USA: the 'Great Divorce'

There comes a point when the members of the military will begin to resent the sacrifices that they are being asked to make by a seemingly ungrateful nation. In late 2005, while researching the demographics of the American military, Kathy Roth-Douquet spoke to the families of a number of Cobra helicopter pilots who had resigned after returning from a tour of duty in Iraq. She concludes that:

> Their reasons for leaving were not the hard work, danger, and distance from their wives and children, but the fact that the same few military personnel were being asked to make sacrifices over and over again, while the rest of society went about their personal business, unaffected by what in any other era of our nation's history would have been a national effort. It seemed that only the military was at war, while the instruction to the rest of the country from the political leadership was to go shopping and travel so the airlines and the economy would keep on cranking.[63]

A telling indication of the sacrifices being asked of soldiers is that by mid-2007 over 170,000 US soldiers had been deployed to Iraq or Afghanistan more than once. Furthermore, in April 2007 it was announced that two army units were to return to Iraq without having had a break of even a year at home. A concurrent announcement extended the tours for active-duty US Army soldiers in Iraq and Afghanistan from twelve to fifteen months.[64]

The manpower demands for fighting the campaigns in Afghanistan and Iraq forced the US military to offer a range of new recruitment and retention incentives. On 3 July 2002 President Bush signed Executive Order 13,269, which made all non-citizens on active duty during the Global War on Terror eligible for immediate US citizenship. Previously non-citizens were required to serve on active duty for at least three years before they could apply for citizenship. Now they could file for citizenship upon entry into the US military. 'In my time with the US military', writes Oliver Poole, a journalist embedded with the 3rd Infantry Division during the invasion of Iraq, 'I met Filipinos, Puerto Ricans, Mexicans, even a British citizen from Glasgow, all of whom hoped their enlistment would help turn their green card into a US passport.'[65]

In September 2003 the Department of Defense assessed that approximately 37,000 active-duty soldiers in the US military were not US citizens (about 3 per cent of the active-duty strength). In addition, there were an estimated 13,000 non-citizen reservists. At that time, 3,000 non-citizens were serving with the US military forces in Iraq. A legal specialist in the US Army observed: 'They are over here for their country. They are over here fighting a war, and it's not even technically their country.'[66] Indeed, the first American service member to die in combat in Iraq was Lance Corporal José Gutierrez, a US Marine and native of Guatemala, who was killed near the southern Iraqi city of Umm al Qasr on 21 March 2003.[67] Gutierrez's family became one of the first benefactors of a Bill passed by Congress in April 2003 awarding American citizenship to non-citizen soldiers killed in combat while serving with the US military during the Global War on Terror. By March 2008 nearly 37,000 non-citizens serving with the US military had been granted citizenship since the war began; 109 of these grants were posthumous. At that time, another 7,300 service members had their requests for citizenship pending.[68]

Pentagon policy states that reservists and National Guard troops can serve on active duty for a cumulative total of 24 months, although this period can be split among multiple deployments that occur over several years. By late 2004 many reservists had reached the 24-month limit and were demobilised. However, such was the need for troops that individuals were being offered a $1,000-a-month tax-free bonus to stay on. The Chief of the Army Reserve, Lieutenant General James Helmly, commented that 'the most likely "volunteers" are those who often enjoy less responsible positions in civilian life', and that 'We must consider the point at which we confuse "volunteer to become an American soldier" with "mercenary".'[69]

Enlistment bonuses for the US Army's active-duty component were also increased in an attempt to arrest the falling number of applicants, along with large cash bonuses to entice serving active-duty soldiers to re-enlist. In 2006 the US Army and Marine Corps spent $1.03 billion on re-enlistment payments, up from $174 million in 2003 when the war in Iraq began. The US Army paid a bonus to two out of every three soldiers who re-enlisted in 2006, while in 2003 the number of re-enlisting soldiers who received a bonus was fewer than two in ten.[70]

In conjunction with bonuses, the US military also raised its upper age limit for enlistment, though the physical standards and the entry medical examination remained unchanged. To overcome declining interest in a military career among the traditional recruitment base, in January 2006 the active-duty age limit for enlistment was raised from 35 to 40, followed by a further increase in June 2006 to allow enlistment up to a maximum age of 42 (by way of contrast, the maximum enlistment age for the British Army is 33). Furthermore, in order to meet enlistment goals the US Army increased the number of high school dropouts and convicted felons recruited, giving further credence to what the journalist Arthur Hadley terms the 'Great Divorce' between the military and civilian America. Hadley defines this as 'the less-than-amicable separation of the military from the financial, business, political, and intellectual elites of this country, particularly from the last two'.[71]

Israel: an egalitarian approach

The freedom to implement voluntary enlistment for their military (the United Kingdom in 1961, the United States in 1973) is not an option available to all nations. Surrounded by hostile Arab nations, Israel was forced to become a 'nation in arms' to ensure its survival. The Israel Defense Forces (IDF) were formally established on 26 May 1948 as the nascent Jewish state fought for its very existence in the War of Independence. The Israeli public view the IDF as the one national organisation that can guarantee the survival of the Jewish state. Defence has remained Israel's highest national priority, though the state has not been directly threatened with invasion since the Yom Kippur War of October 1973. Military service is compulsory for all Israeli citizens and permanent residents.

Israel is the only country in the world that currently conscripts women into its armed forces, although the only occasion when Israeli women have served in combat was during the War of Independence (1948–49), when 108 were killed while serving with the armed forces. Following that war, if women were allocated to an Israeli combat unit they would be withdrawn before the unit went into action, though female conscripts were permitted to serve in Lebanon during the Israeli occupation (1982–85). Since 2001 women have been able to serve in combat positions in the artillery and engineering corps and as border guards. Israeli women do not currently serve in the infantry or armoured corps.

Compulsory military service has been a feature of the Israeli state since its inception, with the initial Defence Service Law being passed in September 1949. Military service is considered the primary rite of passage for entry into mainstream Israeli society, and a favourable report of military service is an important prerequisite for admission into the workforce (particularly for government positions). Reuven Gal, a former chief psychologist of the IDF, emphasises this point:

Military service in Israel is not perceived as compulsory, even though it is. It is not perceived as a penalty, even though it constitutes a major interruption in the life course of Israeli men and women. It is not considered a calamity, even though it is extremely stressful, sometimes even fatal. It is a normative part of the Israeli ethos – an integral phase in the life of any Israeli youth.[72]

The egalitarian basis of compulsory military service is an important social leveller, as citizens from all economic, social and political backgrounds share a common bond by experiencing the same hardships and having the same rights. Compulsory military service also fulfils an important role in the assimilation of immigrants into the Israeli nation (immigrants often begin their military service with a three-month intensive course of Hebrew). Until relatively recently, the military profession was considered one of the most prestigious careers in Israeli society and the military has traditionally enjoyed a high level of public support. As Israel moves away from its socialist origins, however, the financial appeal and greater personal freedom of business and the professions, rather than a military career, are increasingly attracting the Israeli urban socio-economic elite. This factor has been exacerbated by the unpopular role the IDF has been called upon to fulfil in the Occupied Territories.

The central role that the military plays in Israeli society is evident in the number of senior positions in the Israeli government that have been filled by veterans. The state of Israel has had twelve prime ministers (excluding interim prime ministers) since its formation in 1948. With the exception of Golda Meir, all have served in the military in some capacity. Two Israeli prime ministers, Yitzhak Rabin and Ehud Barak, have previously held the rank of lieutenant general and served as the Chief of Staff – professional head of the IDF – while one other, Ariel Sharon, held the rank of major general. Additionally, a number of former Israeli generals have also served as the Israeli Minister for Defense or held other ministerial appointments.

All Israeli citizens and permanent residents receive their initial draft notice at the age of seventeen. They then undergo a series of medical and psychological examinations and have their level of education verified and personal background checked in preparation for their induction into the military at the age of eighteen. Men are required to serve for 36 months (increased from 30 months following the Six Day War of 1967), women for 21 months. A small number of conscripts will choose not to be discharged at the end of their compulsory service and will join the permanent service corps, or *Keva*. The Keva is responsible for the higher command and strategic planning of the IDF, along with the administration, organisation and training of the IDF's conscript and reserve elements. The Keva and the conscripts comprise the standing peacetime military force, which are supplemented by the reserves when required.

After completing their period of full-time service, both men and women become members of the reserve, men serving up to the age of 51 (though those who have served in combat units can be discharged at age 45) and single women to the age of 38. The reserves are structured around formed units up to divisional level. Many reservists will serve in the same unit for a number of years, and thus the reserve element develops strong *esprit de corps* and extremely cohesive teams. Reserve service is set at 39 days per year, which can be extended for a national emergency. In practice, reserve units are activated as required, few women are called up for reserve service, and most men over the age of 35 are not called up because they are considered physically unfit.

In theory, military service is compulsory for all Israelis. However, several exemptions and deferments are available. Orthodox men are granted deferments while pursuing Torah studies at *yeshivas* (religious schools). Orthodox women are exempted on the basis of the Jewish tradition that does not allow unmarried daughters to stray from their father's authority or live in a mixed-gender environment outside the family home. In July 1978 an amendment was made to the Defence Service Law that made it easier for women to be exempted from military service on religious grounds. All that was now required was for a woman to attest before a judge that 'reasons of religious conviction prevent her from serving in the defence service and that she observes the dietary laws at home and away from home, and that she does not ride [travel] on the Sabbath'.[73] Women exempted from military service on religious grounds can perform one to two years' service to the nation in a civil capacity, but few do so. In 2001 women comprised approximately 30 per cent of Israeli military conscripts. About a third of eligible females are granted exemptions (double the figure for men), with an increasing number seeking exemption on religious grounds.

Members of most minority groups (such as Israeli Arabs, which includes Bedouin) are exempted from compulsory military service, though they can volunteer to serve. As the traditional enemies of Israel are the surrounding Arab states, drafting Israeli Arabs has an attendant security risk. Other grounds for exemption are not meeting the military's physical and mental/ psychological standards, though these are relatively low (when compared to other militaries), as the public's expectation is that everyone will share the responsibility for the defence of the nation. To forgo military service on medical grounds it would have to be determined that the applicant would not be able to serve in any capacity in the IDF. Married women and mothers are exempted, and if a woman gets married or pregnant while undertaking compulsory military service she is discharged. Conscientious objection to military service is not grounds for exemption.

Recently the Israeli system of compulsory military service has come under growing pressure to change. An increase in population means there is a manpower (and womanpower) glut. The notion of sharing the burden of military service equally across the nation is still strong in Israeli society, with the result that since the late 1980s the IDF has become increasingly wasteful with its manpower, particularly in the support arms. Many conscripts in non-combat units spend nine out of every fourteen days at home, while others are transferred to the police or border guards. Conversely, the high-technology aspects of military service, particularly in some of the combat arms, require lengthy periods of training to attain the required proficiency. Choosing to serve in certain combat units extends the conscript's service from three to four years so that the IDF receives a better return for the intensive training provided. However, permanent standing armies are expensive – not so conscript armies. Most Israeli conscripts are paid low salaries, while reserve pay is provided by the social security department and therefore not captured in the military budget. Consequently, there is little financial incentive to change the current system of compulsory military service.[74]

* * *

Men either volunteer or are compelled to serve in the armed forces, though there are varying degrees of compulsion, some forms of which, such as economic necessity or the threat of being conscripted, are not always obvious. Additionally, social, parental and peer expectations influence whether an individual accepts or rejects military service. In America, the burden of military service, particularly since the end of the Second World War, falls almost exclusively onto the working and lower-middle classes.

Strange to say in America, those who by reason of accumulation of property have assumed the roles of leisure class and have more or less association with that British element which supplies the scions to the

Army, Navy and the Civil Service, seldom or never consider the propriety of devoting themselves or their sons to the public service unless it be as ambassadors or ministers at foreign courts.[75]

The above comments were made in 1906 but are an accurate reflection of the pattern of military service in the United States in 2006. As at the beginning of the last century, the wealthy and the powerful still generally either avoid or do not even consider military service. To paraphrase a former Vice President of the United States, Dick Cheney, 'they have other priorities'.[76] Contrast this attitude to that of George Washington, who wrote in 1783 that: 'It may be laid down as a primary position, and the basis of our system, that every Citizen who enjoys the protection of a free Government, owes not only a proportion of his property, but even of his personal services to the defence of it.'[77]

Chapter 3

A SOLDIER'S JOURNEY

A range of emotions are manipulated by governments to encourage men to become soldiers and go off to war. Although some of the inducements to enlist are designed to appeal to man's better nature, such as defending the helpless, perhaps the most powerful is the threat of social stigmatisation – in particular, labelling men who refuse to serve in the military as cowards. One of Tim O'Brien's colleagues, who, like O'Brien, had been drafted into the US Army, confronted his drill sergeant during basic training and told him he was opposed to the Vietnam War. The sergeant just laughed and called him a coward.

> He said I was a pansy. It's hard to argue, I suppose. I'm not just intellectually opposed to violence, I'm absolutely frightened by it. It's impossible to separate in my mind the gut fear from pure reason. I'm really afraid that all the hard, sober arguments I have against this war are nothing but an intellectual adjustment to my horror at the thought of bleeding to death in some rice paddy.[1]

Another common method to induce men to enlist is to appeal to their sense of duty, with 'duty' defined as those actions required to fulfil the objectives of the government of the day. British citizens were encouraged to do their bit for 'King and Country' during the First World War, while the Germans they faced across no man's land fought for the 'Fatherland'. During the First World War, appeals targeting feelings of patriotism were often combined with attacks on the masculinity of those who had not yet signed up, the preferred manner of persuasion being the derision of such men by young women. The use of women as proxy recruiting agents was officially encouraged. A poster issued by the Parliamentary Recruitment Committee and addressed to the 'Young Women of London' stated:

Is your 'Best Boy' wearing Khaki?
If not don't YOU THINK he should be?

If he does not think that you and your country are worth fighting for
– do you think he is WORTHY of you?

Don't pity the girl who is alone – her young man is probably a soldier
– fighting for her and her country – and for YOU.

If your young man neglects his duty to his King and Country, the time
may come when he will NEGLECT YOU.

Think it over – then ask him to

JOIN THE ARMY TO-DAY.[2]

Shame and anxiety

Female advocacy of male enlistment in the armed forces was the basis of the 'white feather' campaign of the First World War. Charles Penrose Fitzgerald, a retired admiral of the Royal Navy, established the 'Order of the White Feather' on 30 August 1914, a few weeks after the start of the war. The organisation's aim was to shame men into enlisting in the British armed forces. Its modus operandi was for young women to publicly berate young men not in uniform and then place a white feather – the symbol of cowardice – in their lapel or hat band. This simple tactic was surprisingly effective. The daughter of a man presented with a white feather on his way home from work recalled:

That night he came home and cried his heart out. My father was no coward, but had been reluctant to leave his family. He was thirty-four and my mother, who had two young children, had been suffering from a serious illness. Soon after this incident my father joined the army.[3]

The white feather campaign soon spread throughout the United Kingdom and then to other countries, including Canada and Australia. The movement was notable for its zeal and its confrontational and indiscriminate methods, with little thought being given to the justification or appropriateness of the 'feathering'. There are numerous accounts of boys who were under-age or men who had returned from active duty, either on leave or to convalesce, being presented with a white feather.

I was walking down the Camden High Street when two young ladies approached and said, 'Why aren't you in the Army with the boys?' So I said, 'I'm sorry but I'm only seventeen' and one of them said, 'Oh we've heard that one before. I suppose you're also doing work of national importance.' Then she put her hand in her bag and pulled out a feather. I raised a hand thinking she was going to strike me and this feather was pushed up my nose.[4]

I recall in the early sixties my grandfather, a sergeant in the Lancashire Fusiliers, describing with undiminished indignation being handed a white feather by a young woman on a tram in London while on sick leave and wearing civvies. Time had clearly not dulled his sense of outrage.[5]

Apocryphal tales abounded of a winner of the Victoria Cross being presented with a white feather while home on leave; a coda whereby the lady in question realises her error and falls in love with the young hero was added to appeal to the more romantically inclined. But such tales had a firm basis in reality. Perhaps the most extraordinary documented incident of a feathering was the case of Sergeant Thomas Painting, 1st Battalion, King's Royal Rifle Corps. Having taken part in the retreat from Mons and the First Battle of Ypres he was captured and put in a German prisoner of war camp. Eventually he escaped and made his way back to England via Denmark, only to be presented with a white feather, having arrived home in civilian clothes.[6]

Appeals to support the war effort are most persuasive when a country is fighting a war of national survival, though the government, to suit its purposes, may manipulate the specific nature of the threat. Government propaganda is most effective when it targets a deep-seated anxiety. The 11 September 2001 terrorist attacks, for example, shattered the American people's sense of security; and although no evidence has been produced that Saddam Hussein or any Iraqi intelligence operative played a significant role in the attacks, President Bush's rhetoric clearly identified Iraq as an imminent threat to America. The anxiety that the Bush administration exploited to garner public support for its invasion of Iraq was not that of the recent past but of a generation before. The mental image with which the administration sought to dominate the debate was not an Islamic jihadist but rather a mushroom cloud. On 7 October 2002, President Bush, in a speech in Ohio, stated: 'America must not ignore the threat gathering against us. Facing clear evidence of peril, we cannot wait for the final proof – the smoking gun – that could come in the form of a mushroom cloud.'[7]

The next step is to associate the threatening image with the person or persons you desire to remove from power. Returning to our example, in the same speech Bush commented:

Many people have asked how close Saddam Hussein is to developing a nuclear weapon. Well, we don't know exactly, and that's the problem ... If the Iraqi regime is able to produce, buy or steal an amount of highly enriched uranium a little larger than a single softball, it could have a nuclear weapon in less than a year.[8]

The efforts of the Bush administration were successful, with the majority of Americans initially supporting the invasion of Iraq.

Boot camp: breaking the sense of self

There are two avenues of entry into the military: voluntarily or via compulsion, although, as discussed in the previous chapter, factors such as the so-called 'poverty draft' can blur the distinction. Regardless of their manner of entering the military, the first stage of a soldier's journey is the ordeal of 'basic training' or 'boot camp'.

Basic training seeks to achieve the socialisation and indoctrination of the recruit. Although the specific processes may vary between armies and have been modified over time, the underlying objective of basic training has not changed. This objective is to purge the recruit's civil identity, including any preconceptions he may hold about his rights and personal freedoms, and supplant the civilian value system with that of the military. This is accomplished by various methods, including denigrating those outside the military system and at the same time stressing the virtues of the military community; it is in effect a transformational approach whereby the recruit self-actualises the desire to become part of the military. Other approaches are more individualised, relying on humiliation (including feminisation of the male recruits) and brutality to break an individual's self-esteem, lower their resistance to the values and attitudes that the military wants them to adopt, and reinforce the omnipotent nature of military discipline.

Furthermore, by heavily regulating a recruit's time, including specifying when he will eat and sleep, coupled with an emphasis on there being one right way to accomplish a given task – that being the 'army way' – personal initiative is discouraged. The recruit comes to develop a total dependency on the military to provide his basic needs and direction. The desired psychological state is that the recruit's identity will be so closely tied to his role in the military as to be almost indistinguishable from it. He will define himself by his military role and rank, and his self-esteem will be dependent on how well he can fulfil this role.

This psychological transformation is accompanied by a physical transformation that seeks to reinforce the homogeneity of military service. Jean-Baptiste Barres tells us of his induction into the French Army in 1804: 'I was given my uniform as the day went on, and equipped with the underclothing and the footwear that I should require … We were instructed to let our hair grow in order to make a pigtail and to sell such of our belongings as had not been taken from us.'[9] A century and a half later, although the uniform and haircut had changed, the underlying process had not. Ron Kovic recalled that shortly after his arrival at the Marine Corps Recruit Depot at Parris Island, South Carolina in 1964 his drill sergeant bawled at Kovic's platoon, 'I want you to take your clothes off. I want you to take off everything that ever reminded you of being a civilian and put it in the box.'[10] (Which was subsequently mailed to their families.) The manifestations of the recruit's former civil identity are shed: their hair, their clothes and their personal possessions. These are supplanted by a crew cut and the issuing of uniforms and basic possessions that are common to all recruits. The historian and former US Marine Robert Leckie pointed out that:

> It is the quartermasters who make soldiers, sailors and marines. In their presence, one strips down. With each divestment, a trait is lost, the discard of a garment marks the quiet death of an idiosyncrasy. I take off my socks; gone is a propensity for stripes, or clocks, or checks, or even solids; ended is a tendency to combine purple socks with brown tie. My socks henceforth will be tan.[11]

The recruit is issued an army (regimental) number, which is used to identify him for administrative purposes rather than his name. In this manner the individual's personal identity is subjugated and homogenised to that of a recruit. 'In a barracks, life begins anonymously', wrote Samuel Hynes, who served with the US Marine Corps during the Second World War.

> Every bed is like every other bed, lockers are identical, all arrangements are symmetrical (hang your towel, folded once vertically, on the lefthand end of the foot of your bed; do not display personal photographs; shoes must be placed in the locker in pairs, toes in). There is one common toilet room (no doors on the stalls) and one shower room. No individual is to be distinguished from another.[12]

Once the recruit has internalised the military's value system, the process of rebuilding their confidence and self-esteem begins.

Basic training is designed to be physically exhausting. Sociologists have empirically confirmed that a severe initiation results in entrants placing a high value on group membership, whereas a mild initiation does not engender the same intensity of commitment to the group. The intensity of basic training therefore has a correlation with the desired intensity of organisational loyalty. Yet the apparent severity of basic training is relative to the recruits' limited frame of reference, as most recruits will pass 'boot camp' and go on to attend more physically – and certainly more mentally – demanding training.

The recruits are made to feel that they are constantly in competition with each other as well as with the other sections, platoons etc. Psychologically, this spirit of competition exerts an almost irresistible attraction to adolescents on the cusp of adulthood who are unsure of their status in society. But more subtly, the minor rewards arising from this competition – for example, praise from a mentor-figure such as their drill sergeant, or the 'honour' of leading the company onto the passing-out parade – are a component of the recruits' conditioning in expectation management. Previously, the social conditioning of their early lives had led them to expect material rewards or affection for completing a task; this is not the way of the military.

Intense collective physical conditioning increases the bodily strength and endurance of the recruits, with corresponding growth in individual confidence, but also increased confidence in the physical abilities of fellow recruits. Rigorous and frequent inspections engender pride in the uniform, which, when coupled with indoctrination in the heroic traditions of their particular element of the military, produces pride in their service.

During basic training, emphasis is placed on seemingly archaic close-order drill more tactically relevant to the Napoleonic period than the modern battlefield. While the ability to perform complicated drill manoeuvres fulfils the immediate requirement to march around on their graduation parade, it also builds mutual confidence in the ability of the group to act collectively; but, more importantly, it instils in recruits the need to respond instantly, and without reflection, to verbal commands. Those who consistently fail to carry out drill commands correctly are publicly humiliated by their drill instructors and are usually ostracised by fellow recruits. The contempt in which their peers hold such individuals further reinforces to the group the need to succeed in all tasks assigned to them by their instructors. The overriding desire of most recruits is to be accepted by their peers; so individuality is suppressed by the need to be embraced as a member of the group. Group cohesion develops out of shared hardships, shared punishments and shared triumphs. The aim of basic training is to develop a recruit to the necessary stage of military competence (and socialisation) where he can undergo specialist training.

Making the grade

The completion of basic training is marked by subtle but nevertheless significant differences in the manner in which the soldiers are addressed and treated. They are no longer recruits – the very bottom of the military pecking order – they are now privates. As they are no longer considered to be under probation they receive full salary. The soldiers are also permitted various uniform accoutrements: for example, Israeli armoured corps soldiers can now wear the corps' black beret. The symbology is twofold. The wearing of these accoutrements differentiates the uniform of a private from that of a recruit, but they also serve as a visual reminder that the soldiers have been deemed to have 'made the grade'. 'Even our overseas caps lacked class', observed Paul Fussell when he was undergoing basic training with the US Army during the Second World War. 'They were plain, devoid even of light blue infantry piping. That would come after we'd completed basic training and had earned the right to be designated soldiers, rather than shitheads, assholes and dumb fucks.'[13]

Symbology is also used to connect the new soldier with his country's martial past. During the Second World War, at the completion of their initial training, Soviet soldiers stood before the banners of their regiment and were presented with their weapons in a formal ceremony, which was followed by the assembled men singing martial songs. Likewise, contemporary Israeli soldiers participate in a swearing-in ceremony on the completion of their basic training. With a rifle in one hand and a Bible or Torah in the other, they swear allegiance to Israel and the code of ethics of an Israeli soldier. The ceremony takes place at dusk and is held at one of three symbolic sites: the Western Wall, Masada and Latrun (the site of a major battle during the Israeli War of Independence).

The next stage in the journey of the soldier sees him being sent to various military schools to undergo specialist training. A few comrades from his recruit training may accompany him, but more often than not he will be thrown in with a completely new group of men and the group dynamics will need to be re-established. The completion of specialist training may also be marked by a ceremony, particularly if the soldier is graduating into an elite unit. Guy Sajer tells us that the graduation ceremony into the elite Gross Deutschland Division (motorised infantry) of the Wehrmacht during the Second World War consisted of each soldier marching forward until he stood alone in front of a stand bedecked with flags and holding the assembled officers of the training camp. The soldier then stated: 'I swear to serve Germany and the Führer until victory or death.'[14]

After specialist training is completed, the soldier is considered employable and is sent to a unit. He will once again have to prove his abilities and character to his comrades to gain peer acceptance. If the soldier is fortunate, he will arrive at his first platoon in a time of peace and his learning curve will not have the attendant risk of him being killed. In most cases his fellow soldiers will tolerate him as long as he does not repeat a mistake too many times. If he is unlucky, he will become a reinforcement dispatched to a platoon in combat to make up losses. He will be designated the FNG (fucking new guy) or equivalent, and considered a danger not only to himself but also to his comrades. 'I felt like a kid that just wasn't wanted', recalled Jonathan Polansky, who was sent to Vietnam as an infantry reinforcement for the 101st Airborne Division in November 1968. 'So I struggled over to a few people who were settling down. I introduced myself … nobody wanted to see me because I was a new guy. Nobody wanted a "cherry" out there. Especially this skinny kid who was obviously going to fuck up.'[15] The soldier will either acquire the skills needed to increase his chances of surviving combat or he will be returned home wounded or dead. If a soldier does gain the necessary skills to survive, his psychological well-being will depend on the duration and intensity of his exposure to combat.

Tour of duty: the strain of combat

One of the most perceptive studies of the capacity of men to endure the strain of combat is *The Anatomy of Courage* by Lord Moran. Moran served in the trenches of the Western Front as the medical officer of the 1st Battalion of the Royal Fusiliers, the City of London Regiment, from the autumn of 1914 to the spring of 1917 (a period of 30 months). His book drew extensively upon the diary he maintained during this period and was supplemented by interviews that he conducted with Royal Air Force pilots during the Second World War. In a seminal passage in the preface, Moran wrote:

> Courage is will-power, whereof no man has an unlimited stock; and when in war it is used up, he is finished. A man's courage is his capital and he is always spending. The call on the bank may be only the daily drain of the front line or it may be a sudden draft which threatens to close the account. His will is perhaps almost destroyed by intense shelling, by heavy bombing, or by a bloody battle, or it is gradually used up by monotony, by exposure, by the loss of stauncher spirits on whom he has come to depend, by physical exhaustion, by a wrong attitude to danger, to casualties, to war, to death itself.[16]

Moran's hypothesis was supported by a study conducted by the US Army during the Second World War into the psychological effects of prolonged exposure to combat.

> There is no such thing as 'getting used to combat' ... Each moment of combat imposes a strain so great that men will break down in direct relation to the intensity and duration of their exposure. Thus

psychiatric casualties are as inevitable as gunshot and shrapnel wounds in warfare.[17]

To an extent, the level of combat fatigue among American troops in the Second World War was dependent on the operational theatre in which they served. Troops who fought in the Pacific theatre, in particular those who took part in the fiercely contested but relatively short-duration island-hopping campaigns, could look forward to an operational pause between battles. Not so soldiers in the land-based campaigns fought in Italy, France and the Low Countries, where constant pressure needed to be maintained on the withdrawing Germans.

Despite the lengthy periods that the American troops, particularly in Europe, spent on the front line, the reason why units did not break down was that there was a constant flow of reinforcements to replace the combat-weary men who had become casualties. The historian Gwynne Dyer emphasises in *War* that:

Most units in prolonged combat in modern war, therefore, consist of an uneasy mixture of some utterly green and unsure replacements, some surviving veterans of many months of combat, most of whom are nearing collapse, and a portion of soldiers – the larger the better, from the unit's point of view – who are still in transition from the former stage to the latter.[18]

This is a viewpoint also held by John Baynes, who notes that the British soldiers of the First World War became less and less willing to face the dangers of the trenches as their experience of combat increased. He comments that armies, as well as men, become tired of war and that it was only the influx of reinforcements, occasioned by heavy losses on the battlefield, which maintained the morale of the British formations. 'Keen, eager young men kept pouring out to replace the killed and wounded, and in this way morale was constantly renewed.'[19]

The British military authorities of the First World War were aware of the psychological risk of keeping men in combat for extended periods so, where possible, troops were rotated to the rear to gain respite from the battlefield. This rotation would either be on a unit or individual basis. Units would periodically be brought out of the front line. During this rest period, collective training would be undertaken and the troops might be called upon to support their comrades in the trenches. Throughout the First World War, 'resting' British troops would often be required to undertake fatigues, such as carrying supplies forward to the front-line troops, digging trenches or erecting wire entanglements. On an individual basis, men might depart the front line on leave (either local or home leave) or be sent to a training school; the opportunity for rest provided to NCOs and officers attending these schools being at least as important to the overall functioning of the army as the training received.

In November 1944, the American army in Europe introduced a rotation plan whereby troops who had been wounded at least twice, decorated twice for bravery, or had spent at least six months at the front, would be eligible for rotation back to the continental United States for four months. The initial quota was 2,200 men per month (later increased to 5,500), though passage back to the States was dependent on berths being available on ships. The impact that this rotation had on the manning of the divisions in combat was negligible, but the plan had the desired effect of providing the combat troops with some hope, albeit remote, that there was another alternative way to be released from combat other than by death or wounds. Ernie Pyle commented that only the most optimistic of soldiers believed that he personally would be rotated home. One soldier he spoke to had calculated that under this plan he would be returned to America in seventeen years' time.[20]

The findings of the psychological studies into the effects of prolonged exposure to combat during the Second World War resulted in the US military introducing the one-year tour of duty (thirteen months for Marines) during the Vietnam War. Yet Vietnam was noted for its high proportion of psychiatric casualties – many of the symptoms becoming apparent only after the affected men returned home. A significant contributory factor to the psychological damage caused by the Vietnam War was that the enemy forces were interspersed among the civilian population. The enemy were virtually indistinguishable from the civilians the soldiers were sent to Vietnam to protect – unless they were actually firing at you. There were no completely secure rear areas in-country, a fact that became patently clear after the January 1968 Tet uprising, during which virtually every American base in Vietnam, including the US embassy in Saigon, was attacked. Furthermore, the widespread use of mines and booby-traps by the enemy 'led to a decomposition of the normal, the familiar, the safe', concluded the psychiatrist Jonathan Shay. 'Every familiar item of the physical world could be made to be or to conceal an explosive by the Vietnamese, whether a shiny aluminum rice carrier, a Parker-51 fountain pen, a bicycle, a coconut, Coke cans, C-ration cans, and discarded American artillery-shell casings.'[21] So the US soldiers were never able to completely relax their guard during their entire tour in Vietnam, increasing their exposure to the stresses of combat. For this reason, soldiers left Vietnam for any extended breaks: Bangkok, Taipei, Hong Kong and Sydney being the most popular R&R locations.

The US soldiers serving in the occupation force in Iraq experienced a psychologically threatening environment that was similar to Vietnam. Although Saddam's army was quickly defeated – as expected – what the US planners had not anticipated was a widespread insurrection, where even Baghdad's heavily fortified 'Green Zone' was not immune to attacks by mortars, improvised explosive devices and suicide bombers. An article in the *US News & Weekly Report* in November 2004 opined: 'The Iraq War is a minefield of psychological threats: guerrilla attacks, the uncertain distinction between safe zones and battle zones, the pervasive sense of an enemy around every corner.'[22] A *Time* article, also in November 2004, reinforced the all-pervasive nature of the threat that the American soldiers faced in Iraq: 'Every trip "outside the wire" brings the possibility of attack from any direction, from people who look like everyday citizens and from everyday objects – cars, oilcans, dead animals, even human beings – refashioned into deadly bombs.' The article quoted a Marine serving in the Fallujah region who stated: 'It's relentless, from the moment you arrive until the moment you leave, you're in danger.'[23] In October 2005, the US Army's Center for Health Promotion and Preventive Medicine, on the basis of interviews and surveys completed after soldiers had served in Iraq, revealed that one in four US veterans of the Iraq conflict had physical and psychological ailments, ranging from unhealed wounds to suicidal tendencies.[24]

Prolonged exposure to combat produces an emotional dullness, with soldiers affecting a demeanour that is unmistakeable to experienced observers. Ernie Pyle, who served as a war correspondent in Europe during the Second World War, called it the 'look'.

> It's a look of dullness, eyes that look without seeing, eyes that see without conveying any image to the mind. It's a look that is the display room for what lies behind it – exhaustion, lack of sleep, tension for too long, weariness that is too great, fear beyond fear, misery to the point of numbness, a look of surpassing indifference to anything anybody can do. It's a look I dread to see on men.[25]

A generation later, and a continent away, the journalist Michael Herr would discern the same look, which he termed the 'thousand-yard stare', on the combat-weary US Marines fighting in Vietnam.[26]

The experiences of Eugene Sledge, as recorded in *With the Old Breed*, his memoir of the Pacific campaign of the Second World War, provide a germane example of the downward trajectory of the personal prospects of the combat soldier.

Sledge was advised by his family to stay in college so that he could qualify for a commission in one of the army's technical branches. But he 'had a deep feeling of uneasiness that the war might end before [he] could get overseas into combat', so he enlisted in the Marine Corps in December 1942.[27] He was accepted into an officer training programme and sent off to continue his studies. But finding himself stuck back in college, Sledge, along with half of his detachment (90 men in all), deliberately failed his first semester so that he could enter the Marines as an enlisted man. Following basic training, Sledge completed his specialist training and was taught how to operate the 60mm mortar. With the benefit of hindsight he wrote:

But I don't recall that anyone really comprehended what was happening outside our own training regime. Maybe it was the naive optimism of youth, but the awesome reality that we were training to be cannon fodder in a global war that had already snuffed out millions of lives never seemed to occur to us. The fact that our lives might end violently or that we might be crippled while still boys didn't seem to register. The only thing that we seemed to be truly concerned about was that we might be too afraid to do our jobs under fire. An apprehension nagged at each of us that he might appear to be 'yellow' if he were afraid.[28]

Sledge landed on the Pacific island of Peleliu on 15 September 1944. A few days later he was assigned to a combat patrol that was sent out to make contact with the enemy and remain in contact until artillery and/or air strikes could target the Japanese. 'Slowly the reality of it all formed in my mind: we were expendable! It was difficult to accept. We come from a nation and a culture that values life and the individual. To find oneself in a situation where your life seems of little value is the ultimate in loneliness.'[29]

Sledge escaped being wounded or killed on this patrol, but as the number of casualties among his comrades grew, he became increasingly dejected and began to doubt that he would survive the war.

We merely existed from hour to hour, from day to day. Numbed by fear and fatigue, our minds thought only of personal survival. The only glimmer of hope was a million-dollar wound or for the battle to end soon. As it dragged on and on and casualties mounted, a sense of despair pervaded us. It seemed that the only escape was to be killed or wounded. The will for self-preservation weakened. Many men I knew became intensely fatalistic. Somehow, though, one could never quite visualise his own death. It was always the next man. But getting wounded did seem inevitable. In a rifle company it just seemed a matter of time. One couldn't hope to continue to escape the law of averages forever.[30]

Finally, after 30 days of almost unrelenting combat, Sledge was informed that his unit was to be withdrawn from Peleliu. He poignantly recalled: 'I suppose I had become completely fatalistic; our casualties had been so heavy that it was impossible for me to believe that we were actually leaving Peleliu.'[31]

Dead men walking

The policies of armies, although practical, can actually encourage fatalism among soldiers. During the First World War, British Army Orders specified that a 'Minimum Reserve' of approximately 10 per cent of the battalion's strength was to be sent to the transport lines to wait out the battle. This reserve would form the basis around which a reconstituted battalion could be rebuilt if the remainder was decimated in combat. The need for such a reserve could hardly have filled the soldiers getting ready to 'go over the top' with much confidence in their long-term prospects. Moreover, it is somewhat difficult for a soldier to remain optimistic of his chances of survival when he is repeatedly informed that his nation is engaged in an attritional struggle. The front-line soldier notes the endless calls for volunteers, the huge number of men undergoing training, and the massed replacements being assembled. As he makes his way to the front he may pass by the burial pits prepared for the casualties of the coming assault, and observe the extensive medical infrastructure established on the edge of the battlefield. He therefore adjusts his own life expectancy accordingly. 'It was here that I for the first time became aware that each platoon had a stretcher along – gruesome reminder to timid hearts of the bloody business at hand,' mused Horace Baker, a member of the 32nd Division, American Expeditionary Force, as his unit made its way towards the Meuse-Argonne battlefield in the closing stages of the First World War.[32]

The soldier has no say in where he is sent to fight and why. He has virtually no influence over the intentions and actions of his enemy. While he can master the rudiments of his profession, even the most capable soldier can do little to prevent death in its many varied forms from seeking him out on the battlefield. It is therefore not surprising that many soldiers become fatalistic when subjected to prolonged periods of combat. In the following perceptive passage, James Jones, who served with the US Army's 25th Infantry (Tropic Lightning) Division in the Pacific during the Second World War, sums up the extreme fatalism that affected some American soldiers during this campaign. He wrote that the soldier must accept

> the fact that his name is already written down in the rolls of the already dead. Every combat soldier, if he follows far enough along the path that began with his induction, must, I think, be led inexorably to that awareness. He must make a compact with himself or with Fate that he is lost. Only then can he function as he ought to function, under fire. He knows and accepts beforehand that he's dead, although he may still be walking around for a while.[33]

On the other side of the world, Siegfried Knappe, a German soldier fighting in the Russian campaign, experienced the same fatalism following the death of a fellow officer. Both men had thrown themselves to the ground when a salvo of Russian artillery shells struck some trees beside them. They were lying beside each other: one was killed by shrapnel, the other left unharmed. Knappe recalled that his response to the incident was that

> I had to become fatalistic about it and assume that eventually it would happen to me and there was nothing I could do to prevent it … I knew that I was going to be killed or badly wounded sooner or later. The odds against me escaping unscathed were impossibly high, and I accepted my eventual death or maiming as part of my fate. Once I had forced myself to accept that, I could put it out of my mind and go on about my duties; I would not have been able to function had I not done so.[34]

There were also Soviet officers who embraced the same form of fatalism to allow themselves to continue to function in the surreal world of combat. 'I have told myself that I will be killed whatever happens, today or tomorrow … I go into battle without any fear, because I have no expectations', a Captain Kozlov commented to the famed Soviet war correspondent Vasily

Grossman. 'I am absolutely convinced that a man commanding a motorized rifle battalion will be killed, that he cannot survive. If I didn't have this belief in the inevitability of death, I would be feeling bad and, probably, I wouldn't be able to be so happy, calm and brave in the fighting.'[35]

An acute manifestation of the fatalism of soldiers is the belief that it is best to be killed early in a campaign rather than endure the privations of active service only to be killed in the war's latter stages. Robert Leckie recounted that one of his fellow Marines expressed this sentiment as the 1st Marine Division was being decimated in the attack on the Japanese-held island of Peleliu in September 1944: '"Remember the guys who got it back on Guadalcanal?" Runner asked. "We used to think they were poor slobs – getting it so soon. Maybe they were the lucky ones. They didn't have to go through all of this crap and wind up getting it anyway."'[36]

Such fatalistic thoughts were not restricted to the soldiers who fought in the seemingly endless conflict of the world wars. Philip Caputo, who fought in the Vietnam War, tells us that:

I only knew I had ceased to be afraid of dying. It was not a feeling of invincibility; indifference, rather. I had ceased to fear death because I had ceased to care about it. Certainly I had no illusions that my death, if it came, would be a sacrifice. It would merely be a death, and not a good one either. A good death involved a certain amount of choice, ritual, and style. There are no good deaths in war.[37]

But Caputo survived the jungles of Vietnam and his discharge from the Marines left him feeling as 'happy as a condemned man whose sentence had been commuted'.[38] This is a particularly telling analogy because the fate of a condemned man, like the fate of a soldier, rests with others. Anthony Swofford, who served with the US Marine Corps in the 1991 Gulf War, echoed Caputo's sentiments: 'Just before we engaged the Iraqis, I'd decided that I would soon die and this was okay, and I went forward into battle with the dumb death stare of the dead walking. But after the war I was shocked back to life …'[39]

Evan Wright, a journalist who accompanied elements of the US Marines during the 2003 invasion of Iraq, wrote of a similar mindset that was recounted to him by the commander of the platoon in which he was embedded. The young officer called his coping mechanism for combat the 'Dead Man Walking Method': 'Instead of reassuring himself, as some do, that he's invincible or that his fate is in God's hands (which wouldn't work for him since he leans towards agnosticism), he operates on the assumption that he's already a dead man, so getting shot makes no difference.'[40]

The abandonment of hope and the realisation that they will probably not survive the war has an attendant psychological cost. Robert Graves, who served with the British Army during the First World War, observed that the death of the last of the unwounded survivors (less the rear echelon elements) of the original battalion that came out to France in August 1914 filled the men with a murderous rage.[41] The survival of the veteran soldier, in the face of the great probability of death, had given them hope that they too might survive the carnage. His demise stripped them of this illusion; death was to be the common lot.

The culture of obedience

The efficient functioning of an army requires the subjugation of individual desires and wants and the embracing of a collective perspective. As the collective effort is the sum of individual efforts, armies need to ensure that each soldier carries out his assigned role. The scale of modern armies and the geographical spread of their constituent elements constantly threaten to overwhelm individuals. Soldiers may come to feel that the enterprise is so vast that their minor contribution to the battle could not possibly make any difference, and thus may rationalise away the effect of their non-participation. To avoid this situation, modern armies place great emphasis on ensuring that soldiers understand how their individual role contributes to the overall scheme of battle, and the importance of their individual efforts towards achieving the army's objectives.

The benefit of soldiers embracing this collective perspective is that it decreases the angst that might have otherwise impaired their effectiveness on the battlefield. Philip Caputo observed that some of his soldiers were anxious and solemn the night before his unit's first combat patrol in Vietnam. By the next morning, however, they 'were cheerful in the resigned way of men who know they have no control over what is going to happen to them'.[42]

The necessary conditioning of the soldier began during basic training, when he was first exposed to the omnipotent nature of military discipline and individualism was ruthlessly suppressed. Throughout his career, he will be moved from post to post in accordance with the wishes of the army, his work schedule will be set by the army, and he may be required at short notice to leave his family and be deployed whenever and wherever the army chooses. All these factors reinforce to the soldier that he is a resource of the army to be used as it sees fit. In certain circumstances a soldier may be permitted to resign or retire from the army. However, the social disruption of his frequent moves means that most of his close friends and acquaintances are probably also in the services and, to a greater extent than most other occupations, the soldier defines himself by his job. The decision to leave the army therefore has greater repercussions on a personal level than that which civilians would normally experience when they decide to change occupations. The enormity of the decision to leave the military is reinforced by the extended notice required; for example, all officers and soldiers serving in the British Army are required to give seven months' notice with their resignation.

The desired state of mind is that soldiers will not question assigned tasks, but rather will view them as a mission to be completed without considering the long-term consequences. That is, soldiers will not question whether the tactical value of achieving the objective justifies the personal risk to life and limb. John Baynes notes that the prevailing attitude in the British Army of 1914 was 'orders is orders'; every soldier was instructed from enlistment that any order given by a superior officer was to be instantly obeyed. This ingrained obedience meant that questions could be asked only after obeying the order and not before.[43] The psychologist Norman Dixon postulates in *On the Psychology of Military Incompetence* that the needless slaughter of many of the battles of the First World War was partly due to this crippling culture of obedience, whereby officers did not voice their doubts about a proposed military action, 'preferring to conceal evidence from their superiors rather than be thought wanting in courage or loyalty'.[44]

An army cannot function if soldiers continually question orders and seek justification for their missions. But conversely, the ingrained desire of soldiers to 'get on with the job' and the reluctance to revoke orders once given have condemned many men to pointless deaths.

The value of a military objective is often truly determined only with hindsight. Hindsight allows a cognitive settling of the balance book, but all too often the results achieved do not justify their cost – a cost that is measured in shattered lives. Hugh Dundas' brother, a pilot in the Royal Air Force during the Second World War, was killed in November 1940 shortly after shooting down a German plane. This aerial duel 'certainly contributed nothing, one way or the other, to the course or outcome of the war', lamented Dundas. 'But it affected my life deeply. I think that hardly a day has gone by since then when I have not thought of John.'[45]

The short-term viewpoint of most soldiers throughout history has resulted in many lives being squandered for little appreciable effect on the outcome of the conflict. The historian Joseph E. Persico calculates that the armies involved in the First World War lost 11,000 dead, wounded and missing on 11 November 1918 (Armistice Day) – losses greater than those suffered by both the Allies and the Germans on D-Day (6 June 1944).[46]

The death of Private Henry N. Gunther of the 313th Regiment, 157th Brigade of the American Expeditionary Force best illustrates the futility of these final attacks. Gunther's regiment launched an attack on Ville-Devant-Chaumont at 9.30am on the 11th. At 10.44am a runner informed the commander of the 157th Brigade, Brigadier General William Nicholson, that the Armistice had been signed. Nicholson responded: 'There will be absolutely no let-up until 11am.' Gunther was shot at 10.59am – he was the last American (and indeed Allied) soldier to be killed during the war.[47] What possible difference could Gunther's final fatal charge have made, or for that matter what difference did the final assaults that resulted in thousands of other soldiers being killed and wounded on that winter morning make? The Armistice had already been signed.

Another example of soldiers' lives being squandered for little appreciable strategic gain was the American campaign to wrest the Pacific island of Peleliu from its Japanese defenders during the Second World War. Peleliu is the southernmost isle of the Palau Islands group which lies approximately halfway between the Marianas and the Philippines. The attack on Peleliu had been conceived to protect the eastern flank of General Douglas MacArthur's army when it invaded the Filipino island of Mindanao. When the American objective shifted from Mindanao to the neighbouring island of Leyte, Peleliu became strategically irrelevant and Admiral William Halsey suggested calling the operation off. But Admiral Chester Nimitz, perhaps influenced by intelligence reports stating that Peleliu was lightly garrisoned and could easily be taken, insisted that, even though the overall objective had changed, the Marine force then steaming towards the Palaus should take Peleliu anyway. In fact it was not lightly garrisoned: over 10,000 Japanese soldiers of the 14th Infantry Division held the island. The Japanese defenders awaited the Americans in pillboxes reinforced with concrete and steel and in coral caverns, some of which were six storeys deep. Their orders were to kill as many American Marines as possible. The Americans suffered almost 10,000 casualties, including almost 1,800 dead, in their 30-day campaign to capture Peleliu. William Manchester, who took part in this campaign, wrote in his memoirs of the Pacific war that the enemy garrison at Peleliu 'could have been left to wither on the vine without altering the course of the Pacific war in any way'.[48]

Wars are not bloodless affairs and soldiers being wounded and killed is the necessary cost of achieving a campaign's objectives. To this end, armies require soldiers to supplant individual hopes and aspirations with collective goals. J. Glenn Gray, a philosopher who served in the US Army in North Africa and Europe during the Second World War, wrote in *The Warriors: Reflections on Men in Battle* that 'most soldiers in wartime feel caught in the present so completely they surrender their wills to their superiors'.[49] A British doctor, part of a field ambulance unit during the

First World War, lamented in a letter to his wife in late 1917 that the spirit of the troops had changed since the war began.

> If anything it is more admirable: there is more patience and self-sacrifice: but it is infinitely pathetic to witness. No keen curiosity now, no careless enthusiasm, not even hate to carry them on: but instead a sense of duty, and a bowing down to the inevitable – the inevitable power which drives them on from behind.
>
> The troops have settled down to war as slaves to their task: if they fall short of what is expected they risk discomfort, punishment, even death: whereas if they please their masters there are certain rewards to be won in the form of holiday and rest from the line. It is a sad spectacle to see free citizens of a civilised empire thus degraded.[50]

Not daring to look ahead

Many soldiers will surrender the notion that they are in control of their own fate and will embrace a short-term perspective on life. Walter Downing, who served with the 57th Battalion of the Australian Imperial Force during the First World War, recalled that 'we neither knew nor cared what the morrow might bring. One accepts the immediate present, in the army.'[51] Many of his comrades shared this sentiment, and one would later write that his fellow soldiers considered only the present because 'the probability of seeing their own corpses lying in the mud stifled their desire to look into the future'.[52] Hervey Allen, an American soldier of the First World War, emphatically stated that 'There is no man who is so totally absorbed by the present as the soldier'.[53] David Hackworth wrote of his service in Korea: 'In the infantry, I found, you live for right now. You don't give a damn about tomorrow, because you don't even know if there'll be one.'[54] Colin Sisson, mid-way through his tour of duty with the New Zealand Army in Vietnam, began to notice how 'I only lived for the moment, or perhaps for the next meal or rest stop at most'.[55]

Yet the soldiers who fought on the ground had a much greater probability of surviving the war than the aircrew of the Royal Air Force's Bomber

Command during the Second World War, the majority of whom were killed before they could complete their operational tour. Guy Gibson, himself later killed in a raid on Germany, wrote of his colleagues in Bomber Command: 'The losses in percentage in any one raid are not unduly high – rarely above ten per cent – but remember this used to go on for some sixty missions, and the bare fact remains that out of a squadron of twenty-five crews, not many are left at the end of three months.'[56] Gibson's memoir of his service during the war, *Enemy Coast Ahead*, provides a litany of lost comrades until:

> I was the last one left, the last one out of a bunch of boys who belonged to 83 Squadron at the beginning of the war, to fight until the end of Hitlerism. They had all fought well, but they had paid the price. Some were prisoners of war, I knew, but many were dead. As I lay in bed thinking, I knew I was lucky to survive, but it would come to me any day now. We would go on and on until the whole squadron was wiped out … I did not see any point in living … All my friends had gone now – there were new people – different – with different views on life, different jokes and different ways of living. I was the last one left.[57]

Don Charlwood was an Australian who served as a navigator with Bomber Command. His memoir, *No Moon Tonight*, is an elegant testimony to the courage of the aircrew who continued to fly into the flak and fighters of occupied Europe knowing their chances of survival were remote. His narrative is punctuated by the loss of his comrades. Unclaimed mail would be spirited away, there would be empty beds in the barracks when they returned from a mission and the dead men's personal effects would be removed within a day by the ominously named Committee of Adjustment. Charlwood ruefully recalled that 'we even became accustomed to the idea that to reach thirty ops. was no longer possible, that home was a place for which we could afford no longings'.[58] Instead the men focused on short-term goals. 'From day to day we were seldom far from the station; but every six weeks we were granted six days' leave. To live until the next leave became the greatest hope of each of us.'[59]

Charlwood's fellow pilots in Fighter Command also experienced the loss of many of their companions and embraced a similar fatalism. 'For my part, I looked ahead for a brief week or so, no more; indeed, no further dare a fighter pilot look', wrote Jim Bailey in his memoir of the Second World War. 'For the future that opened up before me promised only many, many years of fighting, and that being the case, there seemed little weary chance that any of us would survive it.'[60]

Over half a century later, Nathaniel Fick also found that the suffocating reality of his immediate future was all he could concentrate on as he contemplated leaving a quiet Californian town for the war in Iraq.

> I looked around at the other tables. There were people my age on dates, whispering and smiling … These people looked forward to Saturday, and Sunday, and the coming months and years of their lives. Mine felt as if it had ended. I didn't have a future. Trying to conjure up a mental image of myself after Iraq, I found that I couldn't. Iraq loomed like a black hole into which all the thoughts and acts and hopes and dreams of twenty-five years were being sucked.[61]

The consolations of faith

The possibility of their imminent death forces soldiers to confront their own mortality. Some will turn to religion. Gottlob Bidermann served with the Wehrmacht's 132nd Infantry Division on the Eastern Front during the Second World War. He observed that as the German forces pressed into Russia, 'Many of the soldiers who had not previously been so inclined began to attend religious services, and with the growing consciousness of our mortality we became more aware of the presence of the chaplain'.[62] Many soldiers' memoirs comment upon a dramatic increase in attendance at regimental church parades immediately prior to an attack; for example, well-attended prayer sessions were held on the densely-packed ships of the Allied invasion fleet as they sailed towards Normandy for the D-Day landing. Moreover, David Cooper, the chaplain of the 2nd Battalion of the Parachute Regiment, recalled that the entire battalion attended the final church parade held on the MV *Norland* before the landing at San Carlos during the Falklands War. As the threat of combat drew near, soldiers who were previously ambivalent about religion sought out what Cooper categorised as an 'insurance policy'.[63] Günter Koschorrek served as a machine gunner with the German 21st Panzergrenadier Regiment on the Eastern Front during the Second World War. He embraced religion as a Soviet T-34 tank spun around above him, attempting to collapse the sides of the trench that Günter and his colleague were sheltering in: 'I haven't prayed since I was a child, believing in my youth that I was strong enough not to need the help of a higher being. But now, facing death and fearing for my life, those long forgotten words come to mind.'[64] Soldiers may find prayer helpful because their belief in divine intervention removes some of the randomness of war and reasserts, at least in their own minds, their value as an individual who is worthy of surviving.

The notion of battlefield absolution may also provide some comfort to soldiers about to enter combat. Cardinal Patrick Francis Moran, the head of the Catholic Church in Australia, stated in a service held shortly after the outbreak of the Boer War that if a volunteer died on the battlefield, heaven's portals would open up to receive him and he would gain salvation in the eyes of his fellow man and God.[65] Prior to the men being committed to battle during the First World War, the Catholic padre of Robert Graves' battalion informed them that 'if they died fighting for the good cause they would go straight to heaven, or at any rate would be excused a great many years in Purgatory'.[66] This sentiment was also evident among Australian troops, one of whom tried to reassure his parents (and most likely himself) in a letter he wrote before moving up into the trenches: 'I am spiritually prepared for death and should such be my fate remember I shall go straight to heaven.'[67]

Religious faith was not widespread among the Soviet troops of the Second World War; however, as Catherine Merridale notes in *Ivan's War*, the trappings of religion fulfilled a deep-seated psychological need in the men: 'There were few believers, but most used prayers and ritual gestures out of superstition, as spells, crossing themselves to ward off death.'[68]

But the devastating effects of weapons can result in soldiers questioning their religious beliefs, their previously sanitised exposure to death having not adequately prepared them for the mutilation and evisceration of the battlefield. 'It is my honest opinion, a very humble one, that the sight of battlefields must always be a great blow to the lingering belief in personal immortality', mused Hervey Allen.[69] Eric Sykes, a British paratrooper of the Second World War, confessed that 'After experiencing the tragedies of war I became more and more detached from religion, and all of the religious learnings and teachings, to a point of complete disbelief in a supreme being ...'[70] A Russian soldier of the Second World War felt very much the same: 'When you see the atrocities that are taking place minute by minute, you just think, God, if you're so omnipotent and just, how can you let so many innocent souls suffer this torment and die?'[71] Soldiers are forced to face the realisation that the human body is only a 'fragile case stuffed full of disgusting matter' (as Philip Caputo puts it). Caputo found it difficult to believe that the bloody eviscerated bodies lying at his feet would be capable of resurrection on Judgement Day and that their souls had passed on to another existence.[72]

Accepting the presence of death

The presence of death all around them forces soldiers to accept the very real possibility of their imminent demise. Siegfried Knappe, who fought with the Wehrmacht throughout the Second World War, was shocked by the first bodies he encountered on the battlefield.

[W]e knew that in war people get killed. But 'knowing' it intellectually was entirely different from seeing and experiencing it ... the word 'killed' still had a clinical connotation about it compared to its meaning when you saw lying on the ground before you a bloodied, mutilated, foul-smelling corpse that had previously been a vital, living human being.[73]

The bodies that Knappe came across were those of some French Moroccan soldiers; coincidentally, they had been killed during a firefight in a cemetery. He remembered 'their limbs in grotesque positions, their eyes and mouths open'. Knappe wrote that:

> The experience was impossible to forget. This was what we were doing to people and what they were doing to us. It was devastating to realise that this was what we had to look forward to every day, day after day, until the war was over. From that moment on, death hovered near us wherever we went.[74]

By focusing on his responsibilities as an officer, Knappe was able to accept the 'ever-present nearness of death' and function on the battlefield.[75] Other soldiers embrace different coping mechanisms. J. Glenn Gray observes that some soldiers believe that while death will strike down others, they themselves have a personal invulnerability and that somehow death will pass them by. 'These soldiers cherish the conviction that they are mysteriously impervious to spattering bullets and exploding shells. The little spot of ground on which they stand is rendered secure by their standing on it.'[76] 'Not for one moment did I contemplate the possibility of anything unpleasant', wrote Robert Crisp, a tank officer with the British 8th Army, when reflecting upon his feelings prior to his initial encounter with Rommel's Afrika Korps. 'And with that went an assumption that there was bound to be a violent encounter with the enemy, that it would end in our favour, and that if anything terrible were going to happen it would probably happen to other people but not to me.'[77] Richard Hillary, who flew Spitfires during the Battle of Britain, was firmly of this belief: 'That I might be killed or in any way injured did not occur to me. Later, when we were losing pilots regularly, I did consider it in an abstract way when on the ground; but once in the air, never. I knew it could not happen to me.'[78] Hillary suffered extensive burns to his hands and face when he was shot down in September 1940. After protracted surgery and a lengthy period of recuperation he returned to flying duties and was killed in January 1943 when he crashed his Bristol Blenheim during a night training flight.

Others will seek to rationalise away the risk. Michael Durant, an American Black Hawk pilot who was shot down and captured in Mogadishu, Somalia in October 1993, admitted: 'I never thought about what might happen if I got killed. I didn't think I *could* be killed. There was an expression for that kind of helo jock [helicopter pilot] denial: *Big Sky, little bullet*.'[79] Hugh McManners, reflecting upon his experiences during the Falklands War, summed up in *Falklands Commando* that:

> The most universal attitude of soldiers is the belief that whatever is going on they will not be the one to get hurt or killed. When others are killed the worry is short-circuited by saying 'if it's got your name written on it, then no matter what you do, it's going to get you'. This optimistic fatalism is the only way to avoid becoming convinced of the inevitability of death, or worse – mutilation.[80]

Some soldiers will succumb to what the psychologist M. Brewster Smith termed the 'strategic abandonment of hope'.[81] In contrast to those who accept the inevitability of death as a means to continue functioning on the battlefield, these soldiers have essentially accepted that they will not survive the war and have become apathetic. An American soldier who fought in the European theatre of the Second World War described such soldiers: 'They don't give a damn whether they get killed or not. They lose courage. They don't aim, can't hit the ground fast. They're scared all right but they don't care. When they're running they run about fifteen yards and then start walking – they don't give a damn.'[82]

A soldier who becomes so fixated on the probability of his death in combat that he directs the majority of his energies towards increasing his chances of survival is an ineffective soldier. Militaries therefore seek to desensitise soldiers to death, so that they will not dwell upon their own mortality and also will remain combat-effective and not lapse into sorrow and mourning when their comrades are killed beside them. This desensitisation enables soldiers to psychologically endure the carnage they encounter on the battlefield. Paul Fussell described the practical application of this coping mechanism in *Doing Battle: The Making of a Skeptic*.

Before we'd finished in Europe we'd seen hundreds of dead bodies, GIs as well as Germans, civilians as well as soldiers, officers as well as enlisted men, together with ample children. We learned that no infantryman can survive psychologically very long unless he's mastered the principle that the dead don't *know* what they look like. The soldier smiling is *not* smiling, the man whose mouth drips blood doesn't know what he's doing, the man with half his skull blown away and his brain oozing onto the ground thinks he still looks OK. And the man whose cold eyes stare at you as if expressing a grievance is not doing that. He is elsewhere. The bodies are props on a set, and one must understand that their meaning now is that they are props, nothing more.[83]

Colin Powell, reflecting on his second tour of duty in Vietnam, expressed a similar sentiment: 'People in combat develop a protective numbness that allows them to go on.'[84]

Often the most effective soldiers, although not necessarily lacking in intelligence, have an unquestioning nature. Lord Moran believed that the soldiers who fared best during the First World War were blank, unimaginative people.[85] 'A man needs many things in war, but a strong imagination is not one of them', concluded Philip Caputo. He remarked that, 'in Vietnam, the best soldiers were usually unimaginative men who did not feel afraid until there was obvious reason'.[86] These sentiments were echoed by James Jones, who wrote that experienced soldiers 'had learned, maybe the most important of all for survival, that danger only existed at the exact place and moment of danger, and not before and not after'.[87]

Alcohol and drugs in war

One of the coping mechanisms frequently embraced by soldiers to withstand the psychological strain of combat is the consumption of alcohol. According to Jakob Walter, a German conscript serving with Napoleon's army during the Austrian campaign of 1809, when the army marched into Bregenz, 'Cellars were broken into, and wine was carried out in buckets everywhere. Even several kegs were left running. Everyone became intoxicated until finally a strict order put an end to all this.'[88] Adrien Bourgogne, a sergeant in Napoleon's Imperial Guard, tells us that during the occupation of Moscow his comrades uncovered a 'great quantity of wine in the cellars, and some Jamaica rum, also a large cellar filled with barrels of excellent beer, packed in ice to keep it fresh during the summer. We found at our boyar's house fifteen large cases of wine and sparkling champagne.'[89] The alcohol store was put to good use by Bourgogne and his comrades to while away the chill of the long Russian nights. Fifty years later, following the battle of Gettysburg in the American Civil War, William Fletcher recounted that one of his comrades discovered several barrels of whisky.

> He came to camp and told some of the boys in the company and they gathered up all the canteens of the company and filled them. It was soon noised around and whisky was soon in the different nearby camps; and in some instances, the camp kettle was filled. There was soon, as far as I could see, drunken soldiers – it was said that half the camp was drunk and the other half doing the little camp duties and keeping the drunken ones straight.[90]

Some armies (as well as navies, for that matter) have traditionally provided alcohol to their troops in the field. The Royal Navy supplied its sailors with the so-called 'rum ration' (approximately a quarter pint) from the mid-17th century up to 31 July 1970, when the practice ceased upon the orders of the Admiralty Board who had concluded that the rum issue was no longer compatible with the 'high standards of efficiency' required of naval personnel.[91] The British Army likewise periodically provided a rum ration to its troops, the practice continuing up to the Korean War. The rum ration was particularly appreciated by the soldiers serving in the cold, water-logged trenches of Belgium and northern France during the First World War. Max Plowman lyrically recounted:

> Rum of course is our chief great good. The Ark of the Covenant was never borne with greater care than is bestowed upon the large stone rum-jars in their passage through the wilderness … I honestly believe

some [men] I know would commit suicide if the rum ration were withdrawn. And in truth the rum is good – fine, strong, warming stuff – the very concentrated essence of army-council wisdom.[92]

The French *poilu* of the First World War would accept abysmal conditions in the trenches as long as he received his ration of *pinard* (rough red wine). The provision of the wine ration remained crucial to the morale of the French soldier for much of the 20th century, indicative of which is that among the 'necessities' provided to the besieged French garrison at Dien Bien Phu were almost 50,000 gallons of wine. Such was the importance of the wine ration that the French Army developed a wine concentrate called 'Vinogel' so as to ensure supplies to mobile bodies of troops. Unfortunately the Vinogel tasted awful, though this fact did not stop a raid by members of the French Foreign Legion on the Viet Minh trenches surrounding Dien Bien Phu on 30 April 1954 to secure two crates of it that had fallen into no man's land.[93]

From August 1941 the Soviet Army provided a vodka ration of 100 grams a day to be issued to all active-duty soldiers. 'Of course, it was seldom that we drank 100 grams only because as we bore heavy losses every day that amount of vodka was meant to cover a bigger number of people than there actually were', Fyodor Sverdlov blissfully recalled. 'I usually drank 200 grams of vodka at breakfast, 100 grams of vodka at lunch, and if there was no combat in the evening, I would drink another 200 gram glass sharing dinner with a company of friends.'[94] Maurie Pears, an officer serving with the 3rd Battalion of the Royal Australian Regiment during the Korean War, tells us: 'Evenings and nights were spent swapping tall yarns on the hillside, while slowly consuming our daily ration of one Asahi [Japanese] beer bottle after a hard day's dig. Oh, how sweet it was! Beer never tasted better.'[95] Likewise, the Italian soldiers of the First World War enjoyed the brandy that arrived at the forward positions with the ammunition. Emilio Lussu, an infantry officer in the Italian Army, observed that as his battalion waited in the trenches to attack the Austrians, 'The only movement to be seen was that of the brandy flasks, up and down, from belt to mouth and mouth to belt, like shuttles on a great loom'.[96]

As recounted in *Storm of Steel*, his memoir of his service with the German Army during the First World War, Ernst Jünger seemed never to want for an alcoholic beverage, consuming in the trenches quantities of brandy, schnapps, grog (rum and water), red wine, cognac, advocaat, and crème de menthe. But even the German Army realised that its soldiers could have too much of a good thing. Following an incident where some of Jünger's comrades became drunk on looted French wine, measures were taken to prevent a reoccurrence. Jünger wrote that 'In similar situations later, we were simply to shoot holes in barrels and carboys and other containers of alcohol'.[97] Alcohol was also readily available to units of the German Army during the Second World War. Guy Sajer was informed by a soldier on the Eastern Front that 'There's as much vodka, schnapps and Terek liquor on the front as there are Paks [a German anti-tank weapon]', and recalled that alcohol 'was freely distributed along with our ammunition'.[98] Johann Voss, who served with the Waffen-SS in Finland, recalled that their monthly rations included bottles of cognac.[99]

During the First World War the American Army attempted to enforce temperance among its soldiers, a policy that reflected the influence of the prohibitionists. By the Second World War, the prohibitionists had been discredited and alcohol was more freely available to the fighting men, with US Army officers periodically receiving an issue of liquor (spirits rather than beer). As for the enlisted men, although they were not permitted to buy spirits from the army canteens, they were allowed to purchase beer when off duty.

For much of the European campaign of the Second World War, soldiers' alcohol consumption was generally restricted to what they could purchase from civilians or 'liberate' – usually locally produced wine or spirits, though alcohol remained plentiful in some units that had had the good fortune of seizing the massive alcohol stores of the withdrawing German Army. For example, the British Guards Armoured Division, which led the advance of the British and Canadian 21st Army Group into Belgium, captured intact the German military's principal champagne store at Louvain. A troop commander in this division recounted that his share of the haul was twenty cases of champagne.[100]

Alcohol was difficult to come by for the American troops serving on the front line during the Korean War. Joseph Owen, who served with the 7th Marine Regiment, 1st Marine Division tells us that after the battalion's first major battle the company's gunnery sergeant called the platoon sergeants together and distributed four gallons of sick bay alcohol (ethanol) that he had obtained in a trade for a bugle taken from a dead North Korean soldier. Owen recalled that 'Pat Burris "slicked" four big cans of pineapple juice, and this was blended with the alcohol in steel helmets that served as punch bowls. "Hawaiian Whoopee," Burris called it.'[101]

When alcohol was not readily obtainable, soldiers would seek to manufacture their own. Edward Costello wrote of his colleagues during the Peninsular War of the early 19th century:

> Some of the men, for want of better pastime, succeeded in constructing a still, with which they managed to make spirits from a quantity of dried grapes, found in the old wine house; a discovery, however, soon took place, much to our chagrin, and the still was destroyed by our old Captain Peter O'Hare.[102]

Evgeni Bessonov noted in his account of his service with the Soviet Army during the Second World War that the company medic was responsible for the production of 'moonshine':

> Moonshine made from sugar beet was strong (it could even burn), but it stank like hell so it seemed that our moonshine-making technology was far from perfect. Battalion commander Kozienko would regularly inspect the companies and destroy the distilling equipment, but we would assemble it again and continue brewing alcohol. We would come from our training, have a half-glass of moonshine, and it would feel great ...[103]

Bill Mauldin remembered that at Anzio 'the boys at the beachhead were fixing up their own distilleries with barrels of dug-up vino, gasoline cans, and copper tubing from wrecked airplanes. The result was a fiery stuff which the Italians call grappa … It wasn't bad stuff when you cut it with canned grapefruit juice.'[104]

The austere environment in which most of the Pacific campaign of the Second World War was fought denied soldiers the opportunity to 'liberate' alcohol because it simply was not available. Bootlegging therefore became endemic. James Jones recalled that:

[I]n my outfit we got blind asshole drunk every chance we got … We made our 'swipe' [a Hawaiian word for bootleg liquor] by stealing a five-gallon tin of canned peaches or plums or pineapple from the nearest ration dump, and putting a double handful of sugar in it to help it ferment, then leaving it out in the sun in the jungle with a piece of cheesecloth or mosquito netting over it to keep out the bugs. It was the most godawful stuff to drink, sickly sweet and smelling very raunchy, but if you could get enough of it down and keep it down, it carried a wonderful wallop.[105]

The consumption of home-made alcohol could produce tragic results. The *Official US Medical History of the Second World War* noted that 'During the period October 1944 to June 1945 … there were more deaths in the European theatre due to a single agent, alcohol poisoning, than to acute communicable disease'.[106] Ed Laughlin, a US soldier who fought in Europe during the Second World War, provided a graphic account of the effects of alcohol poisoning in an unpublished manuscript. Laughlin tells us of two comrades, nicknamed Mutt and Jeff, who had earlier offered him some alcohol.

About 8 P.M. there was a lot of moaning and crying out … from where Mutt and Jeff had their bedrolls. The moans turned into screams and someone ran to get the medics. By the time the medics got there, Mutt and Jeff were having convulsions, tearing at their stomachs … trying to vomit, blood was trickling out of their mouths. It was determined at the hospital that they were drinking wood alcohol [methanol], no one knew where they were getting it, and Mutt and Jeff died horrible deaths shortly after midnight.[107]

Catherine Merridale recounts a similar incident that occurred a few days before the German surrender in May 1945 when some soldiers found a canister of methanol. Eventually ten Soviet soldiers would die as a result of drinking the alcohol.[108] The Soviet war correspondent Vasily Grossman told of the deaths of two Soviet soldiers who had drunk their bottles of anti-yperite (an antidote for chemical warfare attacks), believing the liquid to be a form of alcohol, and noted that a number of others, celebrating the Soviet victory, had drunk out of barrels containing industrial poison stored in the Tiergarten in central Berlin. Grossman poignantly remarked that 'many of those men celebrating were living corpses'. The poison began to take effect on the third day after consumption.[109] Moreover, Colin Brown tells us that during the Korean War, 'it was strictly forbidden to purchase or consume Korean native-made whisky (known as "Lucky 7"), brandy or cider, since it generally contained high levels of methyl alcohol and other dubious ingredients such as boot polish and methylated spirits, which could cause blindness'.[110]

Alcohol also helped to ease the anguish of combat for aircrew, particularly for those squadrons based in England throughout the Second World War. When off duty, aircrew were regular patrons of the bars of the various messes on base or they would visit the pubs in the adjoining towns. Guy Gibson wrote of one occasion when snow blocked the roads off the base for a week: 'The Mess ran out of beer. A long-range Wellesley bomber on the other side of England, in response to urgent requests, came over low one day and dropped us a few crates by parachute, but the boys soon drank it all up.'[111] The loss of a plane and the death of their comrades usually generated a bout of heavy drinking among the surviving aircrew. The custom was to hold an impromptu wake in the officers' mess, with all drinks being charged to the dead pilot's mess account. This account would then be paid out of mess funds.

In subsequent wars, especially Vietnam, the temporary escape from reality provided by alcohol for the soldiers of the Second World War was now sought through the hash pipe or the needle. Illicit drug use was widespread among the American soldiers in Vietnam; by 1971, more soldiers were being evacuated because of drug abuse than for war wounds. That is not to say, however, that the use of alcohol by soldiers for 'self-medication' in the war zone had ceased. Alcohol was readily available in Vietnam, either supplied by the bars and clubs in the Vietnamese cities or in the clubs at the US base camps. Robert Mason wrote of visiting the officers' club at the International Hotel in Saigon: 'The bar served any drink you could name, made with American booze for a quarter ... as the bourbon flowed into my bloodstream, I began to warm to the occasion ... The events of the rest of the night are lost to me. I knew that both of us drank too much.'[112]

To avoid causing offence to local sensibilities, Saudi Arabia was officially an alcohol-free zone for the British soldiers deployed there during the build-up of forces for the 1991 Gulf War. The ingenuity of soldiers and their friends back in the United Kingdom, however, soon devised ways to contravene such proclamations. James Hewitt, who served with the Life Guards during the war, wrote of receiving a 'well-disguised square parcel from Fortnum and Mason containing a bottle of Jameson's Irish Whiskey encased in polystyrene'. Hewitt recalled that the authorities soon grew wise to the Fortnum and Mason whisky packages, 'so whisky found its way to the men in shampoo bottles and, of all things, chocolate boxes.'[113]

During the 2003 Iraq War, the US military in Iraq was also officially 'dry' – that is, the owning, making, drinking or selling of alcohol was forbidden. But as with their British counterparts a decade or so earlier, relief from abstinence was to arrive courtesy of care packages from home. Colby Buzzell, who served with the US Army in Iraq in 2003–04, wrote of receiving two 22-ounce bottles of Guinness in a care package from his wife for his birthday, while a colleague received a care package containing bottles of rum, vodka, tequila and whisky.[114] Evidently, scrutiny of the contents of these packages was stepped up, because by 2007 soldiers were getting friends back home to mail them gin or clear rum in mouthwash bottles, the addition of blue or yellow food colouring completing the subterfuge. For those soldiers who chose not to trust the vagaries of the postal system, locally produced moonshine (known as 'hajji juice') was cheap and easily obtained.

The importance of superstition

The capricious nature of combat enhances the tendency of soldiers to be superstitious. Soldiers seek in the realm of the supernatural some means of exerting control over their fate, believing that they have lost all ability to control the course their lives are taking in the rational world. In *A Passionate Prodigality*, his memoir of the First World War, Guy Chapman observed that in the trenches, 'many men after a period of war took to themselves some mascot on which they pinned the little faith left in them, a guardian ju-ju, a mystic rite, to omit which spelled disaster'.[115] Paul Fussell writes of the British soldiers of the First World War:

> [N]o front-line soldier or officer was without his amulet, and every tunic pocket became a reliquary. Lucky coins, buttons, dried flowers, hair cuttings, New Testaments, pebbles from home, medals of St. Christopher and St. George, childhood dolls and teddy bears, poems or Scripture verses written out and worn in a small bag around the neck like a phylactery, [Siegfried] Sassoon's fire-opal – so urgent was the need that no talisman was too absurd.[116]

Soldiers' superstitious beliefs include attaching special importance to certain numbers, whether they are considered lucky or unlucky. For example, Willi Heilmann, who flew with the Luftwaffe during the Second World War, fell prey to dark forebodings after his thirteenth victory was indicated by being painted onto the fuselage of his aircraft, and 'felt that a line had already been drawn sealing my fate'.[117] Anniversaries would also take on a mystical significance. An American soldier of the Second World War recalled that he was greatly relieved when he realised his birthday was over: 'For some strange reason I had persuaded myself that if I could live through my nineteenth birthday, I could make it all the rest of the way through the war' (he did).[118] Heilmann wrote: 'A birthday is sacred to superstitious pilots. To fly on such a day is blasphemy.'[119] Steve Hughes, the regimental medical officer of the 2nd Battalion of the Marine Division during the Falklands campaign, developed a fixation that he would die on his 25th birthday. He survived. 'My birthday had passed; I felt a great weight off my mind', Hughes later remarked. 'Stupid of course, but although more dangerous times were ahead, I was never as frightened as I had been on 12 June.'[120]

Items of lingerie were considered especially lucky, particularly among bomber crews of the Second World War, many of whom also insisted on wearing articles of clothing associated with a close escape in the past. This practice is confirmed by Richard Holmes, who catalogues in *Firing Line* the talismans carried by a Lancaster bomber crew: a red and blue scarf, a hat, a girlfriend's brassiere, a yellow scarf patterned with red dragons, a silk stocking, a Land Army brooch, a pink chiffon scarf and a tiny bone elephant.[121] Given the high casualty rate among Bomber Command during the Second World War – over half of these men were killed – it is perhaps not surprising that they were among the greatest advocates of charms and talismans. Don Charlwood recalled that when they borrowed another crew's plane for a mission they were required to rub three times the belly of 'Yododhi, a fearsome being with a red and blue complexion' that was painted on the side of the fuselage, lest their guns freeze up, and also had to rub the horseshoe hanging over the navigator's table to ensure they didn't get lost.[122]

A generation later, Michael Herr noticed while accompanying US ground troops on patrol in Vietnam the blind faith many combat soldiers placed in their lucky charms: carrying around in their packs five-pound bibles, relics from enemy soldiers they had killed, girlfriends' under-wear, locks of hair, etc.[123] Tobias Wolff carried a heavy gold pocket watch throughout his tour in Vietnam. The watch had been passed down from his fiancée's grandfather to her father and then from his fiancée to Wolff. 'It went with me everywhere, rain or shine. That it continued to tick I regarded as an affirmation somehow linked to my own continuance, and when it got stolen towards the end of my tour I suffered through several days of stupefying fatalism.'[124] Raleigh Cash, who fought in the so-called

Battle of Mogadishu in Somalia on 3–4 October 1993, carried with him
a small medallion given to him by some Thai Rangers. According to the
Thais, this medallion would protect its wearer from all dangers except a
knife attack. Cash, along with a fellow forward observer, wore the medallion throughout the battle. Out of their team of eight they were the only
men not to be wounded. Cash proclaimed: 'I believed in that medallion
symbol so much that I decided to have it tattooed on the small of my back.
That symbol is with me now for the rest of my life.'[125] Andrew Exum, who
served with the US Army's 10th Mountain Division in Afghanistan in 2002,
would tie a piece of parachute cord around his right wrist at the start of
each mission, removing it only when he had returned safely to base.[126]

The historian John Ellis writes in *The Sharp End: The Fighting Man in
World War II* that many of the lucky charms carried into battle by soldiers
were mementoes of their civil existence, and that:

> They summed up the entire web of a man's relationships, defined him
> as a social being with an established place in the world rather than just
> an expendable unit in the military machine. If a man could provide tan-
> gible evidence that his existence had a reality beyond his own physical
> body, he was the more able to reassure himself that he was too impor-
> tant to be simply blown away.[127]

Another manifestation of soldiers' superstitious beliefs, identifiable in
accounts drawn from a number of conflicts, would be the need for pre-
battle preparations to be carried out in a fixed order. For example, the left
sock would always be pulled on first, then the right. The basis for such
rituals could usually be traced back to a narrow escape from danger on a
previous mission. Soldiers (as did aircrew) would also insist on retaining
articles of equipment or clothing that were associated in their minds with
a similar lucky escape.

Once a soldier had embraced the notion of a talisman/pre-battle ritual,
he would attribute all subsequent missions that he survived to its protection, rather than seeking more rational explanations such as superior
weapons, tactics and/or training. The belief in the power of these talis-

mans and rituals can become self-fulfilling, because when a soldier loses his battle charm he may come to feel that his death in battle is inevitable and therefore takes unnecessary risks. His death, noted by his comrades to have occurred shortly after the loss of his talisman, reinforces their belief in the efficacy of such objects. Mansur Abdulin recalled talking with a despondent Soviet tank commander before an attack during the battle for Stalingrad. The tank commander spoke of his imminent death and when Abdulin tried to raise his spirits by pointing out that they all had an equal chance of staying alive, he replied: 'I've been driving a tank since the first days of the war. This is my fifth machine. I'm finished today and that's it. You're just a kid, you don't understand. I've lost my mascot.'[128] The tank commander was killed in the subsequent attack.

Many accounts of combat tell of incidents where a soldier seemingly has prescience of his impending death, which he communicates to his comrades. A colleague of Edward Costello by the name of Brooks had a recurring dream concerning the headless body of a rifleman. A few days after recounting this dream to his companions, Brooks and Costello took part in the siege of a fortified town during the Peninsular War. Costello, who was a little ahead of his comrades as they made their way forward to a trench, heard the rush of a cannon ball and felt his jacket splattered with something. Turning, he 'beheld the body of Brooks headless, which actually stood quivering with life for a few seconds before it fell. His dream, poor fellow, had singularly augured the conclusion of his own career.'[129] Timothy Gowing wrote of an incident prior to the battle of Alma during the Crimean War where a fellow soldier 'told me he had a presentiment that he would fall in the first action'.[130] Gowing's comrade was shot dead the following day. Rice Bull noted the uncharacteristically melancholy demeanour of a colleague in the Union Army during the American Civil War and asked what was troubling him: 'He answered that for the last few days he had the feeling that he would not survive the next battle; he believed it would be soon and would not be surprised if it came that day.'[131] Later that day he was struck in the forehead by a bullet and died that night. An Australian soldier serving in France during the First World War told of an incident where a man had a premonition that he would be killed, and ... asked to get off that day, but the sergt. major thought that he was trying to get out of going up to the guns and roared him up. He was a quiet decent boy, and ... He said, 'Alright – I know I'm going to die, even if I stop here.' And he gave all his things away and went up, and just as he was leaving after having delivered his shells, I saw the shell land on him.[132]

Hugh Dundas recalled a visit from a fellow pilot who told Hugh that he wanted him to have his personal belongings in the event of his death. Dundas would write in *Flying Start*, his memoir of his service with the Royal Air Force during the Second World War, that he felt the pilot had had a premonition that he would be killed. 'If indeed he had such a premonition, it was fulfilled next day, when, in a dogfight south of Dover, George was shot down and killed.'[133] A similar incident occurred as the men of the US Army's 84th Division advanced into Germany during the Second World War, as described in *The Men of Company K*. 'While we were waiting Bill broke down. He was crying and in bad shape. He told me he couldn't go on, he knew he was going to be killed,' an American soldier recalled. 'Two of us tried to calm him down and reason with him. He finally got hold of himself and moved out with the rest of us. We should have told Bill to stay behind. He was killed a few minutes later out in the middle of that next open field.'[134] Perhaps soldiers, sensing their death, become more fatalistic and less cautious, exposing themselves to greater risk, thereby increasing the probability of their predicted demise becoming a self-fulfilling prophecy.

There is probably an element of selective memory in such reminiscences. On other occasions soldiers had likely predicted that they would shortly be killed, only to survive. Alex Bowlby, who fought with the British Army during the Italian campaign of the Second World War, tells us of a soldier going around informing everyone that he was going to die in a forthcoming attack. The premonition was fulfilled when a mortar shell exploded beside him. The following night, one of the slain soldier's comrades started 'talking about Meredith, how he'd known he was going to be killed. I reminded him that Meredith had said the same thing before

Peschiena [another attack].'[135] Yet it is the occasions when these predictions have been fulfilled that usually persist in memory and are recounted in narratives of combat, implying a mystic connection between soldiers and death.

Michael Herr observed that in addition to inanimate objects, some combat units in Vietnam would also place their faith in a 'charmed grunt', in whom the collective need for safety was personified. This soldier may have had a lucky escape, such as stepping on a dud mine, and now the other men in his platoon would stick close to him in combat, hoping that his 'aura' would protect them. This conviction would continue until this particular soldier became a casualty or rotated back to America, in which case the search would begin for the next 'charmed grunt'.[136] The first lieutenant of HMS *Avenger*, which had just survived an Argentine Exocet attack during the Falklands War, expressed a similar belief. The officer recounted the effect that the captain's calm demeanour had on the crew: 'He had an aura and was much admired. It was not hard to convince people that this ship was going to be all right. His luck was a talisman that seemed to protect the whole ship.'[137]

The closer a serviceman comes to the end of his tour of duty, generally the greater his faith in omens and charms. The US Army required that its soldiers complete a twelve-month tour of duty in Vietnam. A soldier's combat effectiveness noticeably decreased as the end of his tour loomed, the so-called 'Short-Timer Syndrome'. General Tommy Franks, who served as an artillery forward observer in Vietnam, felt that 'it was almost impossible for most line soldiers to concentrate on their jobs when they were short'.[138] 'I got very twitchy,' recalled Robert Mason as he neared the end of his tour in Vietnam. 'Being a short-timer made life difficult. It would almost be better not to know when you were due to return. As the day drew closer – only fifty days to go – the possibility of dying seemed more imminent, like I had already used up my breaks and would be getting it any day now.'[139]

The US Army's *Textbook of Military Medicine: War Psychiatry* states that the 'Short-Timer Syndrome' has been identified in conflicts where exposure to combat was limited to a set period (for example, US soldiers undertook nine-month combat tours during the Korean War) or number of missions (for example, the fixed number of missions for aircrew during the Second World War). In *The American Soldier: Combat and its Aftermath*, the psychologist Irving L. Janis drew attention to the fact that many aircrews commented in interviews after they had finished their tour of duty that the last two or three missions were the hardest of all. Janis noted that 'some men asserted that at times they wished they had not been told how many missions were required so that they could have avoided the intensified "sweating out" of their last missions'.[140] A similar sentiment was expressed by some of the commanding officers of air force squadrons interviewed by the American war correspondent Ernie Pyle during the Second World War:

When pilots got to within three or four missions of the finish, they became so nervous they almost jumped out of their skins. A good many were killed on their last allotted mission. The squadron leaders wished there were some way they could surprise a man and send him home with six or eight missions still to go, thus sparing him the agony of those last few trips.[141]

Esprit de corps: the psychology of small-group solidarity

Putting the belief in luck, talismans and omens aside, one of the key enablers of military performance is to place soldiers in circumstances where they have to fight or suffer the consequences of not doing so. When soldiers are placed in a life-threatening situation they have only two possible responses: fight or flee. If they choose to flee, then the entire military disciplinary apparatus is arrayed against them. They will also have to overcome the intense social conditioning designed to keep soldiers at the front, not least of which is the belief that flight from danger will result in their comrades considering them a coward. There is no greater disgrace for a soldier.

S.L.A. Marshall, who conducted extensive studies into the behaviour of men in combat during the Second World War, pointed out in *Men Against Fire* that fear pervades the battlefield. But if a soldier was serving among men whom he had known for some time, then he would seek to hide his fear from his comrades to preserve his reputation. Marshall claimed that because of this social pressure, 'the majority are unwilling to take extraordinary risks and do not aspire to a hero's role, but they are equally unwilling that they should be considered the least worthy among those present'.[142] These sentiments were echoed by an American soldier of the Second World War who remarked: 'The thing that keeps a soldier going in the face of horrendous violence and unbelievable living conditions is simply self-respect and the psychological need for the respect of your fellow soldiers.'[143] Lord Moran observed that the need to maintain their reputation kept soldiers in the front line during the First World War and that few sought to avoid battle by reporting sick. 'The few sick that I saw in the trenches were old friends', he wrote, 'they came without hope, they had no reputation to lose, in the Company they were well known. These apart the men would not come near me until they were hit.'[144]

Small-group solidarity fulfils a vital role in preventing psychological breakdowns among soldiers. American military psychiatrists during the Second World War concluded that the psychological breakdown of most soldiers was precipitated more by a disruption to their immediate small group than exposure to a particular threat: 'The soldier lost his group relationship and ... forfeited all the strengths and comforts with which it had sustained him. As a member of the team he would have been able to take it, alone he was overwhelmed and became disorganised.'[145] An opinion shared by the senior psychiatrist for the British 2nd Army, who in July 1944 stated that 'the emotional ties among the men, and between the men and their officers ... is the single most potent factor in preventing breakdown'.[146]

Esprit de corps is strongest among soldiers who have endured the strain of combat together and is generally weakest among hastily assembled groups of men. Eugene Sledge observed that 'a man felt that he belonged to his unit and had a niche among buddies whom he knew and with whom he shared a mutual respect welded in combat'.[147] This factor was recognised by the US Marine Corps, whose policy in the Second World War was to return wounded Marines to their old company once they had recovered (the British Army followed a similar policy). By contrast, following the

North African campaign of 1942–43, the US Army instituted a policy whereby soldiers requiring hospitalisation were dropped from their unit's roll. When these men were released from hospital they were absorbed into the Replacement System and sent wherever the need was greatest. The possibility of returning to their old unit and their comrades was small, occurring only when the unit had an appropriate vacancy and had submitted a requisition for the soldier in question. This industrialised approach to combat embraced by the US Army during the Second World War reduced their soldiers to the status of a labour force. An American soldier sent to a replacement depot bitterly recalled: 'We were just numbers, we didn't know anybody, and I've never felt so alone and miserable and helpless in my entire life – we'd been herded around like cattle at roundup time.'[148]

Marshall studied at length the integration of battlefield stragglers into combat units during the Ardennes campaign of 1944. He observed that the individual straggler was of so little value to their new unit that they were considered virtually useless in combat. He concluded that these soldiers displayed an inherent unwillingness to risk danger on behalf of men with whom they had no social identity. Conversely, Marshall noted that three or four men who came from the same unit, and who therefore knew each other, would stay and fight if integrated into a new unit.[149] The historian Marc Bloch, who served throughout the First World War with the French 272nd Reserve Infantry Regiment, observed: 'I believe that few soldiers, except the most noble or intelligent, think of their country while conducting themselves bravely, they are more often guided by a sense of personal honour, which is very strong when it is reinforced by the group.'[150]

During the Second World War, even if a US soldier had come into a unit as a replacement, he would remain with that unit for the remainder of the war unless moved on by wounding or promotion. Over time, he usually bonded with the men of his unit and in most cases developed lasting friendships. What was the exception in the Second World War became the rule during the Vietnam War, with virtually all soldiers arriving in-country as an individual replacement to undertake a twelve-month tour of duty. A soldier's removal from combat was dependent not on victory, as was the case in the Second World War, but upon merely surviving their tour. The lack of experience, and in some cases incompetence, of the newly arrived soldier represented a direct threat to the survival of the other members of their unit. So new arrivals in Vietnam were often shunned rather than welcomed. In time, if he survived, the soldier would attain combat effectiveness; but then, all too soon, the looming end of his tour would focus his attention on personal survival rather than the prosecution of the war. The Australian Army sent formed units to Vietnam rather than individual replacements. The Australian soldiers often served alongside American units and detected the difference in cohesiveness between the two forces. Gary McKay, who served with the Australian infantry in Vietnam, wrote that:

> The Americans suffered from an individual rotation system which weakened teamwork and created disorder. American troops did not train together before they served in Vietnam and the constant replacement of their personnel did not allow teamwork or a cohesive spirit to develop ... Australians, on the other hand, were part of highly trained and coordinated units and knew each other well. Most of my platoon had been together for almost a year before we sailed.[151]

In Vietnam, the collective bonding experienced by American units of the Second World War was noticeably absent and was largely supplanted by a desperate band of individuals focused only on their own survival. A symptom of this attitudinal shift was the 'fragging' (so named because the weapon of choice for such acts was the fragmentation grenade, which was usually rolled into the victim's tent at night) of officers and non-commissioned officers whom the perpetrators considered too eager for combat. In such instances, the officer/NCO was deemed to be a greater immediate threat to the soldiers' survival than the enemy.

The Pentagon, mindful of the inherent problems of the individual replacement system, as borne out by the Vietnam War, sought to replace entire units rather than individuals in subsequent conflicts. Consequently, Congress gave the Pentagon 'Stop-Loss' authority after the Vietnam War, though the first time this authority was applied was in 1990 during the troop build-up for the 1991 Gulf War. At that time, a 'Stop-Loss' order was applied to the entire active component of the US Army. This order banned voluntary separation and retirement, while a corresponding 'Stop-Movement' order suspended reassignment of personnel. In November 2003, 'Stop-Loss' policies were enacted for all units deployed to, or scheduled to be deployed to, Afghanistan and Iraq as part of the Global War on Terror. An army spokesperson commented: 'The bottom line of this is unit cohesion. This way, the units train together, deploy together, fight together and come home together.'[152]

News from home

The desired unit cohesion, however, can be threatened if the military's value system is supplanted by external sources. If family and civilian friends, in particular wives or girlfriends, are ambivalent about or even fundamentally opposed to a soldier's continued participation in a conflict, then a shift in social identity may occur. In such circumstances, the soldier may place greater value upon returning alive and uninjured to his family and friends than upon his reputation among his fellow soldiers. Thus the soldier, not really caring what his fellow soldiers think about him, will become risk-averse and the small-group bonds that collectively provide an army's cohesion will begin to break down. To prevent such a situation, militaries actively encourage family members to be supportive of the cause and their deployed husbands/fathers/brothers – in effect, they should 'rally round the flag'.

One manner in which families can be supportive of loved ones serving overseas is by regularly sending mail. 'You will all think me very "exigent" and a great nuisance, but if you only knew what a comfort and pleasure it is to hear from home, you would pardon my constant demands for a few lines,'[153] entreated Henry Clifford in a letter to his father from the siege of Sevastopol during the Crimean War. 'No men ever hungered more for letters than we did,' wrote Rice Bull. 'What a joy it was to hear from father and mother, brothers and sisters, and friends ...'[154] Guy Sajer, then serving with a German transportation unit, remembered that everywhere they went among the dispersed German outposts on the Russian front they were greeted by the same question: 'Any mail?'[155] Gary McKay tells us that 'mail was an important part of our lives as we were busting our guts in South Vietnam and occasionally getting shot at, and it seemed the only people who really cared, apart from the men with you, were parents, wives and girlfriends'.[156] Hugh McManners commented in his memoir of the Falklands War that:

Everyone knows from the last war just how important the mail is for morale, but you really have to be in a situation like ours to appreciate it. Letters from home become the only truly private thing that you have, and the only proof of your existence as an individual outside the military machine.[157]

Patrick Cordingley, the commander of the British 7th Armoured Brigade during the 1991 Gulf War, tells us that in the lead-up to the conflict, 'the greatest morale booster was the "bluey", the forces aerogramme. These were free and we all made the most of them. Each day the brigade would send out thousands of letters, and would in turn receive just as many.'[158] A decade or so later, when the British soldiers returned to the Gulf for the 2003 invasion of Iraq, Tim Collins, the commander of the 1st Battalion of the Royal Irish Regiment, noted that in addition to surface mail (which he described as 'terribly slow'), each soldier was issued with a twenty-minute phonecard per week to be used for the satellite phones. He observed that 'being able to call home proved a considerable boost to morale'.[159] The British soldiers taking part in the invasion also received 'e-blueys'. An e-bluey was essentially an email sent to a computer server in theatre that was printed out, enveloped and sent forward to the troops, thereby avoiding the delays inherent with surface mail. Moreover, the progression from combat to stabilisation operations in Iraq and the consequent move into semi-permanent garrisons increased the internet access available to the troops, with most major bases having an internet kiosk for use by the soldiers to keep in contact with family and friends.

The morale boost generated by the arrival of mail normally results in mail being given one of the highest priorities for transportation. But mail (or for that matter phone calls home) can also cause problems for military commanders. Paradoxically, by reminding a soldier of his life beyond the military, mail can weaken the bonds that bind soldiers together. This sentiment is evident in the comments of a parachute officer who served in the Falklands War and was interviewed by Richard Holmes for his book *Firing Line*:

[H]e dreaded the arrival of mail because it reminded him that he had another persona: in addition to being merely a cog in a military machine and of little individual value, he was also a husband and a father whose death would have devastating consequences. Remembering his role as a family man made him feel uneasy when the situation demanded that his military role should be dominant.[160]

Providing home leave for soldiers can also be problematic from a military effectiveness viewpoint. If a soldier's family is directly threatened by the enemy, as was often the case for the Israeli soldiers fighting the Yom Kippur War of 1973, then home leave may emphasise to the soldier what he is fighting to protect. But, in the absence of this direct threat, home leave may merely serve to remind the soldier of the discomforts of soldiering, that others are risking less than he, and the effect of his absence on those he loves.

The military unit as family

Militaries tend to define in abstract terms the reasons for soldiers to continue fighting: such as duty, honour and country. This abstraction is particularly useful when the soldier's family, friends and/or home are not directly threatened by enemy action. The ever-present danger for the military is that a soldier may begin to dwell upon what he is being asked to risk – his life – and seek to determine for whose benefit he is risking it.

The young John F. Kennedy, while stationed in the South Pacific during the Second World War, wrote:

We are at a great disadvantage – the Russians could see their country invaded, the Chinese the same. The British were bombed, but we are fighting on some islands belonging to the Lever Company, a British concern making soap … I suppose if we were stockholders we would perhaps be doing better, but to see that by dying at Munda you are helping to ensure peace in our time takes a larger imagination than most men possess.[161]

A British soldier, recalling his initial reaction when the IRA ambushed his patrol in 1975, put it more brusquely than the eloquent Kennedy: 'I started thinking about my wife, family, mother, kids … What the fucking hell am I doing here?'[162] When a soldier ceases to think of society in abstract terms but rather focuses on individuals, he may come to feel that these individuals, safe and comfortable at home, and in the soldier's mind almost certainly oblivious to his privations and suffering, have no right to ask this task of him. Militaries would prefer that soldiers concentrate on the slight matters which they can actually influence, such as the specifics of their assigned task, rather than dwelling upon the larger issues of why they are risking their lives in combat and whether this risk is justified.

An introspective soldier is a hesitant soldier, so militaries emphasise the deeds of those who have gone before to encourage soldiers to submerge their own hopes and desires and embrace the collective spirit of the unit. An officer of the Scots Guards who served in the Falklands War maintained that 'we were confident that we would uphold the traditions of the regiment. That sounds rather pompous, but for a regiment like ours with such a long history, it is an important part of our motivation.'[163] Units with high *esprit de corps* so successfully inculcate regimental pride that individuals develop an ingrained sense of personal responsibility not to 'let the unit down'. Alex Bowlby, reflecting upon why some men could not handle the pressure of combat and deserted, concluded that they 'had been "closed" men, shut off from the tribal spirit that kept the rest of us going. They had fallen because of this. My own immediate source of courage – fear of disgrace – would soon have dried up without it.'[164]

The knowledge that their forebears in the unit were able to withstand the strain of battle gives comfort to those about to face their own baptism of fire. An officer of the 2nd Battalion of the British Parachute Regiment who fought in the Falklands War, reflecting upon the ties that bind a unit together, commented:

[W]e are a body of people welded together by our traditions, by our regiment, by a feeling of togetherness. We're a family of people and you have to remember that. We all know each other, we know each other's families. This is a body of people who would die for each other. If you run away, you're running away from all that. It's like withdrawing the love of your mother, it's that kind of commitment.[165]

This commitment is what sustains the soldier on his journey; which will end with his homecoming or with his death. The ritualisation of death by the military serves to ease the feeling among a soldier's comrades and, if present, family and friends, that the soldier's death was in vain and that he will soon be forgotten. Rituals vary from country to country and from service to service. George MacDonald Fraser recalled that following the death of his section commander in Burma during the Second World War, the dead corporal's military equipment was placed upon a groundsheet. Each member of the section came forward and substituted a piece of his own equipment for an item on the groundsheet. 'While it may have appeared to an outsider to be a callous example of soldiers taking an opportunity to rid themselves of inferior equipment', Fraser poignantly noted, 'of course it had another purpose: without a word said, everyone was taking a memento of Tich.'[166] Fallen paratroopers are honoured by placing the boots of the dead men in formation on the parade ground, while a riderless horse with boots reversed in the stirrups commemorates a dead commander. A killed pilot is commemorated by the so-called 'missing man' formation, whereby a single aircraft veers away during a flypast and pulls up sharply, climbing vertically while the other aircraft continue in horizontal flight. Units may even erect memorials at their home base to serve as a constant reminder of their fallen comrades, an example of which is the Clock Tower at the home base of the British 22nd Special Air Service (SAS) Regiment at Hereford. Engraved on the tower are the names of all members who have lost their lives while serving with the regiment. One of the aims of all such memorials, whether they are a simple shrine in a unit or a focus of national sorrow such as the Arizona Memorial at Pearl Harbor or The Wall (the Vietnam War memorial in Washington, DC), is to reassure the living that a death in battle will be remembered.

* * *

Soldiers undertake this journey on behalf of society. They are thrust into a foreign environment and stripped of their civil identity. They are moulded into instruments of state policy and sent out to kill or be killed for their country. Some will benefit from their combat experience, fondly recalling the intense comradeship of the battlefield. But all who return from conflict will be changed. They will have gained a greater appreciation of the simple freedoms and comforts that the majority of their civilian compatriots take for granted and, for most, their greatest desire will simply be to live in peace. In May 1962, General Douglas MacArthur, in his final public speech, reflecting upon 52 years of military service, told the cadets at West Point: 'The soldier, above all other people, prays for peace, for he must suffer and bear the deepest wounds and scars of war.'[167]

Chapter 4

THE COST OF WAR

Generals (and historians) measure the cost of war in terms of the extent of territory lost, the amount of equipment destroyed and the number of men captured, wounded or killed; while economists measure the cost of war in resources used up and capital expended. But for the soldier the cost of war is not delineated on maps or entered into a balance sheet – the soldier measures the cost of war in terms of comrades killed and maimed. This sentiment was eloquently expressed in an open letter to *Time* by Sean Walsh, a young West Point graduate.

> The passing of the 4,000th service member in Iraq is a tragic milestone and a testament to the cost of this war, but for those of us who live and fight in Iraq, we measure the cost in smaller, but much more personal numbers. For me those numbers are 8, the number of friends and classmates killed in Iraq and Afghanistan, and 3, the number of soldiers from my unit killed in this deployment.[1]

Death can seek out a soldier in many forms. It may be instantaneous and unexpected, delivered via a sniper's bullet or the explosion of a hidden anti-personnel mine, or the soldier might have some warning of its arrival, such as the whine of an approaching artillery shell. But regardless of his level of skill and preparedness, even an adept soldier can be killed in battle.

'In the coming year we would learn how little our decisions determined our futures. Destiny is not born of decision; it is born of uncontrollable circumstances', wrote James McDonough when reflecting upon his tour of duty in Vietnam. 'Throughout the year to come, this point would be driven home time and time again. Rational decision making or technical and physical skills may save you once or twice. But a man in combat is exposed a thousand times.'[2] David Bellavia confronted this stark reality during the battle of Fallujah in November 2004 when he was informed of the death of his battalion's sergeant major.

> If they can kill Sergeant Major Faulkenburg, how have I survived? He was so much more skilled than I, so much more experienced than almost every other soldier out here. Is this more about luck than skill? If it is, we're all only one bullet away from Faulkenburg's undeserved fate.[3]

Pragmatism in the face of death

It is for this reason that soldiers are generally more pragmatic in their attitude towards death than other elements of society. The day before he was due to take part in the opening air offensive against Iraq in 1991, John Peters of the Royal Air Force and his fellow aircrew went back to their rooms, 'to put our affairs in order, which is to say, we prepared for the fact that we might die very soon. It sounds melodramatic, but you have to think this way, it would be wrong to leave it to someone else.'[4]

An example of this pragmatism is that during the First and Second World Wars the British Army paybook contained a will proforma. The pay-book was carried into battle in a soldier's left breast pocket so that it could be easily located in the event of his death. Moreover, Hugh McManners noted that while the British taskforce was sailing towards the Falkland Islands in 1982 the officers coaxed the men to ensure that each had made out a will, the men taking advantage of a provision enabling soldiers en route to war to make out legally binding wills despite being witnessed only by a mate.[5] Likewise, contemporary members of the US military are required to have designated beneficiaries for a death gratuity and unpaid pay/allowances and to have nominated who will decide what will happen with their remains prior to deploying overseas.

Another example of this pragmatic approach to death is that before going into battle soldiers may write a 'last letter'. These letters may be exchanged with a fellow soldier on the understanding that they will be posted only in the event of the author's death. Conversely, the soldier may keep the letter on their body, or with their possessions, so that it will be found and posted postmortem. Before joining battle with the French in November 1813, Edmund Wheatley of the King's German Legion (an element of the British Army) wrote in his diary: 'An order to stand under arms at three tomorrow – a battle for certain. Must pack up. The last time perhaps. Schuck [the battalion adjutant] has promised to send my two letters in case of death. Farewell to all! God help me!!'[6] Just prior to the storming of Sevastopol during the Crimean War, Timothy Gowing wrote in a letter to his parents:

I am determined to try and do my duty for my Queen and Country. I am glad in one sense that this hour has come; we have looked for it for months, and long before the sun sets that is now rising, Sevastopol must be in our hands. I will now say good-bye, dear and best of mothers; good-bye kind father; good-bye, affectionate brothers and sisters. This letter will not be sent unless I fall; I have given it open into the hands of one of our sergeants who is in hospital wounded, and if I fall he has kindly offered to put a postscript to it and forward it. May the God of all grace bless you, dear Parents, and help you to bear the pending blow.[7]

Some soldiers will send a letter home to help prepare their family for what appears to be their imminent demise. On 24 April 1915, the evening before the landing of the Australian forces at Gallipoli, Colonel (later General Sir) John Monash composed a heartfelt letter to his wife.

As this may be the last opportunity I have of talking to you, I want to say briefly that, in the event of my going out, you are to believe that I do so with only one regret, which is, the grief that this will bring to you and Bert. and Mat. For myself, I am prepared to take my chance. While, on the one hand, to win through safely would mean honour and achievement, on the other hand to fall would mean an honourable end. At best I have only a few years of vigour left, and then would come the decay and the chill of old age, and perhaps lingering illness. So with the full and active life I have had, I need not regard the prospect of a sudden end with dismay.[8]

A colleague of Monash, William Malone of the New Zealand Army, also felt the possibility of his death weighing heavily upon him as he prepared to lead his battalion in an attack against the Turkish position on Chunuk Bair at Gallipoli. In a letter to his wife he wrote:

I expect to go thro' all right but, dear wife, if anything untoward happens to me, you must not grieve too much, there are our dear children to be brought up. You know how I love and have loved you, and we have had many years of great happiness together ... I know that you will never forget me, or let the children do so. I am prepared for death and hope that God will have forgiven me all my sins. My desire for life, so that I may see and be with you again, could not be greater, but I have only done what every man was bound to do in our country's need.[9]

Three days after writing this letter, Malone was killed by friendly gunfire when an artillery shell fell short and burst above his trench.

Some 30 years later, Karl Binder, a member of the trapped German 6th Army at Stalingrad, wrote to his wife that:

Time is now so short that I must concern myself about the end of everything ... I have always endeavored to be decent, a comrade, a soldier. I have also tried to be a good husband to you and a good father to the children ... With us, Death is a daily guest. He has lost all horrors for me ... In case I fall, move to Schwäbisch Gmünd as soon as possible. Life is cheaper there ... Throw away my uniforms. The rest is yours ...

I wish you and the children all the best for the future. Let us hope that we shall be reunited in the other world.[10]

Matsuo Keiu, the commander of a Japanese midget submarine sent into Sydney Harbour on the evening of 31 May 1942 to attack Allied shipping, knowing that there was a high probability that he would not return from this mission, wrote a last letter to his parents which he left with a comrade.

I write this letter to you before I depart on the most important mission of my life. For twenty years, I have hoped and prepared for the honour of this task, and I am very happy, for I have achieved everything I wanted to in life. My resignation is set, and if I can do my duty steadfastly, then I will have nothing left to wish for.[11]

Robert Lawrence wrote his last letter home on board the QE2 sailing towards the Falkland Islands. He reflected upon how this practice had almost become a cliché: 'People going to war find themselves acting as they have seen people act in films about people going to war … writing in your letters what you have seen people in films writing in their last letters home.'[12] Cliché or not, the practice has continued up to the present day. 'I love you so much and if anything has happened, know that I was always thinking of you and the kids', wrote Jake Kovco in a last letter to his wife. Kovco was the first Australian soldier to die in Iraq following the 2003 invasion. The letter had been written following a suggestion made by his commanding officer prior to the unit's deployment. The letter continued: 'Please find our children someone who loves them equally and is fair to them. I want you to try and find happiness if anything does happen.'[13]

Although Kovco's last letter was found in his private journal after his death and then sent to his wife, modern communication technology has expedited the passage of such letters, now more likely to be sent via email than in the post. In some circumstances, mobile phones will be used to communicate the sentiments traditionally contained in 'last letters' directly to loved ones. In May 2000, Phil Ashby, serving with the United Nations in Sierra Leone and surrounded in a walled compound by members of the

Revolutionary United Front, made some final phone calls home before he prepared to break out of the rebel cordon with three companions. After going over his preferred funeral arrangements with a friend, he then spoke to his wife. 'I felt relieved to have spoken to her', Ashby wrote in his memoirs. 'I don't know what it feels like to die but I'm sure that in the split second before oblivion you know what's happening and, for me, it would have been unbearable not to have said farewell.'[14]

This pragmatic approach to death extends to soldiers envisaging what would happen with their personal effects postmortem. Wayland Dunaway noted that his comrades in the Confederate Army of the American Civil War discarded packs of cards as they made their way into battle, so as not to be thought a gambler if they were found on their body.[15] Lord Moran once destroyed a little rubber doll taken from a box of crackers because he thought it would look ridiculous if it was found upon his body.[16] Before the US Marines advanced into Kuwait in the 1991 Gulf War they were advised to remove any 'foreign matter' from their packs. Anthony Swofford wrote in *Jarhead* that 'foreign matter' included 'letters from women or girls other than our wives or girlfriends, and also pornography or other profane materials that wives and girlfriends and mothers might not like to receive after our deaths when our personal effects will be shipped to the States, directly to our home of record'.[17]

Perhaps the best example of the pragmatism of soldiers towards the likelihood of their death in combat is an incident recounted by Robert Graves, who recalled one of his soldiers winning £5 in a 'sweepstake' following an offensive. Before the attack began the platoon had pooled all its money. This money was secured in a dugout and was to be divided among the survivors and thus kept out of the clutches of the battlefield scavengers. Graves remarked: 'Those who are killed can't complain, the wounded would have given far more than that to escape as they have, and the unwounded regard the money as a consolation prize for still being there.'[18]

The fate of the fallen

The battlefield scavengers so despised by Graves' men were able to prosper due to the fact that, until comparatively recently, the bodies of slain soldiers, particularly those of a defeated army, received little in the way of postmortem care. Most likely the members of the victorious force would strip the bodies of their foe in the pursuit of booty and would take anything they could sell or use. Benjamin Harris recalled that after the battle of Vimeiro in August 1808, 'I strolled about the field in order to see if there was anything to be found worth picking up amongst the dead'. Harris found an officer from the 50th Regiment whom he recognised. Other battlefield scavengers had already got to the officer and had torn open his clothes and emptied his pocketbook. 'I had moved on but a few paces when I recollected that perhaps the officer's shoes might serve me', recounted Harris, 'my own being considerably the worse for wear, so I returned again, went back, pulled one of his shoes off and knelt down on one knee to try it on. It was not much better than my own; however, I was determined on the exchange, and proceeded to take off its fellow.'[19] Daniel Crotty was tasked with helping to bury the dead following the battle of Seven Pines in May/June 1862 and noted: 'In almost every instance the Union dead are stripped of their boots and shoes, coats and sometimes pants and shirts, pockets turned inside out, by the rebel robbers of the dead, who held that portion of the field before they retreated.'[20]

The absence of concern for the dead was an enduring feature of the Napoleonic Wars. William Wheeler, who fought in Wellington's army, related that after the battle of Salamanca in July 1812 he and his comrades gathered some dead bodies from the battlefield and stacked them on top of each other to make a windbreak, behind which they spent the night.[21] The scant regard accorded the bodies of slain soldiers is perhaps most evident in a short article that appeared in the *Observer* (London) on 18 November 1822. Wedged between paragraphs detailing the cost of oranges imported from South America and a fire affecting some haystacks belonging to a Reverend Dr Lord of Northiam, Kent, is the following missive:

War and Commerce – It is estimated that more than a million of bushels of human and inhuman bones were imported last year from the continent of Europe into the port of Hull. The neighbourhood of Leipsic [Leipzig], Austerlitz, Waterloo, and of all the places, where, during the late bloody war, the principal battles were fought, have been swept alike of the bones of the hero and of the horse which he rode.

From Hull the bones were sent to bone grinders in Yorkshire, where they were granulated and sold to farmers as fertiliser. The article went on to claim that:

[A] dead soldier is a most valuable article of commerce … It is certainly a singular fact, that Great Britain should have sent out such multitudes of soldiers to fight the battles of this country upon the continent of Europe, and should then import their bones as an article of commerce to fatten her soil![22]

Such was the fate of the fallen members of the gallant British squares at Waterloo.

A century later, the bodies of British soldiers would again be left to rot in the fields of Belgium. The nature of trench warfare during the First World War meant that the bodies of the men who fell in no man's land were usually not recoverable and were left out in the open, a fact that was evident and troubling to the men who served in the trenches. Max Plowman in *A Subaltern on the Somme*, his memoir of his service during the First World War, wrote of a fellow officer, Hardy: 'His own, almost his only, fear for himself is lest his corpse should be left unburied. He told me the other day he simply could not stand the thought of his body being left on the wire to rot, and he extracted a promise from me to do what I could if he were killed.'[23] A similar sentiment was expressed by Sidney Rogerson in *Twelve Days on the Somme*: 'Worse than all the anticipation of battle, all the fear of mine, raid, or capture, was the dread of being struck down somewhere where there was no one to find me, and where I should lie till I rotted back slowly into the mud.'[24]

If uncollected, the bodies would turn grey and then become black and bloated as they began to putrefy. Eventually the abdominal cavity would rupture and the decomposing body would be subsumed by the soil so that only the skeleton remained, perhaps to be buried by the subsequent explosions of shells. Thus, in the vivid imagery of the writing of Edmund Blunden, the skulls of the slain men would protrude through the soil like mushrooms.[25]

The construction of new trenches or the repair of existing ones would often expose the detritus of war, the bodies of dead comrades being periodically disinterred. Walter Downing provides a graphic account of the Somme battlefield: 'The dead lay everywhere. The deeper one dug, the more bodies one exhumed. Hands and faces protruded from the slimy, toppling walls of trenches. Knees, shoulders and buttocks poked from the foul morass.'[26] A French soldier, recalling the condition of the trenches at Verdun, stated: 'You found the dead embedded in the walls of the trenches, heads, legs and half bodies, just as they had been shovelled out of the way by the picks and shovels of the working party.'[27] But for the sheer variety of corpses, one would be hard pressed to top Rogerson's account of digging new trenches on the old battlefield at Loos in mid-1916:

Men vomited over the task of digging new trenches, for bodies were unearthed at every yard. The deepening of the front line turned a German officer out of the mud beneath our very feet. A sap led into an overgrown trench full of French skeletons of 1914. Most pitiful, the attempt to straighten a piece of trench broke into a dug-out where sat huddled three Scottish officers, their faces mercifully shrouded by the grey flannel of the gasmasks they had donned when death came upon them.[28]

On occasion, the bodies of the dead were even incorporated into defensive field works. In his memoir of the First World War, *Old Soldiers Never Die*, Frank Richards wrote of some trenches in High Wood on the Somme where the parapet had been built up with the bodies of British soldiers. The British troops had been under fire and were finding the ground difficult to dig. If one of their comrades was killed, his body was placed on the parapet and earth was thrown over him as the trench was deepened. Richards remarked that as he made his way down the trench, 'arms and legs were protruding. In one bay only the heads of two men could be seen; their teeth were showing so that they seemed to be grinning horribly down on us.'[29] German soldiers at Stalingrad, faced with a relentless Soviet advance and the prospect of having to dig into frozen soil, also resorted to constructing defensive field works out of the corpses of their comrades. Mansur Abdulin, a Soviet soldier who fought in this battle, recalled that the Germans opposing him did not dig trenches, but rather created temporary defences by building a low wall consisting of two or three layers of dead bodies and then covering the bodies with snow.[30] A decade or so later, the frozen bodies of Chinese soldiers were used by US Marines to construct a low wall in front of their fighting positions during the campaign at the Chosin Reservoir in the Korean War.[31] A similar measure was employed by the French soldiers defending the Elaine strongpoint against the Viet Minh during the dying throes of the French fortress at Dien Bien Phu. The besieged French piled up the bodies of their comrades in front of their trenches to provide some protection against the enemy's guns.[32]

The bodies of dead soldiers could provide protection from more than enemy fire. Guy Sajer, who served with the Wehrmacht on the Eastern Front, was on his way to the front line when a train loaded with Russian prisoners passed by. The horrified Sajer noticed that the Russians had stacked the dead to shield themselves from the wind.[33] But the soldiers of the Soviet Army of the Second World War, by necessity, became particularly hardened in their attitude towards their dead comrades. In 1941–42, the hard-pressed Soviet Army, reeling under the German onslaught and with new equipment and uniforms reserved for the armies then being raised, could not afford to leave valuable matériel on the battlefield. As with all armies, the weapons of the dead would be collected, but in accordance with an order dated 29 November 1942, the Soviet burial parties were also expected to retrieve 'greatcoats, tunics, hats, padded trousers and jackets, sweaters, gloves, boots, and *valenki* [felt snowboots]'.[34] Thus the soldiers were buried clad only in their undergarments.

These soldiers were buried across the vast Russian steppes as the Soviet armies blunted the German advance and then pushed the Germans back towards Berlin. The more static nature of the battles of the First World War, however, meant that at the war's end there was a much greater concentration of bodies. The Menin Gate memorial, located in the Belgian town of Ieper (Ypres), provides stark testimony to the number of slain British soldiers subsumed by the soil of Belgium. The memorial displays the names of the 54,896 Commonwealth soldiers who were killed in the Ypres Salient up to 15 August 1917 and who have no known grave. But it was found that even this enormous memorial did not contain sufficient space to list the names of all those who fell in the area with no known grave, and so those killed from 16 August 1917 until the end of the war are commemorated on a further memorial, located at the rear of the nearby Tyne Cot Cemetery, displaying the names of another 34,957 soldiers. Yet the number of dead represented by these memorials pales in comparison to the Douaumont Ossuary at Verdun, which contains the remains of approximately 130,000 unidentified soldiers (both French and German) collected from the battlefield after the war. Many of the soldiers commemorated by these memorials were probably buried by their comrades during the war, but were likely placed in an unmarked grave or the record of their burial was lost. Max Plowman pointed out the temporary nature of many field burials: 'an inverted bottle with a bit of paper in it: a forage-cap hung on a stick: a rough wooden cross bearing the pencilled inscription, "To an Unknown British Soldier." These signs recur: pathetic, temporary memorials ...'[35]

The ebb and flow of the battle would often dictate whether or not a soldier's body was identified and placed in a marked grave. Martin Middlebrook notes that only 978 of the 7,485 British soldiers killed on 21 March 1918 (the first day of the German spring offensive) had identifiable graves. He writes that:

It was common practice for the victor in a big battle to recover his own dead from the battlefield first, identify the bodies where possible, and then bury them. The bodies of the enemy were usually buried without identification or ceremony in shell holes and other forms of multiple graves.[36]

Initially the French dead at Dien Bien Phu in the Vietnam War were placed in marked individual graves, but as the casualties mounted, each night a bulldozer excavated a shallow trench that was filled with bodies and covered over by the next morning. Eventually some 8,000 Viet Minh and 2,000 French troops were interred in the reddish earth of Dien Bien Phu. Post-war efforts at graves registration and a proposal to create a vast ossuary floundered in a climate of political mistrust. The bodies of the soldiers were left where they had fought and died, to be covered by the dirt brought by the monsoonal floods and the land reclaimed by the jungle or the crops of the villagers. With the exception of a small Viet Minh cemetery containing 500 bodies, there is no memorial to indicate the resting place of the thousands of others who fought and died in this remote valley in the north-west of Vietnam.

Conversely, a soldier's body may have been identified and placed in a marked grave, but this grave may have been obliterated by the ravages of nature or by deliberate action. For example, the Allied forces withdrew from the Gallipoli peninsula in December 1915 and did not return until after the Armistice. By then many of the temporary grave markers had been washed away or had become illegible. Consequently, less than one sixth of the Commonwealth dead at Gallipoli are buried in identified graves. On the Eastern Front during the Second World War, in an effort to deny Soviet intelligence the identity of formations involved in specific campaigns, as well as to keep the personal details and overall number of fallen soldiers from the enemy, military cemeteries were often stripped of grave markers when the German forces withdrew. Johann Voss bitterly recalled the graves of his fallen comrades being covered over as the German Army prepared to withdraw from Norway in late 1944. He commented: 'They said it was a measure of safeguarding the peace of the dead ... but watching the scene, our hearts were full of grief and despair, soothed only by the promise to those left unnamed in foreign soil: *You shall never be forgotten*.'[37] In some other theatres of the war a deliberate policy was enacted by the enemy to erase the identity of the buried soldiers. The German 6th Army established battalion cemeteries at Stalingrad, with each grave individually identified. After the war the Soviets dug up the German bodies and reinterred them in mass, unmarked graves (as they did for the vast majority of German graves on Soviet soil).

For other forces fighting the Second World War there would often be no body left for the enemy to disinter. In accordance with Buddhist custom, the practice among Japanese soldiers was for the bodies of their fallen comrades to be cremated and the ashes sent back to their family for burial. If the tactical situation did not permit the cremation of the full body, then the little finger would be removed and cremated and the rest of the body buried. If Japanese soldiers thought that their body might not be recoverable from the battlefield, they would cut their hair and nails, wrap them in paper inscribed with their regimental details and send this package to the rear prior to the battle, so that their family would at least have some part of their physical body to bury in the family grave.

The need for identification

For the wars of the late 19th and 20th centuries, the identity of slain soldiers could usually be established. But for much of the history of human conflict there were no practical measures to identify bodies that had been so disfigured by injury, or became so decomposed, as to preclude visual identification. If a soldier's body remained unidentified then his family might endure the emotional agony of being informed that their loved one was 'missing in action'; not knowing if he was dead or alive.

Before going into battle during the American Civil War, some soldiers would write their name, regiment and home address on a slip of paper that would be stuffed into a pocket or pinned to the back of their coat. They hoped that in the event of their death this information would prevent them from being buried in an unmarked grave. This was a realistic concern, as 42 per cent of the Civil War dead remain unidentified; though a major contributing factor here was which side held the field at the end of the battle. Hamlin Alexander Coe, who served with the 19th Michigan

Volunteer Infantry Regiment, wrote of visiting the site of the battle of Chickamauga and being appalled by the manner in which the Confederate troops had treated the bodies of his Union comrades.

> They buried their own men decently, putting a board and inscription at the head of each, but the Union forces they covered so slightly that their hands, feet, and their skulls are now uncovered and exposed to the open air. They burned a great many and their bones are now lying with the ashes above the ground.[38]

Other methods of assisting postmortem identification during the Civil War included soldiers stencilling their name and regiment onto their clothing and equipment; or the men might be identified by the notebooks, letters and diaries they carried in their pockets. The more affluent soldiers could purchase commercially manufactured silver and gold pins inscribed with their details that they attached to their coats. A less expensive option was provided by the sutlers who followed the campaigning armies and who would hand-stamp a soldier's personal information onto a solid brass or lead tag. These tags, like the 'dog tags' of the 20th century, had a hole in the top for attaching a string or chain and were worn around the neck. There were several variations, but most featured an eagle, a shield and the inscription 'War of 1861' on one side and the soldier's name, company and regiment on the other. There was no official policy on battlefield identification. It was left to the individual soldier to decide what form of identification he would use, if any.

Official recognition of the need for standardised battlefield identification of soldiers' bodies would not occur until some 40 years after the end of the Civil War. Although a metal identification disk was first officially recommended for issue to US soldiers in 1899 during the Philippines Insurrection, it was not introduced until December 1906, and not made mandatory for US Army personnel until 1913. Initially a single circular tag, a second tag was added in July 1916. The second tag hung beneath the first on a separate small chain so that one tag could be interred with the body while the other was turned over to the Graves Registration personnel to record the death. During the Second World War, the oblong shape, still used today, replaced the circular identification discs and the information stamped on the dog tag was standardised. One tag was to be tied into the laces of the left boot while the other was to be worn around the neck to increase the probability of at least one tag surviving the mortal injury. In Vietnam the practice was to collect one tag and clamp the other tag in the dead man's jaw.

The German Army first used identity discs (*Erkennungsmarke*) during the Franco-Prussian War of 1870–71. The German disc was oval in shape and had two holes through which a cord was threaded, allowing the disc to be worn around the soldier's neck. During the First World War the German soldiers wore a single identity disc, which was divided in two by a perforated line. The front of the disc contained the name and residence of the soldier, followed by the details of the depot unit they had joined. The back of the disc detailed the soldier's regiment, company and roster number (equivalent to a serial number). The information was repeated on each section of the disc, enabling the bottom half to be snapped off if the soldier was killed. The bottom half was sent back to the depot (who would notify the family), while the top half was buried with the remains. The Wehrmacht also used a tag that could be snapped in two, but the information was restricted to the details of the soldier's depot unit and roster number, along with the soldier's blood group (the details were stamped on both halves of the disc).

Not all nations required their soldiers to carry identity discs, the absence of which doomed the fallen to unmarked graves. Addison Terry, a US artillery officer attached to the famous 'Wolfhounds' (27th Regiment of the 25th Infantry Division) during the fierce fighting of the initial months of the Korean War, tells us that 'because the Reds wore no dog tags and had no identification there was no point in separate graves and it had become SOP (Standard Operating Procedure) to bury them in groups of thirty-five to fifty'.[39]

The usefulness of a dog tag is predicated on the hope that it will survive, reasonably intact, the mortal wound. While this may be true for deaths caused by small-calibre weapons, this may not be the case when bodies are fragmented beyond recognition by blast weapons such as artillery shells, aerial delivered munitions or, as often occurred in Iraq following the 2003 invasion, an improvised explosive device (IED). Colby Buzzell, who served with the US Army in Iraq in 2003–04, observed that a number of his fellow soldiers were getting 'meat tags' prior to deployment, ostensibly to aid postmortem identification of their bodies. Buzzell explained that:

> A meat tag is basically your dog-tag information (name, Social Security number, blood type, and religion) tattooed on your side, usually under your armpit. Soldiers get the meat-tag tattoo so that when an IED blows them into a million fucking pieces, there's a better chance for their carcass to be identified.[40]

Since the beginning of the 20th century, forensic identification of human remains has principally relied on fingerprint and later dental records. But such methods have limitations; not enough remains may be present to identify by conventional means, or records may be incomplete or missing. The US Department of Defense, anticipating a large number of US casualties during the 1991 Gulf War, investigated using DNA for battlefield identification. By the beginning of Operation Desert Storm in January 1991, a system for the collection and storage of DNA samples was in place and DNA was used to establish the identity of two American soldiers killed during the war.

This validation of battlefield identification using DNA during the 1991 Gulf War prompted the Department of Defense to embark upon an ambitious project in 1992 to collect and store a DNA sample from every member of the US armed forces. By 2007, the military's DNA Registry contained over 5 million samples, providing a DNA record of all current US military personnel (active and reserve), as well as Department of Defense civilians and contractors deployed to operational areas. These DNA records were used to identify military victims of the 11 September 2001 terrorist attack on the Pentagon, as well as US casualties suffered during operations in Afghanistan and Iraq. The intention behind the establishment of the DNA Registry was to ensure that there would never again be another US military casualty buried under the inscription, 'Here Rests In Honored Glory An American Soldier Known But To God'.

Sending the remains home

Until relatively recently (in a historical sense), most men killed in battle were buried where they fell – more often than not in a mass grave at the site of the battle. The bodies of the slain were generally not returned to their family for burial unless the campaign occurred close to their village or town. The decomposition of the body, coupled with primitive means of transportation, usually precluded the movement of bodies over great distances.

Naval forces have a longstanding tradition of burial at sea, particularly if the ship's mission made it impractical to return the remains to shore. The deceased were sewn up in their hammock or a piece of canvas (traditionally the last stitch was inserted through the nose of the corpse to ensure that the sailor was not merely cataleptic), some cannonballs were added for additional weight, and the body was tipped over the side. Both the US Navy and the Royal Navy conducted wartime burials at sea up to and during the Second World War (the US Navy currently provides burial at sea during peacetime for active duty members and veterans upon request), with the Royal Australian Navy conducting at least one burial at sea during the Korean War. The contemporary naval practice is to store the remains until return to port or to airlift them ashore.

In the past, the return of a warrior's body to their home for burial was generally because of socio-political considerations based upon the appointment held by the deceased. Three examples of this practice are the funerals of Vice Admiral Viscount Nelson, Field Marshal Lord Raglan and Major General Sir William Throsby Bridges.

Vice Admiral Viscount Horatio Nelson was the commander-in-chief of the British Mediterranean Fleet that defeated the Franco-Spanish fleet off Spain's Cape Trafalgar on 21 October 1805. A French sharpshooter shot Nelson in the closing stages of the battle, the bullet entering his chest and shattering the spine. Nelson was taken below deck, where he bled to death.

His body was placed inside a cask of Spanish brandy to preserve it for the journey back to England. Nelson was laid to rest in the crypt of St Paul's Cathedral on 9 January 1806 (some two and a half months after his death). The other British casualties from the Battle of Trafalgar were buried at sea; the practice in the Royal Navy at the time was to throw the bodies overboard so that they did not clutter up the decks during the battle. Field Marshal Lord Raglan was the commander-in-chief of the British forces during the Crimean War. He died from dysentery on 28 June 1855. His was the only body to be shipped back to England from this campaign. Raglan was buried at the family seat in Badminton on 26 July 1855.

Major General Sir William Throsby Bridges was the commander of the Australian Imperial Force that landed on Turkey's Gallipoli peninsula on 25 April 1915. He died on 18 May 1915 after being shot by a Turkish sniper. Bridges was buried in a military cemetery in Alexandria, Egypt a few days later. A few months later, following a suggestion raised in the Australian parliament, Bridges' remains were exhumed and returned to Australia for a state funeral and reburial at a site overlooking Australia's Royal Military College at Canberra (Bridges was the founding commandant of the college). Of the 60,000 Australian soldiers who died overseas in the First World War, Bridges' were the only remains returned to Australia for reburial for some 80 years. It was not until 11 November 1993 that another of the fallen would return home, when an Australian soldier killed in France was reburied in the tomb of the Unknown Soldier in the Australian War Memorial's Hall of Memory.[41]

Conversely, socio-political considerations could also prevent the bodies of high-ranking military officers from being returned home for burial. General George S. Patton Jr., commander of the US 3rd Army during the Second World War, was mortally injured in a road accident near Mannheim, Germany, on 9 December 1945. Although paralysed from the neck down, Patton survived for a further twelve days, dying on the evening of 21 December. The original intention of his widow, Beatrice, who had flown from the United States to be at his bedside, was for Patton's body to be returned to West Point (the US Military Academy) for burial. At that time, the bodies of all American soldiers killed in Europe during the war remained overseas; none had been returned to the United States for burial. Lieutenant General Geoffrey Keyes, a close friend of Patton's, was fearful of the potential adverse reaction from the families of the other deceased American servicemen if Patton's body was returned to the US. It was left to Patton's doctor to put the matter to Beatrice who, when apprised of the situation, agreed at once that her husband would be buried in Europe. Patton was buried on 24 December in the American military cemetery at Hamm, Luxembourg (near the site of his famous victory during the Battle of the Bulge).[42]

Technological developments in the 20th century, in particular advances in transportation such as powered flight, largely negated some of the practical considerations that had previously restricted the movement of dead bodies. Although it was now possible to return home the bodies of all slain soldiers, official policy on the repatriation of war dead varied from country to country. The families of Americans killed during the First and Second World Wars were given the option of having the bodies of their kin returned to the United States for reburial (bodies were not repatriated until after the end of hostilities). If they chose to forgo this option, the body was removed from its temporary burial site, normally located on the periphery of the former battlefields, and reburied in one of the official American war cemeteries scattered throughout Europe, North Africa and the Pacific. Approximately a quarter of all American servicemen killed during the First and Second World Wars are buried overseas. The remainder were returned to the United States for reburial, less those listed as 'missing in action' or buried/lost at sea. The recovered bodies of all American servicemen killed during the Korean, Vietnam and Gulf Wars were returned to the United States for burial.

Conversely, it was the policy of the countries of the British Commonwealth to bury their war dead in the operational theatre in which they died. (For example, up until the Vietnam War, the vast majority of Australian war dead were buried overseas. It is only since Vietnam that the bodies of Australian soldiers have been repatriated as a matter of official policy.) This Commonwealth policy initially reflected the impracticality of returning the bodies of soldiers slain in far-flung imperial campaigns to their home country for burial, though this was less of an impediment for the soldiers killed in France and Flanders during the First World War. Initially, despite a ban on exhuming remains during the war, some wealthy and/ or influential British families successfully repatriated the bodies of fallen soldiers (usually officers) back to the United Kingdom. Major General Sir Fabian Ware, then head of the Graves Registration Commission, vehemently opposed this practice, emphasising that: 'The one point of view that seems to me to be often overlooked in this matter is that of the officers themselves, who in ninety-nine cases out of a hundred will tell you that if they are killed [they] would wish to be among their men.'[43] Ware, by now the vice chairman of the Imperial War Graves Commission (renamed the Commonwealth War Graves Commission in 1960), reinforced this view in a press statement that was released following the Armistice:

[T]o allow removal by a few individuals (of necessity only those who could afford the cost) would be contrary to the principle of equality of treatment ... The Commission felt that a higher ideal than that of private burial at home is embodied in these war cemeteries in foreign lands, where those who fought and fell together, officers and men, lie together in their last resting place, facing the line they gave their lives to maintain.[44]

The principal reason for returning the body of a slain soldier to their family is that the presence of the body at the funeral provides a focus for the family's grief. Furthermore, the body will most likely be buried locally and thereby allow regular visits from family members, unlike the bodies of soldiers interred in a distant official war cemetery that is rarely, if ever, visited by their loved ones. Yet the preference of many soldiers, as noted by Ware, is to be buried among the men with whom they fought and died. William Manchester wrote that his section was unanimously of the view that if killed in combat their bodies should be buried on the Pacific island on which they fell.[45] In July 1943, Patton indicated in a letter to his nephew that, 'If I should conk, I do not wish to be disintered [sic] after the war. It would be far more pleasant to my ghostly future to lie among my soldiers than to rest in the sanctimonious precincts of a civilian cemetery.'[46] The journalist Peter Arnett recalled being told that the dead soldiers of the French *Groupement Mobile* (Mobile Group) 100, ambushed at Mang Yang Pass by the Viet Minh in June 1954, as per their prior request, were buried upright where they fell, facing France.[47]

The relatives of the British dead of the Falklands campaign were given the option of having the bodies repatriated to the UK or reburied in the official Commonwealth War Graves Commission cemetery at San Carlos on the Falkland Islands. The officer in charge of establishing the cemetery would later comment that when he chose the site he felt that:

[N]o one should be shipped back to the UK. It seemed ridiculous, to be dug up and reburied in Aldershot military cemetery where you'd be forgotten in ten years' time. Better to stay here where nobody would ever forget you, with your mates in this beautiful spot, with the War Graves Commission taking care of everything.[48]

From a psychological perspective, the presence of a body helps bring about a 'sense of closure' in the grieving process. To help ease the grief borne by the families of fallen servicemen, governments will go to great lengths to recover the remains of those killed overseas. Since the early 1990s, improved political relations between the United States and the Socialist Republic of Vietnam have facilitated joint expeditions aimed at recovering the remains of US servicemen killed during the Vietnam War. The US Department of Defense has stated that 'achieving the fullest possible accounting for these Americans is of the highest national priority'. By mid-2003, the remains of nearly 700 Americans had been located, identified and returned to their families for burial since the end of the Vietnam War, though approximately 1,800 American servicemen are still listed as 'missing in action' from this war. The US Defense Force's Joint POW/MIA Accounting Command is currently attempting to locate and identify the 88,000 American servicemen still unaccounted for from the Second World War, the Korean War, the Cold War, the Vietnam War and the 1991 Gulf War.[49]

The US government is not alone in its desire to recover the bodies of the nation's war dead. In the period following the signing of the Armistice up to September 1921, Graves Concentration Units belonging to the British Army's Directorate of Graves Registration and Enquiries exhumed 204,650 bodies from the battlefields of the First World War and reburied them in the official war cemeteries. Despite the systematic and methodical search of the former battlefields, in the three years following the cessation of the general search in 1921, a further 38,000 bodies were uncovered by farmers and various battlefield scavengers. In the mid-1920s the number of soldiers' bodies being uncovered dropped to between twenty and 30 per week. Some 90 years after the end of hostilities, soldiers' remains were still being discovered at a rate of around 30 per year.[50] For example, in September 2006 Belgian gas workers, while laying a pipe, discovered the bodies of five Australian soldiers killed during the battle of Passchendaele. The bodies were reburied in a Commonwealth War Cemetery in October 2007. The lengthy delay between discovery and reburial was to allow for extensive research, followed by DNA testing, in an attempt to identify the bodies (two were identified). Furthermore, in 2007 the Australian government, aided by the volunteer organisation Operation Aussies Home (founded by three Vietnam veterans) whose efforts led to the discovery of the remains, repatriated the bodies of three Australian soldiers killed during the Vietnam War. In August 2008, the remains of the last Australian soldier listed as missing in action in Vietnam, Private David Fisher of the Special Air Service Regiment, were discovered (complete with identification tags). The identification of Fisher's remains left just two Royal Australian Air Force officers (Flying Officer Michael Herbert and Pilot Officer Robert Carver) still unaccounted for from Vietnam. In April 2009 the wreckage of their Canberra Bomber was discovered near the border with Laos and their bodies identified three months later, finally bringing closure to Australia's involvement in the Vietnam War.

The expectation that the body of their loved one will be repatriated provides some measure of comfort to the families of contemporary soldiers. This factor influences government policy with regard to the repatriation of war dead. In early 2003, Pentagon officials were considering the on-site cremation of American battlefield casualties arising from an Iraqi chemical or biological attack, in order to protect the health and safety of other military personnel, particularly those tasked with handling contaminated remains. This option was eventually discounted because of religious objections to cremation and the envisaged emotional effect that such an act would have on the deceased's family. A Pentagon spokesperson stated that returning soldiers' remains to families continued to be 'a top priority'. Accordingly, specially designed 'contaminated remains' body bags were trialled.[51]

Fear of mutilation

Although repatriation policies may assure soldiers of the return of their bodies to their loved ones, the condition that their body will be in is an enduring concern. In particular, the mutilation arising from the explosion of artillery shells causes a great and abiding fear. Lord Moran commented:

> There were men in France who were prepared for [death] if it came swiftly and decently. But that shattering, crude, bloody end by a big shell was too much for them. It was something more than death, all their plans for meeting it with decency and credit were suddenly battered down.[52]

This sentiment was shared by the French soldier Paul Dubrulle, who recorded in his journal the horror of the prolonged German bombardment of Verdun: 'To die from a bullet seems to be nothing; parts of our being remain intact; but to be dismembered, torn to pieces, reduced to pulp, this is a fear that flesh cannot support and which is fundamentally the great suffering of the bombardment.'[53]

In *A Brass Hat in No-Man's Land*, Frank Crozier provided a vivid, first-hand account of the dismemberment caused by the impact of artillery shells. He recalled that while showing a fellow officer around the trenches during the First World War they encountered a soldier carrying a sandbag filled with something. Crozier, suspecting the theft of rations, challenged the soldier to reveal what he was carrying in the bag. "'Rifleman Grundy,'" comes the unexpected answer. He is carrying down the only mortal remains of Grundy for a decent burial in a bag which measures a few feet by inches.' Crozier continued his inspection of the front line and encountered a soldier carrying a human arm:

> 'Whose is that?' I ask. 'Rifleman Broderick's, Sir,' is the reply. 'Where's Broderick?' is my next question. 'Up there, Sir,' says my informant, pointing to a tree top above our heads. There sure enough is the torn trunk of a man fixed securely in the branches of a shell-stripped oak. A high explosive shell has recently shot him up to the sky and landed him in mid air above and out of reach of his comrades.[54]

A generation later on the plains surrounding Stalingrad, Alexei Petrov looked out into a maelstrom of exploding German shells and was amazed to see a tiny figure, no more than three feet high, waving his arms. When he looked closer Petrov saw that it was the upper torso of a Russian soldier, whose legs and hip lay beside him on the ground, having been sheared off by a shell burst. The man looked at Petrov and tried to communicate but Petrov heard only the sucking in of air. The arms grew still and the man's eyes glazed over. The torso remained upright next to the rest of the soldier's body.[55] On the other side of the world in the Pacific theatre, William Manchester witnessed the evisceration caused by the artillery bombardments on Iwo Jima. He graphically recounted that 'there seemed to be no clean wounds; just fragments of corpses … You tripped over strings of viscera fifteen feet long, over bodies which had been cut in half at the waist. Legs and arms, and heads bearing only necks, lay fifty feet from the closest torsos.'[56] The impact of rounds fired by tanks had a similar effect on the human body. 'Those who lie here are not just dead bodies, with one wound in them or possibly with one part missing', wrote Günter Koschorrek, describing the aftermath of a round fired by a Soviet T-34 tank hitting the crew of a German 88mm gun. 'Here are individual lumps of flesh from arms, legs and buttocks, and in one instance from a head, on to which part of a damaged helmet still clings.'[57]

In later conflicts, the mutilating effects of mines and high-explosive booby-traps would generate the same dread among soldiers as did artillery fire for those who served in the First and Second World Wars. In the Vietnam War, the percentage of casualties killed or wounded by mines and booby-traps was almost four times greater than that of the Second World War. Gary McKay wrote in *In Good Company*, his memoir of the Vietnam War, that 'I think what frightened me more than anything was the danger from mines. I had known two platoon commanders in 2 RAR, Bill Rolfe and Pat Cameron; Pat had lost a leg and Bill had lost both. Mines were a real bastard and you could never tell where or when you would hit them.'[58] Tim O'Brien, who served as a private with the US Army in Vietnam, recalled:

You look ahead, a few paces and wonder what your legs will resemble if there is more to the earth in that spot than silicates and nitrogen. Will the pain be unbearable? Will you scream or fall silent? Will you be afraid to look at your own body, afraid of the sight of your own red flesh and white bone?[59]

O'Brien decided to be ultra-careful, trying to second-guess where the Viet Cong may have laid a mine, weighing up carefully where to put each foot down. He tried to step in the footprints of the soldier in front of him until this soldier turned around and cursed him for following too closely. One of O'Brien's comrades commented: 'It's an absurd combination of certainty and uncertainty: the certainty that you're walking in minefields, walking past the things day after day; the uncertainty of your every movement, of which way to shift your weight, of where to sit down.'[60]

A similar sentiment was expressed in Philip Caputo's memoir of the Vietnam War, *A Rumor of War*. Caputo, who served as an officer with the US Marine Corps in South Vietnam in 1965–66, pointed out that 'the infantryman knows that any moment the ground he is walking on can erupt and kill him; kill him if he's lucky. If he's unlucky, he will be turned into a blind, deaf, emasculated, legless shell.'[61] Caputo noted that the Viet Cong preferred to use command-detonated mines and booby-trapped high-explosive shells. The mines consisted of hundreds of steel pellets packed around a few pounds of C-4 explosive and often resulted in the amputation of arms, legs and heads. But even more dismembering were the booby-trapped shells. Caputo graphically recounted that 'Lieutenant Colonel Meyers, one of the regiment's battalion commanders stepped on a booby-trapped 155-mm shell. They did not find enough of him to fill a willy-peter bag, a waterproof sack a little larger than a shopping bag. In effect, Colonel Meyers had been disintegrated.'[62] Robert Mason, who served as a helicopter pilot with the US Army in Vietnam in 1965, graphically described in *Chickenhawk*, his memoir of the war, an incident where he was sent to pick up the casualties from a jeep destroyed by a booby-trapped howitzer round. The round had been buried in the road and was remotely triggered when the jeep passed by. When Mason arrived there were two survivors out of the six men travelling in the jeep. After the wounded (both of whom died on the flight back to base) were loaded on board the helicopter, the remains of the dead men were collected:

> The man that had lost his leg had also lost his balls. He lay naked on his back with the ragged stump of his leg pointing out the side door … Only the torn skin from his scrotum remained … The scurrying grunts tossed a foot-filled boot onto the cargo deck. Blood seeped through the torn wool sock at the top of the boot. The medic pushed it under the sling seat. I turned around and saw a confused-looking private walking through the swirling smoke with the head of someone he knew held by the hair.[63]

The overwhelming battlefield superiority of the American military at the beginning of the 21st century, coupled with the unlikelihood of opponents choosing to fight the US Army on a conventional battlefield, means that the greatest source of casualties for US soldiers on operations will continue to be command-detonated or booby-trapped explosive devices, rather than indirect or direct fire. In mid-2007, a report to Congress noted that improvised explosive devices (which includes roadside bombs and suicide car bombs) were responsible for over 60 per cent of all American combat casualties (both killed and wounded) in Iraq and half of all combat casualties in Afghanistan.[64] 'By the law of averages, whichever road you choose, there's eventually going to be a bomb waiting for you at some point', wrote Chris Hunter when reflecting on his service with the British Army in Basra. He continued:

> What makes it so damn hard is that there's so little you can do to defend yourself against the bastard things. You could be the most highly trained soldier on the planet, but no amount of training, skill or judgement can protect you from the random and indiscriminate effect of a bomb. It's nearly all down to luck.[65]

Desensitised to death

As a consequence of their continued exposure to death, many soldiers become accustomed to the sight (and smell) of bodies on the battlefield. Benjamin Harris recounted an incident where he came upon the bodies of three Frenchmen killed during the Peninsular War:

> War is a sad blunter of the feelings I have often thought since those days. The contemplation of three ghastly bodies in this lonely spot failed then in making the slightest impression upon me. The sight had become, even in the short time I had been engaged in the trade, but too familiar. The biscuits, however, which lay in my path, I thought a blessed windfall, and, stooping, I gathered them up, scraped off the blood with which they were sprinkled with my bayonet, and ate them ravenously.[66]

Colin Campbell, taking part in the siege of Sevastopol during the Crimean War, confessed in a letter home that 'The vast amount of fatigue, misery and death which is always before me here only begets a callousness which, I suppose, will disappear when one returns to a proper mode of life. The sight of a dead body at present excites no more emotion in me than the sight of a dead beetle ...'[67] John Casler wrote in his memoirs of the American Civil War that the soldier 'becomes familiar with scenes of death and carnage, and what at first shocks him greatly he afterwards comes to look upon as a matter of course'.[68] An Australian soldier, following the battle of Lone Pine at Gallipoli, wrote in a letter to his family: 'The dead were 4 & 5 deep & we had to walk over them: it was just like walking on a cushion.' Anticipating the reaction of his family, he added: 'I daresay you will be surprised how callous a man becomes.'[69] Gottlob Bidermann, although initially shocked when he first encountered the bodies of enemy soldiers as the German Army advanced across Russia, ruefully recalled that:

[I]n the months and years to come I would become benumbed to death on the battlefield and that such scenes would be commonplace to us all. In the months to come our reaction to the deaths we had witnessed would become callous and accepting. We would have searched the corpses for documents, collected weapons, and gathered equipment for our own use.[70]

But it is somewhat less common for soldiers to become oblivious to wounded men. Perhaps it is easier for them to develop an emotional detachment from corpses; after all, they can always look away. Bidermann confessed that 'throughout the long years of the Russian campaign, helplessly witnessing the badly wounded soldiers in their agony always profoundly affected me far more than when a comrade met an immediate and painless death'.[71] The cries of the wounded reach out to soldiers and remind them of their own mortality. 'In the few days we've been here we've heard the awful screams of the wounded – how terrible it must be to die lying on the frozen ground', recounted Günter Koschorrek, one of Bidermann's comrades on the Eastern Front. He continued: 'The thought fills us with horror – we might lie there, with nobody to help us.'[72] J. Glenn Gray postulates that the reason why soldiers have a greater aversion to wounded rather than dead comrades is that soldiers have a benchmark for pain, as pain is 'very real in the memories of everyone'. Personal experience of death, however, is abstract for all. Gray argues that:

[M]ost combat soldiers have witnessed enough gaping wounds and listened to the agonised cries of the wounded often enough so that they cannot consciously endure the thought of the same thing happening to them. Though the dread of death may be at the bottom of conscious processes at such moments, the fear of being painfully injured is much in the foreground.[73]

This sentiment was shared by Philip Williams, who fought with the Scots Guards during the Falklands War. 'As the injured were carried in on stretchers nobody looked at them. I didn't either. I suppose we were all scared we might see images of ourselves lying there all maimed and bloody. Not that I was afraid of dying', Williams wrote in his memoir, *Summer Soldiers*. 'When you're my age you don't even think that it's possible, not even when people are shooting at you. But I didn't want to be injured. That was the thing all of us dreaded, I think. Injury was far worse than death in our minds.'[74]

Witnessing men being wounded and killed on the battlefield changes the nature of the fear felt by soldiers. The abiding fear among soldiers untested in combat is that they will turn out to be a coward and will let their comrades down, as this accords with their limited frame of reference. But once soldiers have experienced their first battle, realised that they can cope with its dangers and are not a coward, the fear of social disgrace is largely supplanted by the fear of being permanently crippled or disfigured. For such men, having seen first-hand their fellow soldiers dismembered and maimed, the possibility that the same may happen to them is all too real. 'I was not especially afraid of fire from infantry weapons, because in my experience, if one is hit, chances are that one will either recover or will die quickly', wrote Georg Grossjohann, a German veteran of the Second World War. 'My Achilles heel was artillery fire, because I had seen the effects of shrapnel often enough to know that it could tear off a limb. I could only imagine how awful it would be to live without arms, legs, or with other cruel mutilations.'[75] Evgeni Bessonov, a Soviet veteran of the Second World War, expressed the same sentiment more succinctly: 'The worst thing is to become a cripple, it is better to die right away.'[76]

The most feared wound

The US Army's *Textbook of Military Medicine: War Psychiatry* states that: 'Wounds of the external genitalia are the most feared combat injuries.'[77] David Hackworth pointed out that soldiers being deployed by helicopter into an operational area in Vietnam would sit on their steel helmets, as 'few trusted the Huey's thin underbelly to keep them in full possession of their most vital organs'.[78] Dan Schilling recalled that when he was travelling through the streets of Mogadishu in October 1993 crammed into the back of a Humvee:

> It wasn't my head and torso that I was worried about, it was my groin. The way I was situated, with my back towards the front of the vehicle and my crotch facing the tailgate, I was scared to death that I would get shot there … in the back, man, there was nothing between your future family and high velocity rounds but a tin tailgate.[79]

So Schilling took off his pack, which contained two radios, and placed it between his legs.

Stephen Ambrose writes in *Band of Brothers* of an incident where the primary concern of a soldier pierced by shell fragments during a mortar attack in Normandy was not the large hole in his left buttock, nor the wound to his right wrist, but rather the blood seeping into his right trouser leg at the crotch. Fearing the worst, he asked a fellow soldier to take a look. He was OK. The soldier was greatly relieved. The two shell fragments had lodged in the top of his leg and, according to the wounded man, had 'missed everything important'.[80] Frank Hunt was severely wounded by the explosion of a mine while serving with the Australian Army in Vietnam in July 1969:

> I can still remember lying there and feeling the blood running down my knee on to my thigh. I immediately thought, 'Oh no,' and stuck my hand down my trousers and felt the old fellow: it was all full of blood. I thought to myself, oh fuck. I might as well be dead. I was more worried

about my balls than I was about my legs or whether I was going to die or not, because if they got me in the balls I wanted to be dead.[81]

A similar incident involved Simon Weston, a soldier with the Welsh Guards, who was aboard the *Sir Galahad*, a landing ship belonging to the Royal Fleet Auxiliary, when it was hit by Argentine bombs during the Falklands War. Despite sustaining horrific burns to almost half his body and seeing the skin of his hands come off in layers, his greatest concern was for another part of his anatomy. Weston turned to a medic:

'For God's sake, man,' I blurted, 'give us a situation report on the wedding tackle, will you?'

He lifted the waistband of my underpants and had a good eyeball. 'All present and correct,' he said. 'Looks in perfect working order.'

'Thank God for that.' At least one per cent was left. If the family jewels had been missing I don't think I'd have bothered coming back.[82]

A member of David Bellavia's company was wounded in the genitals during the US Army's assault on the Iraqi city of Fallujah in November 2004. The force of an improvised explosive device had embedded a dead-bolt lock from an exterior gate in the man's penis and some shrapnel had torn his scrotum. Reflecting on the incident, Bellavia wrote: 'Getting hit in the crotch is every soldier's worst nightmare. We can either dwell on it and drive ourselves crazy, of make fun of it. Laughter is our only defense.'[83]

The pervasive fear among soldiers of genital mutilation means that some of the most feared weapons are the anti-personnel mines designed to explode at waist height. The German 'S-mine' (*Schrapnellmine* 35) of the Second World War, colloquially known as a 'Bouncing Betty', was usually buried with just its igniters (three prongs) protruding above the ground. When triggered, usually by a foot depressing the igniters, a propelling charge would fling the projectile about a metre into the air, where it would detonate. Contained within the mine were 350 small steel balls that accelerated outwards at high speed and became the main wounding agent. The mine was designed to eviscerate and/or, as Stephen Ambrose comments in *Citizen Soldiers*, to inflict 'the wound that above all others terrified the soldiers'. Ambrose quotes a Lieutenant George Wilson (US Army), who claimed to have seen every weapon used by the Wehrmacht and who declared that the 'S-mine' was 'the most frightening weapon of the war, the one that made us sick with fear'.[84]

In recognition of this fear of genital wounds, the latest-generation body armour, the Interceptor body armour system worn by American troops in Afghanistan and Iraq, has a detachable groin protector. The groin protector is clipped on to the front of the flak vest and contains a ballistic insert (a ceramic plate designed to provide protection from small arms fire and fragmentation).

What becomes of the wounded?

Although measures such as the provision of body armour can provide some protection, the likelihood of soldiers receiving crippling or disfiguring wounds on the battlefield has never been well publicised, and is an issue that most militaries would rather avoid. War memorials invariably depict physically complete men standing resolute. Seldom do memorials portray a corpse and rarely, if at all, do they depict a crippled or severely wounded soldier. Soldiers killed in action are commemorated on 'rolls of honour' but the crippled, the infirm and the maimed are discharged from the service and pensioned off. Wounded soldiers hanging around barracks are considered bad for morale – as was discovered by Lieutenant Robert Lawrence of the 2nd Battalion of the Scots Guards.

Lawrence was severely wounded during the Falklands War (he was shot in the head). When he was discharged from the army after an extensive period of medical treatment he went back to his former barracks to bid farewell to his old platoon. Before he could do so, a new company commander asked him why he was visiting the barracks. When Lawrence told him that he had come to say goodbye to his men, the officer looked up briefly from his paperwork and remarked: 'You know, I don't think it's very good for morale for the boys to see you limping around the barracks like this. So if I were you, Robert, I'd hurry up and get out of camp.'[85] From a military effectiveness perspective, it is preferable to have soldiers worrying about whether they will be a coward in battle rather than dwelling upon how they would cope if they lost a limb. This psychological diminution of the inherent risks of combat helps make possible the necessary short-term perspective vital for the efficient functioning of armies.

But it seems that the military is not alone in this desire to avoid high-lighting the struggle of wounded soldiers to cope with a reduced quality of life. Although war films are one of the staples of mainstream cinema, rarely is the public exposed through film to the physical aftermath of life for a severely wounded soldier. One of the first films to broach this subject was the 1946 melodrama *The Best Years of Our Lives*. The film focused on a trio of Second World War veterans and their readjustment to civilian life. Harold Russell, a real-life double amputee, portrayed one of the trio, Homer Parish, a young seaman who had lost both hands in a torpedo explosion. Russell, a Second World War veteran, was a former demolitions instructor who had been severely wounded when a defective blasting cap exploded. He subsequently appeared in an army documentary that depicted the rehabilitation of an amputee. The documentary was seen by the director of *The Best Years of Our Lives* and resulted in Russell being cast. In the movie, Parish, whose hands have been replaced by prosthetic, articulated hooks, worries that his sweetheart since adolescence will feel only pity for him now rather than love. Although the movie concludes with Parish's wedding, the ongoing emotional and physical burden of his wartime service is still apparent. He is reduced to living off his monthly disability cheques and although he displays considerable dexterity with his hooks, one of the movie's most poignant scenes reveals how helpless he has become when he removes them at night. After he has been tucked into bed by his fiancée, she departs, leaving the door slightly ajar in case he needs to go to the bathroom during the night. The scene ends with a shaft of light revealing Parish crying in the darkness – with no hands to wipe away his tears.[86]

Later films, notably Oliver Stone's biopic of Ron Kovic, *Born on the Fourth of July* (1989), have also portrayed the struggle faced by severely wounded veterans as they are reintegrated into society. In January 1968,

Kovic, a sergeant in the US Marine Corps, was paralysed from the chest down after being shot during his second tour of duty in Vietnam. One of the film's secondary themes is Kovic's feeling of emasculation arising from his physical disability. He bitterly wrote in his autobiography that 'I have given my numb young dick for democracy ... Oh God Oh God I want it back! ... Nobody ever told me I was going to come back from this war without a penis. But I am back and my head is screaming now and I don't know what to do.'[87]

But the portrayal of combat, rather than its aftermath, has continued to be a much more popular topic for directors (and their audiences). The public tends to prefer wars neatly packaged for their consumption. There is a defined beginning and end; the scores (number of people killed) are tallied up for each side, the results posted and the winner declared. This simplification obscures the true and ongoing cost of war.

Of the 3 million soldiers who fought the American Civil War, approximately 622,000 were killed. This figure represents the loss of 2 per cent of the US population at that time. Yet even this enormous figure does not capture the true extent of the lives destroyed by the war. Another 500,000 men survived the war but were left permanently disabled. During the war, approximately 80,000 amputations were performed on soldiers. Such was the extent of lost limbs that one fifth of the revenue for the state of Mississippi in 1866 (the Civil War ended in 1865) was spent on artificial limbs.[88]

Three quarters of all operations performed by military doctors during the Civil War were amputations. This is a reflection of the nature of the wounding (approximately 70 per cent of wounds were to the limbs) and that surgical intervention was rarely attempted when the wound was to the abdomen or the head (approximately 90 per cent of such wounds were fatal). 'They [the surgeons] found many that required amputation', noted Rice Bull of the 123rd New York Volunteer Infantry Regiment in the aftermath of the battle of Chancellorsville (April–May 1863); 'the only treatment they had for [the] others was to give them a cerate with which to rub their wounds'.[89] Bull observed that as each amputation was completed, the removed limb was thrown onto an ever-increasing pile in full view of the wounded awaiting the surgeon's attention. Because of the absence of anti-septics and antibiotics, the primary concern of the Civil War surgeon was to remove the destroyed tissue before the onset of gangrene. Therefore all limbs with open fractures were amputated, usually within 24 to 48 hours of wounding.

The number of casualties suffered by the British Expeditionary Force of the First World War exceeded the losses of the American Civil War. Approximately 6 million British soldiers served in this force during the war, of which 750,000 (representing 12.5 per cent) had died on active service by the time of the Armistice in November 1918. In March 1930, twelve years after the end of hostilities, 1.6 million veterans were still receiving a pension to compensate them for a disability arising from or aggravated by their war service. The extent of the ongoing suffering wrought by the war is evident from the fact that approximately 42,000 of these former servicemen had undergone an amputation. Of those who served, almost 2.5 million (approximately 40 per cent) were either killed in action, died of illness or wounds, or suffered some form of disability for which state compensation was awarded.[90]

The economies of the other major combatants also had to bear the ongoing financial burden of rehabilitating and providing for the enormous number of men wounded during the First World War. For example, of the 13 million men who served in the German Army, approximately 2.7 million were permanently disabled by their wounds, 800,000 of whom were awarded invalidity pensions; while France had at least 1.1 million war wounded, of whom over 100,000 were assessed to be totally incapacitated.[91]

The United States military suffered almost 60,000 deaths in Vietnam. In addition, over 300,000 American soldiers were wounded, approximately half of whom needed hospitalisation. Of those hospitalised, approximately half (75,000) were classified as severely disabled upon their release, of whom 23,000 were classified as 100 per cent disabled. Among the American servicemen wounded in Vietnam, 5,000 lost limbs and over 1,000 underwent multiple amputations.[92]

The widespread availability of aeromedical evacuation and the sophisticated treatment facilities established in Vietnam by the US military meant
that many wounds that would have almost certainly been fatal in earlier
wars could now be treated and the patient stabilised. An example of a
soldier who survived horrific wounds was Lewis Puller Jr., who fell victim
to a booby-trapped howitzer round in October 1968. The explosion tore
off his right leg up to the torso and only a six-inch stump remained of
his left thigh. He lost the thumb and little finger of his right hand, while
his left hand retained only the thumb and half a forefinger. Additionally,
Puller suffered severe wounds to both buttocks, a dislocated shoulder, a
ruptured eardrum, his scrotum had been split and his body was peppered
with smaller wounds from shell fragments. He had been in Vietnam for
less than three months. He would be in and out of hospitals for the next
two years.[93]

Since the Second World War the lethality rate of wounds has been
steadily decreasing. The ratio of the number of deaths to the number
wounded was one in three for the Second World War, falling to one in four
for the Korean, Vietnam and 1991 Persian Gulf Wars. For the Iraq War, one
in eight wounded soldiers died from their wounds.[94]

The deployment of mobile surgical teams with combat forces, faster
medical evacuation from the battlefield and continuing improvements in
treatment, including quick-clotting agents applied at the point of wounding to control haemorrhaging, mean that soldiers routinely survive horrific
wounds, but the press largely ignores their plight. A Desert Storm veteran
recounted in an opinion piece, carried by the *Army Times*, that the media
coverage of Operation Iraqi Freedom seemed to overlook the number of
wounded and the extent of their wounds:

While the number of killed in action was carefully updated about every
other day, the number of wounded remained a dark figure – and probably still is … When it came to our wounded in action, Americans were
made to believe these numbers didn't really matter. The wounded were
OK – all would have a good future.[95]

A comment made by a pilot with the US Army's 159th Medical Company (Air Ambulance), who was responsible for battlefield evacuation, provides a stark contrast to this presumption: 'People say, "Well, he didn't die", but a lot of these guys have an arm blown off or their leg blown off below the femur. Their lives are still going to suck.'[96]

Ironically, recent advances in body armour (the Kevlar vest currently worn by US soldiers is equipped with ceramic plates that slip into pockets in the front and rear of the vest and is capable of stopping high-velocity rifle rounds) have increased the percentage of wounded soldiers who suffer crippling or disfiguring wounds. The body armour used by American troops in Iraq and Afghanistan ensures that the torso is well protected; the head is also partially covered by a Kevlar helmet. Therefore, the majority of wounds (over two thirds) suffered by US soldiers serving in Afghanistan and Iraq are to the unprotected arms and legs. By January 2007, over 500 American soldiers had had limbs amputated after being wounded in Iraq or Afghanistan.[97] The amputation rate for US soldiers wounded in Iraq was 6 per cent (as at November 2004), compared with an average rate of 3 per cent for previous wars. Additionally, about one fifth of the soldiers wounded in Afghanistan and Iraq and then evacuated to military hospitals in Europe suffered injuries to the face or neck that occurred below the protection of the helmet, many of whom may now require lifelong medical care. Injuries of this nature often involve irreversible brain damage, breathing and eating impairments, blindness and severe disfiguration. In the first eight months of 2004, the US Army's Walter Reed Hospital in Washington, DC treated 355 soldiers for traumatic brain injury.[98]

The number of American soldiers wounded in Iraq is greater than the cumulative total of American soldiers wounded in all other US conflicts since Vietnam. On 1 May 2003, President George W. Bush declared that 'major combat operations in Iraq have ended' and that 'in the battle of Iraq, the United States and our allies have prevailed'. The US forces then migrated from a combat to a stabilisation role. This new role required greater interaction with the Iraqi populace and necessitated American soldiers moving through Iraqi towns and cities in thin-skinned vehicles or on foot, rather than in armoured vehicles. This increased the vulnerability of soldiers to ambushes by Iraqi irregular forces. By late summer 2003, there were about twenty separate attacks on US forces daily, resulting in an average of ten American soldiers being wounded, many of them grievously. In addition to this background rate of wounding, spikes occurred during major US offensives. For example, the battle of Fallujah began on 8 November 2004 and that week 455 wounded US servicemen were evacuated to Europe for treatment.[99] By January 2009, over 31,000 American service personnel had been wounded in Iraq since the start of the war. Approximately half of the wounded men and women were unable to return to duty.

American soldiers who are severely wounded while serving in Iraq are evacuated to the Landstuhl Regional Medical Center in the south-west of Germany, which is the largest American military hospital outside the United States. The chief of neurosurgery at Landstuhl commented in April 2003 that the war in Iraq was 'disgustingly sanitised on television':

We have had a number of really horrific injuries now from the war. They have lost arms, legs, hands, they have been burned, they have had significant brain injuries and peripheral nerve damage ... These are young children; 18, 19, 20 with arms and legs blown off. That is the reality.[100]

In addition to physical wounds, a large number of American soldiers serving in Iraq were affected by the psychological strain of combat. Combat stress teams were deployed throughout the country and regional 'fitness centres' were set up to allow soldiers suffering from combat stress (previously known as 'battle fatigue') to relax for three days. These centres were equipped with proper beds and provided hot food and 24-hour counselling services. About a quarter of the soldiers examined by the combat stress teams were prescribed anti-depressants to enable them to continue to function in the field. Despite the preference to treat combat stress cases in the field and return them to their units as soon as possible, by January 2004, over 500 US soldiers suffering from psychological problems had been returned to America from Iraq.[101]

The relief of getting wounded

Yet such is the twisted reality of combat that wounds may be welcomed rather than feared. When soldiers have enlisted 'for the duration', a non-fatal wound provides the tantalising prospect of an 'honourable release' from a seemingly never-ending conflict. Soviet infantry of the Second World War served in the front line until seriously wounded or killed, or as they quipped, they ended up in 'the Department of Health (*zdravotdel*) or the Department of the Earth (*zemotdel*)'.[102] A similar sentiment was expressed by General Omar Bradley, who movingly evoked the plight of the American infantry soldier of the Second World War in the following passage from his memoirs:

> [T]he rifleman trudges into battle knowing that statistics are stacked against his survival. He fights without promise of either reward or relief. Behind every river, there's another hill – and behind that hill, another river. After weeks or months in the line only a wound can offer him the comfort of safety, shelter, and a bed. Those who are left to fight, fight on, evading death but knowing that with each day of evasion they have exhausted one more chance for survival. Sooner or later, unless victory comes, the chase must end on the litter or in the grave.[103]

A British soldier who served in the Ypres sector during the First World War recalled:

> The future seemed to be an endless vista of battles, each one worse than the last ... All our discussions ended by complete agreement on one point: that whatever might be the end for the nations, our destinies were clear enough. We would all be hit, and if we recovered we would return and be hit again, and so on till we were either dead or permanently disabled. The ideal was to lose a leg as soon as possible.[104]

Robert Graves noted that an infectious pessimism abounded among the 'old hands' of his battalion who had been at the front since the beginning of the First World War. For these men, surviving one battle merely offered the opportunity to be killed in the next. Graves claimed that 'to get a cushy one is all the old hands think of ... They look forward to a battle because a battle gives more chances of a cushy one, in the legs or arms, than trench warfare. In trench warfare the proportion of head wounds is much greater.'[105] Graves himself later succumbed to this pessimism, often volunteering to go out on night patrols because he felt that 'the best way of lasting the war out was to get wounded. The best time to get wounded was at night and in the open, because a wound in a vital spot was less likely.' He reasoned that 'Fire was more or less unaimed at night and the whole body was exposed. It was also convenient to be wounded when there was no rush on the dressing-station services, and when the back areas were not being heavily shelled.'[106]

A common expression among the soldiers who fought the American Civil War was that when a man was wounded he was said to have received a 'furlough', with the length of the furlough rated according to the severity of the wound. John Casler recounted that: 'A soldier who received a moderate wound was considered in luck, as he could go to the rear and get a rest and nurse his wound, wounded soldiers being the only ones furloughed.'[107] Similarly, among the British and Dominion soldiers of the First World War, a wound serious enough to get the soldier evacuated to Britain was known as a 'Blighty'. An Australian serving with the field ambulance in France recorded in his diary the reaction of the wounded soldiers:

Many a man smiles when he is told he will never be able to fight again or that he won't be right again for some months. Its [sic] Blighty for a spell anyhow, and probably back to Australia again, he may casually remark. The fellows shake hands with and congratulate their mates and brothers when they find that they have a wound that ... will most likely keep them away for a few months.[108]

The opportunity to be *hors de combat* was equally attractive to the soldiers fighting the Second World War. A Soviet veteran recalled that 'there was only one thought, to be wounded quickly, to get it over with, to get to a hospital, at least for a convalescence, for a rest'.[109] A non-permanently disabling wound sufficient to keep the recipient out of combat was known to German soldiers as a *Heimatschuss* (homeland shot). 'It is only a matter of weeks since I was dreaming of glory and heroism and was so full of élan that I was almost bursting', wrote Günter Koschorrek during the retreat from Stalingrad. 'Now I long for a *Heimatschuss* – because it appears to me to be the only way that I can, with any sort of honour, say goodbye to this soul-destroying environment and, for at least a few weeks, get away from this awful country and its gruesome winter.'[110] For the American soldiers of this war it was known as a 'million-dollar wound' and, as William Manchester proclaimed, was 'the dream of every infantryman'.[111] A US soldier of the Second World War recalled: 'Sgt Glisch came walking by me, heading rearward. There was a hole in his helmet and blood running down his face – a face that was covered with a boyish grin. That million dollar wound! I felt left out, and I wished I had a bullet through an arm or leg.'[112]

In the latter stages of the Second World War, the pressing need for infantry in the European theatre of operations curtailed the number of wounded soldiers who were permanently excused from combat. By late 1944, unless they had lost a limb, wounded soldiers would usually be returned to combat. In *Band of Brothers*, Stephen Ambrose quotes a soldier who claimed in a letter home that the 'gayest spot' in an American general hospital in England was the amputation ward, 'where most of the lads, knowing the war was over for them, laughed and joked and talked about home'.[113]

Self-inflicted wounds

Even when a soldier has not enlisted for the duration (as was usually the case for the world wars), there still may be a desire to be *hors de combat*. Michael Herr observed that as the casualties mounted among the Marines fighting for control of the Hue citadel during the Vietnam War – the Marines had roughly one casualty for every metre of the citadel taken – everyone wanted to get wounded.[114] But not all soldiers are willing to leave such matters to chance, and some will actively seek to be wounded. Frank

Richards, who served with the Royal Welch Fusiliers during the First World War, wrote of an incident where a soldier had his thumb shot off while bailing out a waterlogged trench.

Every one of us volunteered to take his place, but one man got the bucket first. He deliberately invited a bullet through one of his hands by exposing them a little longer than was necessary; but the bucket got riddled instead and he spent the remainder of the day grousing and cursing over his bad luck. Everyone was praying for a wound through the arm or leg, and some were doing their level best to get one.[115]

In a similar incident, an Australian officer of the First World War encountered a soldier walking along the top of the parapet. He later found out that the soldier was 'looking for a "blighty", was fed to the teeth with things and wanted to get out'.[116] An officer in the Royal Army Medical Corps during the First World War recalled that in one of his wards were 30 Sikhs, all of whom had been shot through the palms of their hands. The doctor considered it extraordinary that the only part of their body exposed to the enemy's fire was their hands. He concluded that they must have held their hands up above the parapet so that they could be shot and invalided home.[117] Some 90 years later, Joshua Key, while serving with the US Army in Iraq, also courted such a wound: 'While we took cover from flying bullets and shrapnel, I sometimes stuck out my arm, hoping that an enemy bullet might smash into it.'[118]

If the enemy was not so obliging to furnish a non-fatal wound, some desperate soldiers would take matters into their own hands. On the eve of the battle of First Bull Run in July 1861, John Casler, serving with the 33rd Virginia Infantry Regiment of the Confederate Army, was woken by a single gunshot in his regiment's line and wandered over to investigate.

I found one of the men had shot himself through the foot, supposed to have been done intentionally, to keep out of the fight, but the poor fellow made a miscalculation as to where his toes were, and held the muzzle of the gun too far up and blew off about half of his foot, so it had to be amputated.[119]

Almost a century later, General George Patton recalled encountering some American soldiers in a hospital with self-inflicted wounds, two of whom had been shot in the left foot and one in the left hand. Patton wrote that:

> It is my experience that any time a soldier is shot through either of these extremities there is a high probability that the wound is self-inflicted. I got out an order that, from then on, soldiers so wounded would be tried, first for carelessness, and then for self-inflicted wounds. It is almost impossible to convict a man for self-inflicted wounds, but it is easy to convict him for carelessness, for which he can get up to six months.[120]

To avoid disciplinary action, as well as the stigma associated with self-inflicted wounds, soldiers would seek to disguise the origin of such wounds. During the First World War some men wrapped sandbags or their overcoats around their arms or legs to avoid the telltale scorch marks on the skin from the rifle being fired so close. In the closing stages of the battle of Stalingrad, German soldiers, desperate to be placed on one of the last flights out, also sought to disguise their self-inflicted wounds. They did so by firing through a loaf of bread to eliminate the incriminating powder burns and instead of shooting themselves in the arm or leg – the favoured sites for self-inflicted wounds – they shot themselves in the stomach or the chest, thereby avoiding the suspicion of the doctors. Less drastic measures that soldiers would employ to gain a brief respite from combat included placing faeces under the eyelids or alcohol in the ears to induce inflammation, while the chewing of cordite would result in an increased temperature. In the freezing, waterlogged battlefields of Europe, perhaps the most expedient method to ensure evacuation was to induce the onset of 'trench foot'. Soldiers seeking to be evacuated in the tropics would increase their chances of contracting malaria by neglecting to carry out the prescribed preventive measures. David Hackworth recounted that among the soldiers he served with in Vietnam:

> Some slackers saw these illnesses [gastroenteritis and malaria] as a way out of the war, and … went out of their way to drink rice-paddy and well water without halazone tablets, or did not take their antimalaria pills and then went around inviting mosquitoes with their sleeves rolled up and shirts unbuttoned when the sun went down.[121]

The favoured site for self-inflicted wounds among Soviet soldiers of the Second World War was the left hand. Wounds to the left hand became so endemic that, regardless of the circumstances, such a wound was automatically considered self-inflicted and an attempt by the soldier to evade his duty on the battlefield. If a self-inflicted wound was suspected, the soldier faced summary execution at the hands of the NKVD (People's Commissariat of Internal Affairs) Special Departments personnel, or at the very least was sent to serve with a penal battalion. A few Red Army surgeons took the drastic measure of amputating the left hand to remove the incriminating wound before inspection by the NKVD.[122] Self-inflicted wounds were also a great temptation for the German soldiers retreating towards Germany in the face of the counter-attacking Soviet armies. Henry Metelmann, a German soldier of the Second World War, tells us that: 'Especially during the retreat there had been suspicions about self-inflicted injuries, for which the penalty was death, and all injured going back to the *Lazaretts* [hospitals] had to have a special report by their commanding officers as to the genuineness of their injuries.'[123]

Yet despite the prospect that a self-inflicted wound offered of surviving the war, the overall number of such wounds in any conflict has remained small. For example, the official history of the Australian Army Medical Services during the First World War noted that only 701 incidents of self-inflicted wounds were recorded for the 295,000 members of the Australian Imperial Force who served in France during the war.[124] The ties of comradeship and the demands of personal honour stayed the hand of many soldiers who might otherwise have been tempted. In *The Men of Company K*,

Harold Leinbaugh and John Campbell commented that the soldiers who were evacuated with self-inflicted wounds were notably absent from their otherwise well-attended company reunions after the war. They pondered: 'Are feelings about that particular past still too strong – for them, and for us?'[125] Eugene Sledge proudly stated that there were no self-inflicted wounds among the Marines of his unit. 'What was worse than death was the indignation of your buddies. You couldn't let them down.'[126]

* * *

Advances in science and technology have largely negated the concerns of earlier generations of soldiers in relation to the postmortem recovery and identification of their remains. Efficient medical evacuation and improvements in surgical and treatment procedures mean that soldiers now routinely survive wounds that only a generation or so ago would almost certainly have proved fatal. But the nature of many of these wounds and the soldier's ongoing struggle for recovery, combined with their readjustment to a diminished quality of life, are often not understood by a public who prefer their wars distant and defined. The true cost of war is only fully appreciated by those who must bear it.

Chapter 5

LOVE, SEX AND WAR

In the mythology of ancient Greece, Aphrodite, the goddess of love, had an affair with Ares, the god of war, and bore him three children. One of the children arising from their union was Eros, who fired arrows of passion into the hearts of mortals. According to J. Glenn Gray:

> [W]e spent incomparably more time in the service of Eros during our military careers than ever before or again in our lives. When we were in uniform almost any girl who was faintly attractive had an erotic appeal for us. For their part, millions of women find a strong sexual attraction in the military uniform, particularly in time of war.[1]

For much of the last few hundred years, soldiers have usually been incarcerated in a regimented environment that denied them the regular social interaction with women experienced by the rest of society. An example of this regimentation is that in the pre-1914 British Army a soldier needed his commanding officer's permission to marry. If permission was granted, his wife would be brought 'on the strength', would be moved with the unit and could obtain quarters abroad. Because of the financial liability incurred, commanding officers allowed only a limited number of soldiers to marry. A member of the rank and file (corporal and below) needed to have completed seven years' service, been awarded at least two good conduct badges and have at least £5 in savings before he could apply for permission to marry. This segregation from women, when coupled with the fact that soldiers are generally fit, young and in their sexual prime, produced an environment characterised by the objectification of women and a preoccupation with the sexual act.

In such an environment many soldiers readily believed rumours alleging that the authorities had added chemicals to the rations to reduce the soldiers' sex drive. One of the most persistent rumours among British soldiers during the First World War was that the army was adding bromide to the tea to reduce their sexual longings. Among the doughboys of the US Army the chemical culprit was identified as saltpetre (potassium nitrate). These rumours were usually harmless enough until soldiers refused to take prescribed medicine because they believed it would have an adverse effect on their sexual performance. During the Second World War, a number of Allied soldiers contracted malaria when they stopped taking their Atabrine/Mepacrine tablets because they were rumoured to make them impotent.

But soldiers' preoccupation with sex should not be seen merely as a consequence of the need of hormonally charged young men to satisfy their lust. Numerous studies have shown that the sexual act fulfils psychological as well as physical needs. Among these is the need to feel valued as an individual, to feel worthy of affection. The psychologists Robin M. Williams Jr. and M. Brewster Smith noted that soldiers are cut off from the normal sources of emotional support, such as family, close friends and their local community. Furthermore, the anxiety and insecurity generated by combat results in many soldiers adopting a short-term perspective on life. Soldiers exposed to combat therefore tend to embrace a more hedonistic lifestyle than their civilian contemporaries.[2]

Various strategies have been tried by military authorities to help slake the sexual longings of soldiers. The Roman Army had military brothels, staffed by enslaved women, attached to every campaigning army or garrison. Later, campaigning armies would be followed by hordes of camp followers, some of whom were the wives of the soldiers, though many others were prostitutes. For example, '400 mounted whores and 800 on foot' accompanied the Duke of Alva's army of 10,000 men that invaded the Netherlands in 1567.[3]

A traditional method for relieving soldiers' sexual longings was allowing them free rein to rape their enemy's women. The belief that the conqueror had the right to rape the women of a defeated foe has a lengthy historical pedigree. Homer wrote in *The Iliad* that Nestor, the Greeks' senior statesman, urged the warriors attacking Troy to 'let no man press for our return before he beds down some Trojan wife'. The ancient Greeks considered captured women as legitimate booty of war, as did the Hebrews of the Old Testament. The Book of Deuteronomy proclaimed that captured females could become the slaves, concubines or even wives of Jewish men. Even warriors on religious quests, such as the Crusaders of the Middle Ages, raped the women of the territories they conquered. One notable example of this behaviour was the sack of Constantinople (a Christian city) in 1204 by the knights of the Fourth Crusade. In the widespread looting and destruction that followed the fall of the city, few of the women who remained inside the city's walls escaped being raped. Not even the nuns were spared this desecration.

For thousands of years, raping your enemy's women was socially acceptable behaviour for soldiers and permitted by the 'rules of warfare'. Wartime rape also served to intimidate and demoralise a defeated enemy. In *Against Our Will: Men, Women and Rape*, Susan Brownmiller emphasises that:

> Men of a conquered nation traditionally view the rape of 'their women' as the ultimate humiliation, a sexual *coup de grâce* ... Apart from a genuine, human concern for wives and daughters near and dear to them, rape by a conqueror is compelling evidence of the conquered's status of masculine impotence.[4]

The yearnings of sailors for female companionship were traditionally even more acute than those of soldiers, as sailors were often sequestered on board ships for months at a time and camp followers were non-existent. According to Admiralty regulations, women were not permitted on Royal Navy ships of the late 18th and early 19th centuries when under sail, though they were allowed to remain on board when the ship was in port. As British seamen were rarely granted shore leave in home ports, because of fears of them deserting, boatloads of prostitutes would be rowed out to every ship of war when it sailed into harbour. A boatman would select the best-looking prostitutes from among those on the docks, each boat normally carrying up to ten women. Once the warship was secured at anchor, the boats were permitted to come alongside. The sailors of the lower deck would then come onto the boats and pick out a woman, paying the boatman a shilling or two for her passage to the ship. The woman would then descend to the lower decks with her 'husband'. It was observed that for a port visit of HMS *Revenge* in 1805, 450 women came on board for a crew of 600. While aboard, the women slept with the men, despite the total lack of privacy on the decks and each sailor being allowed only fourteen inches' breadth for their hammock.[5] In 1822, Admiral Hawkins of the Royal Navy commented:

> Let those who have never seen a ship of war picture to themselves a very large low room (hardly capable of holding the men) with 500 men and probably 300 or 400 women of the vilest description shut up in it, and giving way to every excess of debauchery that the grossest passions of human nature can lead them to; and they see the deck of a 74-gun ship the night of her arrival in port.[6]

War and the spread of venereal disease

Historically, the promiscuous tendencies of soldiers and sailors often entailed a lasting social cost. The sedentary nature of agrarian societies meant that few people left the confines of their village unless called upon to undertake a military campaign. Soldiers therefore became a vector for the spread of diseases, including venereal diseases, throughout Europe. Herodotus, writing in the 5th century BCE, recounted how the Scythian soldiers who plundered the temple of Aphrodite in the Syrian city of Ascalon 'were punished by the goddess with the female sickness', which also afflicted their descendants.[7] Although Herodotus' meaning is not clear, this is probably the first historical account of soldiers contracting and spreading a venereal disease (VD). Less speculative is the prevailing explanation of the spread of syphilis throughout Europe.

Syphilis was originally called the 'French disease' because its first recorded outbreak occurred in 1494 in the army of Charles VIII of France, which had been forced to break off its siege of Naples and was returning home. The prevailing theory accounting for the arrival of syphilis in Europe is that infected crew-members of Columbus' first voyage brought it back from the Americas in 1493. A number of Columbus' crew later joined the army of Charles VIII. Syphilis was initially spread by the returning soldiers but by around 1500 had reached epidemic levels in Europe. At the time, there was no cure, though mercury was widely used to treat the disease and did have some success in slowing down the onset of its latter stages. Unfortunately, the side effects of treating syphilis with mercury included loosened teeth, kidney damage, anaemia, tremors and various mental disorders. If syphilis reached its tertiary stage it could lead to dementia, crippling and disfiguring tumours on bones, and death. Death normally occurred because of the disease's weakening of the cardiovascular and nervous systems.

The suffering of the infected individuals aside, what was of greater concern to the authorities was that the spread of syphilis and other venereal diseases could lead to significant manpower wastage among soldiers and sailors. Venereal diseases have therefore been of continuing concern to military authorities, though the measures they have enacted to control the spread of such diseases have rarely been successful. The 1862 report of the Royal Commission on the Health of the British Army noted the army's high level of venereal infection (almost three times that of the French Army, which regulated prostitution), prompting the British Parliament to pass the Contagious Diseases Acts of 1864, 1866 and 1869. These Acts represented an attempt by the British government to regulate prostitution in the manner of most other European countries, whereby prostitutes were entered onto a register and expected to reside in designated parts of town and undergo a regular medical examination to detect the presence of sexually transmitted diseases.

The Contagious Diseases Acts provided for the policing of prostitutes in a number of specified ports and garrison towns with the intent of reducing the number of soldiers and sailors infected with venereal diseases. All prostitutes in these towns were required to be registered and have obligatory periodic genital inspections. Any woman suspected of being a 'common prostitute' could be apprehended by the police and forced to undergo a genital inspection to determine whether she was infected with a venereal disease. The police had wide discretionary powers in relation to whom they could accuse of being a prostitute. If a woman was so accused, it was up to her to prove her virtuousness; she was guilty until proven innocent. Additionally, if a woman did not acquiesce to the genital inspection, she could be arrested and brought before a magistrate and imprisoned if she continued to refuse the examination. If found to be infected, the woman would be imprisoned in a 'Lock' hospital (a hospital dedicated to the treatment of venereal disease) until she was no longer infectious (or became pregnant).

The Contagious Diseases Acts generated fierce opposition from feminist groups. These groups particularly opposed the Acts' underlying assumption of female culpability for the spreading of venereal diseases. Tellingly, the Acts did not require soldiers or sailors to be routinely inspected for the symptoms of sexually transmitted diseases. After an extended protest campaign by a broad coalition of groups – including Quakers, prostitutes, working-class men (whose wives and daughters were specifically targeted by the Acts) and feminists – the Contagious Diseases Acts were suspended in 1884 and repealed in 1886.

Sex and empire

While social mores handicapped attempts to regulate the prostitutes servicing the soldiers stationed in the United Kingdom, soldiers serving in the outposts of the British Empire faced a somewhat different situation. In 1883, the House of Commons conducted a Commission of Inquiry into the use of prostitutes by the British Army in India. The inquiry found that an Indian 'bazaar' was attached to each of the British Army's cantonments and that each bazaar contained a prostitutes' quarter. Reflecting the provisions of the Contagious Diseases Acts, which the authorities were then struggling to implement in the UK, the Indian prostitutes were registered and required to undergo a periodic medical examination. This system continued up to 1888, when the formal arrangements were abolished because of a public outcry in Britain. Henceforth, many of the prostitutes continued to follow regiments between cantonments, though now without the oversight of the periodic medical checks.

The end of the regulation of Indian prostitutes led to an increase in the incidence of venereal disease among British soldiers, prompting a memorandum from the Commander-in-Chief, India, Lord Kitchener, to his soldiers in 1905. The memorandum warned that the common women, as well as the regular prostitutes throughout India, were 'almost all more or less' infected with venereal disease and cautioned the soldiers that only by 'avoiding altogether the many facilities for indulgence which India affords' could they remain safe from infection. The memorandum went on to state that 'syphilis contracted by Europeans from Asiatic women is much more severe than that contracted in England' and provided a graphic description of the physical wastage caused by the disease.

It assumes a horrible, loathsome, and often fatal form through which in time, as the years pass by, the sufferer finds his hair falling off, his skin and the flesh of his body rot, and are eaten away by slow kankerous [sic] and stinking ulcerations; his nose falls in at the bridge and then rots and falls off; his sight gradually fails, and he eventually becomes blind; his voice first becomes husky, and then fades to a hoarse whisper as his throat is eaten away by foetid ulcerations which causes his breath to stink.[8]

Loss of manpower arising from venereal diseases was also a concern of the British military hierarchy during the Boer War of 1899–1902. Of the approximately 450,000 British and Dominion troops who fought in this war, over 20,000 other ranks (no figures are available for officers) were admitted to hospital for treatment of a venereal disease. The most troubling aspect of these infections was that almost half the men were infected with the then incurable syphilis.[9] But the prevailing Victorian-era sexual morality meant that sexually transmitted diseases were not discussed in polite society. Little had changed a decade or so later, as public fastidiousness still prevented an open discussion of the impact of venereal disease on military manpower. Consequently, the forces that fought the First World War suffered preventable manpower losses. A case in point is provided by the experiences of the Australian troops sent to Egypt in 1915.

All soldiers found to be infected with a venereal disease were discharged from the 1st Australian Imperial Force (AIF) prior to its embarkation for Egypt. Fear of a public outcry, however, hampered the efforts of medical officers to mitigate the spread of venereal diseases once the force was deployed overseas. Regimental medical officers were unofficially permitted to give lectures on the dangers of venereal disease but were unable to distribute condoms to the troops. So the Australian soldiers were unleashed on Cairo with a general warning about the dangers of venereal disease, plenty of money and a liberal amount of leave. The official medical history for the war noted that within a fortnight of the Australians' arrival in Egypt a 'startling outburst of venereal disease occurred'. Furthermore, the official history commented that although 'the outbreak came as a surprise to the military command, it was not unforeseen by medical officers who had studied the history of armies placed in a similar situation'.[10]

The previous policies were reversed in an attempt to reduce the incidence of infections, with regimental medical officers now encouraged to address the prevention of venereal disease with the men (chaplains addressed the associated moral issue), and condoms became freely available. The most punitive measure adopted was the stoppage of pay for men hospitalised with a venereal disease. The average period spent in hospital for a venereal infection was 35 days. The fact that the men had contracted such a disease would be evident by the entries in their pay book and would prompt embarrassing questions upon their return to Australia. Despite these measures, during the four months that the 1st Division of the AIF was training in Egypt, VD incapacitated over 2,000 men and resulted in 3 per cent of the force constantly being in hospital. Treatment for VD was described by the official history in these terms: 'the staff inadequate, the conditions deplorable, methods and equipment rudimentary.'[11]

Sex in the world wars: attempts at regulation

During the First World War there were approximately 417,000 hospital admissions for venereal disease (VD) among the British and Dominion troops (though some of these admissions were for relapses/reinfection). VD accounted for almost 19 per cent of all hospital patients. To quantify the manpower wastage this represented, each day in 1918, on average, 18,000 British and Dominion troops were in hospital in France because of a sexually transmitted disease – equivalent to seventeen battalions of men. The average time spent in hospital was four weeks for a gonorrhea infection and five weeks for syphilis.[12]

The primary source of venereal infection was the unregulated prostitutes that frequented the designated rest areas for the British troops. Frank Crozier recalled that when his battalion was using the recreational centre at Bailleul, France, 'despite the greatest care, our "other rank" casualties from venereal [infection] give greater cause for anxiety than our losses in the line. At last we catch the culprit – an infected girl who hops from camp to camp and ditch to dyke like the true butterfly that she is. Then all is well.'[13]

In addition to the unregulated prostitutes of France, another concern for the military authorities was the influx of whores who converged on army camps in Britain and the prostitutes who sought out soldiers returning to London on leave from the front, in particular those who congregated around Waterloo and Victoria railway stations. Reflecting the intent of the Contagious Diseases Acts of the mid-19th century, the Defence of the Realm Act was amended in March 1918 to make it an offence for a woman to pass on a sexually transmitted disease to a serviceman. The prescribed penalty was either a fine of £100 or six months' imprisonment.

To minimise manpower losses from venereal infection, the British Army initially permitted its men to visit the French-operated mobile field brothels. These brothels operated out of purpose-built vans that were driven around the rear areas of the battle zone. Brothels also flourished in some of the undestroyed towns near the front. The prostitutes in the brothels patronised by the British soldiers were given regular genital inspections by the British Army's medical officers in an attempt to restrict the spread of venereal infection.

Business boomed for the brothels of wartime France. Many young soldiers, knowing they had a good chance of being killed in combat, and not wishing to die a virgin, were eager to sample the brothels' wares. For the rest of the men, the absence of the usual social restraints eased their passage to the brothel's door. Frank Richards remembered a queue of about 300 men lined up outside a brothel in Béthune in northern France, patiently awaiting a few minutes with one of the four prostitutes inside,[14] while a brothel at Rouen serviced 171,000 men in its first year of operation.[15] From April 1918, however, following an outcry from the British clergy, the French brothels were 'officially' placed out of bounds.

During the war, British officers frequented the 'Blue Lamp' brothels, while the men were patrons of the 'Red Lamps'. The rank-based delineation of brothels was a prudent measure, as the absence of such a system could jeopardise the necessary social distance between officers and their men. By way of a postscript, during the Vietnam War, Philip Caputo and a fellow officer were embarrassed when they met up with some of their soldiers in a brothel. Caputo pointed out that 'officers are not supposed to be saints, but they are expected to be discreet'.[16]

The British were not the only combatants concerned about the spread of VD during the First World War. On the other side of the Atlantic, the American military authorities were also keen to control the spread of sexually transmitted diseases among the 'Doughboys' bound for the battlefields of Europe. In August 1917, the Secretary of War and the Secretary of the Navy jointly issued orders forbidding open prostitution within five miles of army or navy posts.

The most infamous of the American brothel districts shut down by the US military during the war was Storyville in New Orleans. Storyville was America's first legally designated prostitution district and had been created by city ordinance in 1897 in an attempt to limit and regulate prostitution in the city. Contained within Storyville's 38 blocks were numerous saloons and bordellos, which provided employment for hundreds of prostitutes. The mayor of New Orleans was ordered to shut down Storyville by the Department of the Navy in late August 1917. Following a message from the Secretary of the Navy to the mayor threatening, 'you close the red-light district or the armed forces will', Storyville was officially disestablished by city ordinance in November 1917. This action resulted in prostitution once again becoming criminalised and hence an underground and unregulated activity in New Orleans. A similar sequence of events resulted in the closing of the prostitution districts of other large American cities, such as New York.

The efforts of the US military to control prostitution near its major bases during the First World War met with limited success. The venereal disease rate among American soldiers stationed in the continental USA throughout the war was almost four times higher than that of the soldiers sent to Europe with the American Expeditionary Force (AEF). But even the relatively low rate of venereal infection among AEF personnel resulted in the loss of 6.8 million man-days and gave rise to 10,000 medical discharges.[17]

Venereal infections were also responsible for a steady drain on the manpower of the protagonists during the Second World War. Various measures were trialled to reduce the incidence of sexually transmitted diseases during the war – most were unsuccessful.

Venereal infection rates for the British Army varied between the different theatres of war and largely reflected the men's access to prostitutes. In the Middle East, the number of men hospitalised with VD in 1941 and 1942 was higher than the number of battle casualties. The British Army also had a large number of men hospitalised with VD during the Italian campaign. Indeed, VD was the most prevalent disease contracted by British soldiers in Italy in 1945. But the highest rate of VD was among the 14th Army in Burma, the so-called 'Forgotten Army'. In 1942, 1944 and 1945, the annual VD infection rate was approximately 70 British other ranks per thousand men. But in 1943, the annual rate was 158 men per thousand – the highest infection rate for the British Army in any theatre during the Second World War. The American forces also had high VD rates. Among the men of the 3rd Army during its campaign in Europe, the annual infection rate was 144 per thousand men, while the annual infection rate among the American forces in Italy during the latter stages of the war was 150 per thousand men.[18]

The high venereal infection rate in the Middle East theatre arose because the soldiers had access to its cities and towns while on leave. The generals in charge of the British troops in the Middle East initially conceded that as the men were likely to seek out and use the services of prostitutes, the most effective method of minimising the spread of venereal infection was to control the brothels. In Tripoli there were several designated 'official' brothels, whose prostitutes underwent regular medical inspections. These brothels were open from 1 to 6pm and were under the control of a Royal Army Service Corps NCO. Patrons were segregated, with separate establishments provided for 'coloureds', white privates and corporals, senior NCOs and warrant officers, and officers.[19] An 8th Army soldier recalled:

The army, with its detailed administrative ability, was able to organise brothels in a surprisingly short time and a pavement in Tripoli held a long queue of men, four deep, standing in orderly patience to pay their money and break the monotony of desert celibacy. The queue was four deep because there were only four women in the brothel. The soldiers stood like units in a conveyor belt waiting for servicing ... Brothels for officers were opened in another part of town, where a few strolling pickets of military police ensured that the honoured ladies were not importuned by those who did not have the King's commission.[20]

But the unrelenting opposition of the chaplain general eventually brought about the abandonment of this policy. In 1940, General Bernard Montgomery had issued a directive concerning the prevention of venereal infection among the British troops in France. This directive suggested to his subordinate officers that rather than taking a disciplinarian approach and advocating chastity, the best method of preventing the spread of venereal disease was to encourage the men to seek rapid medical attention following intercourse. The directive so upset the chaplain general and the commander-in-chief that Montgomery was almost sacked. Montgomery, perhaps mindful of the official response to his previous efforts in this area, was an enthusiastic closer of brothels when he arrived in the Middle East. He placed the brothel district in Cairo out of bounds, and in co-operation with the Canadian and American military authorities, he closed down all known brothels in the operational area of north-west Europe.

As noted, the closing of official brothels served only to drive the prostitutes underground, where they no longer underwent periodic medical checks, and certainly did nothing to slake the desire of soldiers for their services. As usual, market forces prevailed. Despite the brothels in Naples being officially closed, in April 1944 the American Bureau of Psychological Warfare estimated that out of 150,000 nubile females in the city, 42,000 were either part- or full-time prostitutes.

In *The Sharp End: The Fighting Man in World War II*, John Ellis provides a vignette concerning the practical effects of closing official brothels during the Second World War. In Delhi, the British Army operated an official brothel known as the 'regimental' brothel, with a corporal manning the entrance. A soldier wishing to visit the brothel gave the last three digits of his army number to the corporal and paid five rupees. He was then handed a chit with a room number on it and a condom. The prostitutes were paid for their services and were given a weekly medical examination by members of the Royal Army Medical Corps. Once the existence of these 'official' brothels became known to certain sanctimonious elements of the British public, the resulting furore brought about their closure. This action drove the prostitutes out onto the streets or into illicit establishments. Within three weeks of the closure of the official brothel, every bed in the previously almost deserted VD ward of the army hospital was full, with patients overflowing onto the surrounding verandahs.[21]

The British and American governments also enacted widespread measures at home in an attempt to control the spread of venereal disease during the Second World War. In 1942, the British government launched the largest public health campaign then mounted, consisting of several hundred thousand posters distributed throughout the country warning of the hazards of casual sex. These posters were intended to shock. One of the designs consisted of a skull topped by a woman's veiled hat asking, 'Hello boy friend, coming MY way?', and warned: 'The "easy" girl-friend spreads Syphilis and Gonorrhea, which *unless properly treated* may result in blindness, insanity, paralysis, premature death.' More graphic measures were employed in the overseas theatres of war. A British officer serving in Cairo recalled how soldiers were forced to walk through a hall decked out with photographs of men and women showing the grisly effects of venereal infection, with their sexual organs eaten away by the disease.[22]

Other measures used to combat the increase in prostitution and the spread of venereal diseases on the home front included similar legal provisions to those put in place during the First World War. In December 1942, Regulation 33B of the Defence of the Realm Act enacted measures designed to control the spread of sexually transmitted diseases. The language of the regulation was deliberately non-gender-specific to encompass protection for all those engaged in essential war work, including women working in factories, though few doubted that its main targets were female prostitutes who infected servicemen. The regulation permitted doctors to question patients infected with a venereal disease about their sexual partners. If someone was identified as a source of venereal infection by two or more patients, a Medical Officer of Health was empowered to require him or her to submit to a medical examination and to remain under medical treatment until pronounced free from VD. Failure to comply with a 'treatment notice' was considered a contravention of the Act and was punishable by a fine or imprisonment. Regulation 33B was not repealed until 1947.

The policies enacted by the US government to control 'organised' prostitution near military bases during the Second World War were similar to those employed some 25 years earlier during the First World War. In July 1941, Congress passed the May Act. This Act gave federal authorities the legal power to suppress prostitution near military and naval establishments when local authorities were unable to do so. Under the provisions of the Act, base commanders were empowered to summon the FBI to shut brothels. Offences against this Act were considered misdemeanours and were punishable by a fine of US$1,000 and/or imprisonment for up to a year. Many base commanders were reluctant to enforce the provisions of the Act, as they regarded access to prostitutes as an essential element for maintaining the men's morale. But the threat of federal intervention caused the mayors and town councils of base towns to take the lead in the attempts to curb prostitution.

The May Act, as were all other official efforts banning prostitution, was largely ineffective. Easily targeted organised brothels were shut and prostitutes were forced to operate more surreptitiously. Many resorted to street-walking, while others solicited soldiers in bars. Some operated out of taxis or caravans (trailers), which could be towed away from camps and city limits when the authorities began to take too close an interest. As the prostitutes were no longer subject to any form of regulation, the venereal disease rate among the soldiers and sailors greatly increased.

The ineffectiveness of the May Act prompted the army to try a different tack, that of education. Posters warning of the dangers of VD were prominently displayed around bases. Every six months, GIs were forced to sit through graphic films showing the effects of VD (one such film was called *She May Look Safe, But ...*) and were issued a pamphlet entitled 'Sex Hygiene and Venereal Disease'. Other measures included lectures by medical officers on how to avoid contracting VD and the issuing of eight condoms a month to each soldier – 50 million condoms were distributed by the end of the war. To avoid a public outcry, it was claimed that the condoms were used to keep moisture out of rifle barrels. Soldiers also had monthly inspections of their 'short arm' (penis) by the medical officer. Paul Fussell, who served with the US Army during the Second World War, recounted the shock of his first short arm inspection:

It began with the ominous order, 'Put on raincoats and shoes!'
'Is that all?'
'Shut up. Do as you're told, and don't ask questions.'
In a nearby recreation 'hall' we lined up and passed before medical officers, who commanded each man to open his raincoat and then, unbelievably, to 'skin it down'. No yellow matter appearing, we were allowed to return to barracks and to change into our normal proletarian work clothes.[23]

Another preventive measure was the establishment of 'prophylactic stations' in the red-light districts. Soldiers who had engaged in sexual intercourse were 'officially' required to report to these stations, where their genitals would be subjected to a painful treatment – whereby an antiseptic solution was forced up the urethra by means of a tube, retained for five minutes and then expelled – that supposedly reduced the risk of contracting a venereal disease. The military also developed a preventive ointment that could be applied by soldiers after sexual intercourse if unable (or unwilling) to visit the prophylactic station. Known as PRO-KITs, this ointment was issued free to the men. These measures and treatments were not very successful, and large numbers of servicemen continued to report to hospitals with venereal infections.

Various punitive measures were also enacted. The British Army deducted 1s 6d per day from the pay of soldiers hospitalised with VD. Officially this deduction was for 'hospital charges', though it would also have raised questions from soldiers' wives or girlfriends about why their pay was reduced. The Americans implemented much tougher measures. In Italy they placed soldiers infected with venereal disease in barbed wire-enclosed stockades and made them wear uniforms with the letters 'VD' emblazoned across the back of their jackets and on their trouser legs.

Sex, infidelity and propaganda

But for those soldiers who remained faithful to their wives and girlfriends, their war service entailed other hardships. The length of the Second World War, coupled with the fact that many soldiers were serving in remote theatres and so denied home leave, resulted in widespread separation anxiety among the men, particularly in relation to the suspected infidelity of their wives and fiancées. This concern was exploited by enemy propaganda, which suggested to the front-line troops that, while they were enduring the squalor and danger of combat, their wives or girlfriends were living it up with other men. Normally the men cuckolding the combat soldiers were identified as civilians who had avoided the draft and were getting rich at home, or the soldiers of their allies who were enjoying the delights of the home country while others fought abroad. This particular form of propaganda reached its zenith during the Second World War and was mainly resorted to by the Axis powers. A German propaganda leaflet produced in 1944 showed an illustration of an English girl and an American serviceman re-dressing next to a rumpled bed with the caption, 'While you are away'. The reverse of the card stated:

The Yanks are 'lease-lending' your women. Their pockets full of cash and no work to do, the boys from overseas are having the times of their lives in Merry Old England. And what young woman, single or married, could resist such [a] 'handsome brute from the wide open spaces' to have dinner with, a cocktail at some night-club, and afterwards ...

Anyway, so numerous have become the scandals that all England is talking about them now.[24]

Likewise, the Japanese produced a propaganda card with an illustration of an Australian soldier standing in New Guinea, above which is the caption, 'Australia Screams. The Aussie: What was that scream. Something up?' The rest of the illustration portrays an American officer holding a struggling young lady. Above the American officer is the caption, 'The Yank: Sh... Sh... Quiet, girlie. Calm yourself. He'll be on the next casualty list. No worry.'[25]

Soldiers' concerns about their wives' fidelity are not unfounded. When the 32nd Regiment of Foot reached Lahore, India in April 1848 it found the lonely wives of their comrades. Robert Waterfield, a soldier in the regiment, wrote:

The regiments to which their husbands belonged was up the country with Sir Walter Gilbert, and not having any one to watch over them, or to keep them within bounds, they came out in their true colors, and proved false to their plighted vows. There were some few exceptions, and I am afraid but few, and the scenes enacted by the false ones was, in some cases, disgusting in the extreme.[26]

A hundred years later, the wives of the British soldiers stationed in India were sent to Quetta for the duration of the Second World War. John Masters blissfully recalled that while there, 'Good girls grew lonely, naughty girls grew naughtier'.[27] Paul Fussell tells us that the concept of the 'Dear John' letter – whereby a soldier was informed by his wife or girlfriend that henceforth they would be with someone else – originated among the American soldiers of the Second World War.[28] So it is perhaps not surprising that one of the most popular songs of the war years was the subtle plea for fidelity entitled 'Don't Sit Under the Apple Tree with Anyone Else but Me'. Likewise, one of the most popular Russian poems of the Second World War, which was soon set to music, was the similarly themed 'Wait for Me'.

The official history of the British Army during the Second World War concluded that the psychological strain of a prolonged absence from home became apparent in most men after being away for two years. The policy then in force was that the longest period men should be kept overseas, in all but exceptional cases, was three years. The official history noted that a woman's fidelity usually broke down after two to three years of separation, commenting: 'The number of wives and fiancées who were unfaithful to soldiers serving overseas almost defies belief.' The official history went on to state: 'Nothing did more to lower the morale of troops serving overseas than news of female infidelity, or the suspicion of it.'[29]

Again, these suspicions are often well-founded. David Hackworth and his classmates encountered 'the lonely Army wife, whose husband was on an unaccompanied tour', when they attended an officers' training course at Fort Benning, Georgia during the Korean War. Hackworth fondly reminisced: 'These were temporary widows looking for company; if the magic was there, you had an odds-on first-night score, and guaranteed good meals and high times for the rest of your Benning experience, with no strings attached.'[30] Indeed, Hackworth, then a first lieutenant, had an affair with the wife of a major serving in Korea.

The suspected infidelity of their wives/girlfriends was also a source of angst for the American soldiers deployed to Saudi Arabia in preparation for the 1991 Gulf War. Anthony Swofford recounted how his Marine regiment had a 'Wall of Shame', where photographs of wives or girlfriends suspected of having affairs were affixed with duct tape. He noted that 40 or more photographs were attached to the Wall (actually a six-foot-tall post), many of which had narratives of the men's cuckoldry emblazoned across the duct tape.[31] Later, when reinforcements arrived in Saudi Arabia from the United States and joined Swofford's platoon, one of them revealed the extent of the adultery being committed by the wives of the deployed men. He commented that he had noticed cars pulling into the driveways of the 'war wives' late at night and leaving early in the morning. The Marines called the men cuckolding them 'ghostpeckers', which Swofford explained was 'a pecker that's fucked your old lady, but you'll never know'.[32]

A decade or so later, Joshua Key was taunted by one of his sergeants the day before he deployed to Iraq: 'It's a known fact that your wife is gonna start fucking some other guy, the moment you're out of the country. You wait. You'll see. It happens to them all.'[33] While Key's wife did remain faithful, one of the fellow soldiers in his company was not so fortunate. Key recalled overhearing a conversation as he waited in line for his turn on the phone. 'I could hear her saying that she was leaving him, and then I could hear a man bark into the line: "Your little bitch wife is my princess now." And then the line went dead.'[34] James Pritchard, a US Army chaplain based in Baghdad, recounted that in the first five months of 2008 he had 38 soldiers come to him to discuss marital problems, over a quarter of whom had found out that their wife was leaving them or having an affair. 'It's a big issue, especially with younger soldiers who've married somebody they haven't known very long', commented Pritchard. 'They suddenly have extra money coming in and the lifestyle of a spouse at home lends itself to extra-marital affairs. We've had soldiers go home and find the house empty, the wife and kids gone.'[35]

Given the abounding suspicion of infidelity, it is not surprising that wartime separation causes the divorce rate to spike. The number of divorces granted in Britain in 1918 – 1,000 – was unprecedented. During the Second World War, Compassionate Posting Boards were established by the British military in all major overseas theatres to cope with the incessant flow of applications for home postings because of actual or impending marriage break-ups. Furthermore, 90 per cent of the work of the Legal

Aid stations throughout the war was generated by soldiers petitioning for divorce.[36] In 1947, shortly after the end of the Second World War, another unprecedented total of 60,000 divorces was reached in Britain, while in America, between 1940 and 1946 the divorce rate doubled and reached 1 million divorces per year by the end of the decade. In the majority of cases, the grounds for divorce were men citing their wives for adultery.

However, this seems somewhat hypocritical given the well-documented patronage of brothels and the large number of casual affairs had by soldiers serving abroad, many of whom were married or engaged. A study commissioned by the US military in 1945 found that half the married men stationed overseas for more than two years had had an affair. A book on the soldiers of Patton's 3rd Army claimed that: 'The average soldier who landed at Utah Beach and survived to take Germany, the man who was neither stud nor sissy, probably slept with something like twenty-five women during the war – and few of them were, I might add, prostitutes.'[37]

Russell Braddon wrote fondly of the brothels located on the aptly named Lavender Street, which greeted the Australian soldiers of the 8th Division upon their arrival in Singapore in 1941 with a huge banner slung across the street proclaiming 'Welcome to the AIF'. The ardour of the Australians to sample the delights contained within was only temporarily diminished by the medical officer's announcement that 96 per cent of Malayan women had VD; a figure that was raised to 99.9 per cent two days later.[38] Likewise, American soldiers descended en masse on the brothels of Paris during the Second World War, while their British and Canadian counterparts provided a steady stream of patrons for the bordellos of Brussels. As had occurred in Italy, the devastation wrought on the economies of France, Belgium, Holland and Germany by years of war and occupation had produced severe shortages of necessities (as well as other items, such as cigarettes and petrol). These shortages meant that the soldiers' access to consumer items, particularly cigarettes, virtually guaranteed them female companionship, as sex was a commodity openly traded by the impoverished women of occupied Europe.

Fraternisation

But not all of the sexual encounters between British, American and Canadian soldiers and European women had such crass commercial overtones. Occupied Europe was largely bereft of young men, and many of the young women, long denied male companionship and caught up in the euphoria of liberation, were all too willing to satisfy the sexual longings of the advancing soldiers. The soldiers themselves, many of whom were apprehensive about whether they would survive the war, and knowing that they would shortly be moving on to another town or village, were only too happy to engage in a series of brief sexual encounters as they advanced into Germany.

Although fraternisation with the newly-liberated citizens of occupied Europe was encouraged, orders were issued prohibiting fraternisation with the Germans. British and Canadian troops caught doing so were liable for a stoppage of pay of seven to fourteen days for the first offence, up to 28 days' stoppage of pay for the second offence, and a court martial for the third.[39] American soldiers were subject to a US$65 fine if caught fraternising with Germans. Similar measures were enacted to restrict fraternisation between Soviet soldiers and German women. Orders were promulgated stating that any Soviet soldier seen entering or leaving a private home in Germany was to be arrested and punished.

The British, American and Canadian soldiers were initially hesitant in their dealings with German women. But the soldiers soon found them to be as sexually willing as the women they had left behind in France, Belgium and Holland, particularly as the collapse of the German economy meant that many German women were left with the choice of trading sex for goods with the Allied soldiers or living in poverty. Widespread fraternisation soon made a mockery of the no-fraternisation policy, which was not officially repealed until September 1945, some four months after the end of hostilities in Europe.

Market economy or command economy?

The Soviet approach to slaking the need for female companionship among their soldiers was somewhat different to that of their Western allies. A large number of women served with the Soviet military in the field, specifically as signallers and clerks on headquarters staff and as medical personnel (though others served as pilots, snipers and drivers). The ready availability of women enabled senior officers to take 'campaign wives'. Senior officers would arrange for their mistresses to be posted to their headquarters staff, even to the extent of creating a position, whereby they would accompany them throughout the campaign. This practice was widespread throughout the Red Army during the Second World War.

The Allies were largely content to let prostitution operate under the principles of a market economy, that is, in accordance with the principles of supply and demand. The approach of the Axis powers to satisfying the sexual needs of their soldiers varied considerably; in many cases they preferred a command economy model more analogous to their totalitarian political systems.

The German Army maintained a number of brothels during the Second World War in an attempt to control the spread of venereal disease among its soldiers. These brothels were officially designated as auxiliary military installations and the local military commander was responsible for their supervision and operation. Hans Richter provides an account of the functioning of such brothels in *The Time of the Young Soldiers*, his memoir of service with the German Army during the war. Richter wrote of an incident where one of his soldiers did not return from a task to fetch some water from a French town. Richter led a search party into the town and came across a large number of German soldiers patiently standing in line outside a house. 'At regular intervals the door opened, and a medical corps sergeant would let one soldier out and another in', observed Richter. 'Those leaving grinned broadly at the men in the queue, pointing back at the house with unmistakable gestures of satisfaction as they put away their paybook in their breast pockets.' He found the missing soldier near the front of the queue. 'Well, how was I to know it would take so long?', the soldier offered up by way of explanation.[40]

The German Army applied strict hygiene standards for the operation of its brothels. Local doctors, supervised by army medical officers, conducted regular genital inspections of the women. German soldiers were required to use a condom and military police were on duty to ensure the orderly operation of the establishments. In Western Europe, the German Army reserved existing brothels for its exclusive use, while in Eastern Europe, women were often seized and forced to work in the Wehrmacht's brothels.[41]

Many of the famous brothels of Paris (such as No. 12 rue Chabanais and No. 6 rue des Moulins) were reclassified as 'lodging for officers in transit', in accordance with the prudish nature of Hitler's Germany. To ensure that the services provided by these houses of 'lodging' were not misinterpreted, officers arriving in Paris on leave were handed a leaflet explaining that the hotels in question were under German sanitary control. With typical Teutonic thoroughness, the leaflet also provided the name of the brothel's local underground station and the address of the nearest sexually transmitted disease clinic. To prevent officers straying from the officially controlled brothels, the leaflet warned that '99.5 per cent of all venereally infected cases have caught their disease from uncontrolled prostitutes'.[42]

Japan and the 'Comfort Women'

The German coercion of the women of occupied Eastern Europe into prostitution has a correlation with the Imperial Japanese Army's use of 'Comfort Women'. However, to better understand the context in which the widespread use of Comfort Women arose, it is necessary to briefly examine one of the most shocking and barbaric incidents of warfare in the 20th century.

On 9 December 1937, the Japanese Army launched a massive attack on the Chinese Nationalists' capital of Nanking. The city fell four days later. The six-week orgy of violence that followed is generally known as the 'Rape of Nanking'. During this period it was estimated that 260,000 civilians were killed and 20,000 women were raped, most of whom were brutally killed afterwards. The Japanese soldiers raped girls less than ten years old, women over 70, pregnant women and nuns. An estimated one third of all rapes occurred during the day, often on the street in front of witnesses. Women were gang-raped by up to twenty soldiers. Those who resisted were killed immediately.[43]

The murderous rampage of their soldiers at Nanking shocked the Japanese government, but they were even more shocked by the international reaction to news of the massacre. To avoid further international condemnation, the Japanese Army established a network of military-controlled brothels. The primary aim of these was to provide Japanese soldiers with a regulated sexual outlet in the hope of preventing them from randomly raping the women of the occupied territories. Secondary aims of the so-called 'comfort system' were to enhance troop morale, to relieve the tension of combat by providing opportunities for recreational sex, and to minimise the incidence of sexually transmitted diseases among Japanese soldiers by the weekly medical inspection of the women employed in the comfort stations. The Japanese authorities were fastidious in the requirement to use condoms, though occasional supply shortages would result in the women contracting a venereal infection or becoming pregnant. Women found to be infected with VD, or who for some reason missed their weekly examination, were forbidden to work until cured or passed as fit (though there are reports of infected women being abandoned or even killed).[44]

The Japanese Army had first-hand experience of the manpower losses that could arise from venereal disease. From 1918 to 1922, elements of the Japanese Army took part in the intervention, alongside the Western democracies, against the Bolshevik forces in Russia. During this period the Japanese had the equivalent of one division (seven were committed to Russia) incapacitated due to VD.

In addition to its military-run prostitution network, the Japanese Army also held 'short arm' parades of their soldiers each month to check for signs of venereal infection. The penalties for infection included a drop in rank or a spell in the guardhouse. The women who provided sexual services for the

Japanese soldiers were officially known as Comfort Women, with the first 'comfort' house being established near Nanking in 1938.

The Comfort Women were drawn from various nationalities and social circumstances, though mainly from the lower classes. The women were mostly unmarried seventeen- to twenty-year-olds. Some were deceptively recruited by middlemen, initially being told that they would be working in restaurants or laundries, or that they were to form part of the labour draft for war industries. Others were abducted from their villages by Japanese soldiers, while some, such as the Dutch women, were forcibly recruited from civilian internment camps. The Japanese women were drawn from the lower classes and had previously worked as prostitutes in Japan; these women were reserved for Japanese officers and high-ranking civilian officials. The restricted access to Japanese women in the Comfort Women network was a reflection of the Japanese attitude of racial superiority over the other nations of East Asia. In terms of preference, Japanese and Okinawan women were the most highly regarded, then Koreans, Chinese and lastly the south-east Asians. These preferences were reflected in the fee that soldiers were charged for a visit to a Comfort Woman, which usually differentiated between ethnic groups.[45]

Estimates of the number of Comfort Women range from 50,000 to 200,000. The United Nations Human Rights Commission estimated that approximately 80 per cent of the Comfort Women were Korean, and that only about a third of these women survived the Second World War. A large number died of illness or enemy action (such as aerial bombing), or were killed by the Japanese Army – many being caught up in ritual mass suicide when defeat was inevitable. Others, unable to face the social stigma that a patriarchal south-east Asian society accorded to rape victims, remained in the former occupied areas and were employed in the post-war version of the comfort system. Of those who did return home, many found that they had become sterile because of their brutal treatment and the medical procedures they were subjected to in order to prevent venereal disease. Exacerbating the tragedy suffered by these women was the fact that many of their societies placed great emphasis on the production of heirs.[46]

Initially the comfort stations were operated by the Japanese military. Later, as the system expanded, direct military management was considered unnecessary and inappropriate and private operators took over. These men were required to meet the operating standards specified by the military and were given paramilitary status and rank. The armed forces maintained overall supervision of the Comfort Women system and provided health services and transport as required (the women were categorised as 'canteen supplies' on transport requisitions).

The women were usually paid for their services, though there was considerable variation in pay and working conditions depending on locality and the period of the war. They were expected to service up to 30 men a day, given little time off, and had their movements outside their comfort stations strictly controlled. The Comfort Women were accorded paramilitary status and deployed right up to the front lines, some even servicing soldiers in pillboxes and accompanying troops on mobile operations. George Hicks tells us in *The Comfort Women* that 'accounts are given of how the women would arrive with the ammunition – and sometimes even before essential military equipment'.[47]

The military occupation of Japan following the end of the Second World War merely resulted in a change of clientele for many of the Japanese Comfort Women. They were told by representatives of the Japanese Home Ministry that they would need to continue as Comfort Women to 'preserve the chastity of Japanese women from the foul hands of the Occupation forces'. Furthermore, other women, who had been mobilised in preparation for the final defensive battles for Japan, were directed into the Comfort Women system instead. Many others were forced into the system by economic necessity arising from Japan's wartime devastation and resulting economic stagnation.[48]

A network of brothels was established by police officials and Tokyo businessmen under the auspices of the Recreation and Amusement Association (RAA), which was provided with government funding. An advance wave of American troops arrived at Atsugi, just south of Tokyo, on 28 August 1945 and by nightfall had begun to congregate in their hundreds (a RAA executive estimated there were up to 600 soldiers in a line on the street) outside the first of the RAA brothels to open for business – *Komachien* (The Babe Garden). The initial 38 women were quickly increased to 100, with each woman servicing between fifteen and 60 clients a day. The cost to the American GIs for a short session with one of the prostitutes was 15 yen, equivalent to the cost of half a pack of cigarettes.[49]

By the end of 1945, around 350,000 US troops were in Japan serving with the Allied occupation force. At the same time the RAA was employing 70,000 prostitutes in its brothels. Within a few months the majority of the prostitutes were infected with VD, as were a large number of American servicemen, one American unit having an infection rate of 68 per cent. A Japanese academic later claimed that more than a quarter of all American soldiers in Japan became infected with a venereal disease. The high incidence of VD occurred because, unlike with the Japanese servicemen, the operators of the comfort stations had no authority to enforce the use of condoms by their American patrons (and others). The American military authorities tried various measures to control the spread of VD, such as providing penicillin to the Comfort Women and establishing prophylactic stations near the brothels, but to little avail. Once the extent of the venereal infection among the occupation troops became known, and concerned about the potential embarrassment to the occupation force if their widespread use of the brothels became known back home, on 25 March 1946 General Douglas MacArthur declared all brothels off-limits, bringing about the collapse of the RAA. MacArthur's edict served only to drive prostitution underground and a network of mainly small brothels, located close to the major camps of the occupation armies, replaced the official comfort stations. Many, if not all, of these brothels were managed by former comfort station operators and were staffed by former Comfort Women, but they were now bereft of medical oversight – consequently the incidence of VD among the occupation troops continued to rise.[50]

The reduction in the size of the Allied occupation force in the late 1940s brought about a corresponding decrease in the clientele of the Japanese brothels, though they received an influx of new clients with the coming of the Korean War and the use of Japan as a major staging base. W. Watson, a sailor with the Royal New Zealand Navy, tells us that:

While we were at sea we had very bad news in so much that there was a signal telling all the British ships, or everybody in the area, that the Japanese had done a search through Sasebo [site of a major US Navy base on the Japanese island of Kyushu] and had transported 25,000 prostitutes out of the area. That was pretty bad news until we got there and found we couldn't notice any difference. It was an unbelievable situation – you couldn't walk twenty yards down the street without being accosted.[51]

Vietnam and South Korea: 'recreation areas' and 'camptowns'

This interrelationship between love, sex and war was indeed evident in many of the post-Second World War conflicts. Accounts of the campaigns fought by Western armies in the Indochinese conflicts reinforce the traditional ties between whores and soldiers. An extreme example of the symbiotic relationship between fighting men and prostitutes is provided by the siege of Dien Bien Phu.

The French Army base at Dien Bien Phu was besieged by the Viet Minh from March to May 1954. Inside the fortress were two *bordels mobiles de campagne* – French mobile field brothels – one staffed by five Vietnamese women and their madam brought from Hanoi, the other consisting of eleven Algerian prostitutes, which had been moved to Dien Bien Phu when the French abandoned their outpost at Lai Chau. The women had been caught up in the battle and endured the privations and dangers of the siege alongside the French soldiers. Among other duties, they assisted the medical staff as auxiliary nurses. The fate that befell the Vietnamese prostitutes is not known, though it is suspected that the puritan Viet Minh would have sent them off for 're-education'. The surviving Algerian women (four were killed during the battle) accompanied the remnants of the French garrison into captivity.[52]

Many US soldiers serving in Vietnam patronised brothels and, due largely to the efforts of film-makers, the Vietnamese prostitute has become an indelible image of that conflict. In Stanley Kubrick's *Full Metal Jacket* (1987), two of the three speaking parts for Vietnamese women are

rostitutes (the other is a sniper). Additionally, having sex with prosti-
.tes is a ubiquitous feature of soldiers' memoirs of the Vietnam War. In
hickenhawk, Robert Mason recalled that the commanding officer of his
.attalion was so troubled by the escalating VD rate among his men that he
athered the officers together:

> The medics say the VD rate has quadrupled. This behaviour is against
> the code of the American officer, immoral and disgusting. I've decided
> to do something about it … Gentlemen, I know what you're going
> through. I'm human too. But what kind of example do you think we're
> setting for the enlisted men? The girls downtown are all disease-ridden,
> a very tenacious form of VD … So for the time being, I'm holding
> every man here duty-bound to exercise discretion and stay totally away
> from those women … Men, severe situations require unusual solutions.
> I know you may think of it as self-abuse, but I, and the commanders
> above me, think that m-masturbation is now justifiable.[53]

.Iason was somewhat sceptical of the colonel's plan, surmising: 'At that
.ery moment, An Khe [the closest village to the US camp] was filled with
.undreds of enlisted men understandably jumping every female in sight.'[54]
.n American general later placed An Khe off-limits because of the spiral-
ng rate of venereal disease. While an American-regulated brothel was
.eing built, the prostitutes relocated to some islands just downstream from
.he site in the river where Mason's battalion washed out their helicopters.
.Iason tells us that his crew chief would invariably go for a 'short swim'
.hile the helicopter dried out.[55] Johathan Shay recorded in *Achilles in
'ietnam* the assertion of an NCO who informed him that no matter how
.olated a landing zone, after it had been established a few days:

> [W]omen would appear, apparently out of nowhere, and stand quietly
> outside the wire, waiting for their mute solicitation to be taken up. If
> the 'Old Man', the senior officer at the site, did not specifically forbid
> this, the women would gradually be brought inside the perimeter, and
> the men would purchase sex from them.[56]

Philip Caputo recalled that prior to the initial landing of Marines in Vietnam in March 1965, the English-language newspaper on Okinawa reported that 60 prostitutes had relocated themselves from Saigon to Danang in anticipation of the Marines landing there.[57] The journalist Jonathan Schell pointed out that 30 bar-brothels had sprung up outside the village of Phu Loi in South Vietnam in response to the local build-up of American troops, particularly the arrival in the area of the US Army's 1st Infantry Division.[58] Moreover, the recreational centre at Vung Tau on the coast of Vietnam had approximately 178 bars and 3,000 bar girls to entertain the troops from Australia, New Zealand, Korea, Thailand and the USA.[59]

Susan Brownmiller notes that by 1966, the 1st Cavalry Division at An Khe, the 1st Infantry Division at Lai Khe and the 4th Infantry Division at Pleiku had all established official military brothels within the perimeters of their base camps. By way of example, the Lai Khe 'recreation area' was a one-acre compound surrounded by barbed wire and guarded by American military police. In addition to shops selling fast food and souvenirs, there were two concrete barracks, each of which contained 60 curtained cubicles used by the prostitutes and their clients. The girls were checked for venereal disease once a week by US Army medics. Each sexual service cost around US$2, of which the girls would keep approximately 75 cents, the remainder siphoned away in pay-offs to various officials. Brownmiller remarks: 'By turning eight to ten tricks a day a typical prostitute earned more per month than her GI clients.'[60] Further testament to the extent of the American efforts to control prostitution was provided by Frederick Downs, who recalled being briefed by his driver on the operation of 'Sin City' upon his arrival at Pleiku:

> The army tries to control prostitution and V.D. by keeping an area over there where you can go and fuck your eyeballs out if you're an enlisted man, or, being an officer, can sneak in. It's guarded by MPs and everything. The American doctors check the whores three times a week for V.D., but you got to be standing first in line after they get through to be sure you don't get the clap, you know?[61]

he incredulous Downs commented: 'You mean the army supports some-
iing like that?' 'Sure', replied the driver. 'They know they can't stick your
ss over here for a year and not expect you to fuck something, so they must
ave decided since they couldn't stop it, the least they could do was control
, you know? They even make sure the whores don't overcharge you!'[62]

The availability of penicillin to treat syphilis from 1943 greatly cut down
ie period of hospitalisation required for venereal infections. By the time
f the Vietnam War, advances in the field of antibiotics had reduced the
mount of lost duty time for most venereal infections to just a few hours,
fact appreciated by both the American soldiers and their Australian allies,
ho also frequented the Vietnamese brothels, particularly those around
ung Tau. Barry Kelly, who served with the Australian Army in Vietnam,
onfessed that: 'I copped the jack twice over there. Out of fifty-four guys
i our unit there were only two that didn't. One was the adjutant and the
ther was … the son of a pastor …'[63]

For most of the Vietnam War, the venereal disease rate remained
round 260 infected men per thousand American servicemen in-country.
y way of comparison, the infection rate in the United States for the same
eriod was 32 cases of infection per thousand adults, while during the
orean War the average rate among US servicemen was 184 cases per
iousand men. The peak period of venereal infection among US troops in
ietnam was from January to July 1972, when the disease rate increased
) an incredible 700 cases per thousand servicemen. This period coincided
ith the implementation of President Nixon's policy of 'Vietnamisation',
hereby the South Vietnamese troops assumed the responsibility for
ombat and virtually all US troops in Vietnam served in support roles.
uring this phase of the war the US troops generally had poor morale,
5 American involvement in Vietnam was drawing to an inglorious end
nd they resented being forced to risk their lives for a lost cause. The US
oldiers were generally bored and apathetic and had ready access to urban
reas. All of these factors made them more likely to have sex with prosti-
ites, resulting in the spiralling rate of VD.[64]

The US military's widespread use of Asian prostitutes continued
fter the Vietnam conflict. In the towns of Olongapo and Angeles in the

Philippines, the respective sites of the US military's Subic Naval Base and Clark Air Force Base (both of which were closed in the early 1990s), there was virtually no local industry other than the 'entertainment' business. These towns had over 2,000 registered R&R establishments, which provided employment for approximately 55,000 prostitutes. The extensive use of prostitutes is also apparent among the US military stationed in South Korea.

Of the 41 major US military camps in Korea, the twelve largest are serviced by nearby 'camptowns'. An example of the symbiotic relationship between US military bases and their adjoining 'camptowns' is the town of Dongduchon, located some twelve miles from the demilitarised zone separating North and South Korea. A seedy mile of restaurants, shops and bars stretches out from the main gate of Camp Casey, the major US Army base at Dongduchon and home of the 2nd Infantry Division (approximately 13,000 soldiers). The most prominent establishments in the camptowns are the bars, whose owners are licensed by the Korean government to sell tax-free alcohol to American soldiers. As the bars are off-limits to Korean customers, the camptown residents are economically dependent on the patronage of the US servicemen.[65]

The majority of the 37,000 US troops serving in Korea undertake a one-year unaccompanied posting. The patronage of brothels by these soldiers is widely accepted and condoned by their peers. Social reproach is therefore minimised, making the soldiers less discreet when visiting prostitutes than they would be back in the United States.

Homosexuality in the military

Most 20th-century militaries accepted, or at least officially ignored, their soldiers patronising brothels for heterosexual intercourse. But homosexual relationships were another matter. The 1914 *Manual of Military Law*, issued by the British War Office, stated:

> The offence of sodomy is when a male has carnal knowledge of an animal or has carnal knowledge of a human being 'per anum'. Penetration

is required, as in the case of rape, to constitute carnal knowledge. A person over the age of fourteen allowing himself or herself to be known in this manner is guilty of the same offence.[66]

The manual specified that the maximum penalty for the offence of sodomy was imprisonment for life, while for those convicted of 'attempt to commit sodomy', the maximum penalty was imprisonment for ten years.[67] Of note is that sodomy was not a military offence per se, that is, one specified in the Army Act. Rather, these provisions reflected the extant civil law of the United Kingdom, though the military courts had jurisdiction to try any soldier accused of this offence. If convicted, the soldier or officer would also be cashiered. During the period 1914–19, the British Army court-martialled 22 officers and 270 other ranks for sodomy.[68]

Throughout the First World War the German High Command periodically conducted investigations to uncover homosexuals serving in the trenches. The rise to power of the National Socialist (Nazi) Party in the 1930s greatly increased the persecution of German homosexuals. Convicted homosexuals were sent to concentration camps, where they were identified by a pink triangle arm patch. In 1941, Hitler, prompted by the homophobic head of the SS, Heinrich Himmler, issued a directive stating that any member of the SS or Gestapo who engaged in homosexual behaviour would be executed. By the end of the Second World War, the German military had court-martialled 8,000 men for homosexual offences.[69]

Homosexuals were also persecuted (and prosecuted) among the Allied forces. William Manchester recalled that during the Second World War the findings of US Navy courts martial arising from sexual indiscretions, specifically homosexual acts, would be read out to every man in the navy. The punishment for such acts was typically draconian; Manchester noted that the usual sentence was 85 years in prison. 'As unsubtly as possible, we were being warned that no matter how horny we got, we couldn't go down on each other.'[70]

The US Army enacted similar measures during the Second World War. In *Citizen Soldiers*, Stephen Ambrose concludes that homosexual activity among the combat infantry was almost non-existent, attributing this to the fact that:

> [T]he US Army in World War II was brutal in its punishment of known homosexuals. They were stripped of all insignia, drummed out of the service, given long sentences in the stockade, disgraced without mercy. They were also beaten up by their fellow soldiers, who were strongly homophobic.[71]

Though the penalties imposed by the US military on those caught engaging in homosexual acts may seem harsh, they were moderate when compared to the punishment inflicted for such acts in the British Navy of the 18th century. Article 28 of the Royal Navy Articles of War of 1757 stated: 'If any person in the fleet shall commit the unnatural and detestable sin of buggery and sodomy with man or beast, he shall be punished with death by the sentence of a court martial.'[72] This provision remained in force, though seldom was the full extent of the penalty imposed, until the Royal Navy Articles of War were superseded by the Naval Discipline Act of 1860. This Act, in the section entitled 'Offences Punishable by Ordinary Law', stated: 'If he shall be guilty of Sodomy with Man or Beast he shall suffer Penal Servitude.'

There is a distinct similarity between the wording of Article 28 of the Royal Navy Articles of War of 1757 and Article 125 of the Uniform Code of Military Justice, which currently applies to all US military personnel. Article 125 reads:

Sodomy

(a) Any person subject to this chapter who engages in unnatural carnal copulation with another person of the same or opposite sex or with an animal is guilty of sodomy. Penetration, however slight, is sufficient to complete the offence.

(b) Any person found guilty of sodomy shall be punished as a court-martial may direct.[73]

Since the end of the Second World War, the US military has struggled with the issue of whether homosexuals should be allowed to serve in the armed forces. In 1945, responding to the urging of the psychiatric profession, the Pentagon declared that homosexuals were mentally ill and should either be denied enlistment or, if discovered after enlistment, be discharged from the military. In 1949, a Pentagon policy stated that lesbians and gay men constituted a security risk and mandated their discharge from the military. The incompatibility of homosexuality with military service was reinforced by a new policy implemented in 1970, though some discretion was given to commanders to retain members with critical skills. In 1981, following a court decision that questioned the legality of a commander's discretionary powers to retain homosexual service members, a new policy mandated discharge without exception. In 1994, the 'Don't Ask, Don't Tell, Don't Pursue' policy became law. This law represented a compromise between President Bill Clinton, who had publicly stated his opposition to the military's ban on homosexuals serving in the armed forces, and Congress, a majority of whose members opposed repealing the ban. Clinton wrote in his autobiography that: 'On paper, the military had moved a long way, to "live and let live", while holding on to the idea that it couldn't acknowledge gays without approving of homosexuality and compromising morale and cohesion.'[74]

The 'Don't Ask, Don't Tell, Don't Pursue' law retained the prohibition on military service by homosexuals as enunciated in the 1981 policy. The law mandated the discharge of homosexuals for statements, acts and marriage: that is, service members could not state they were homosexual, engage in sexual or affectionate conduct with a member of the same sex, or enter into a gay marriage. The law suggested homosexuals should be clandestine about their sexual preference and that they could continue to serve as long as they kept their sexual orientation private. The main purpose of the law was to put an end to intrusive formal and informal investigations aimed at determining the sexuality of service members. But following its passage, policy violations and the number of homosexuals discharged from the military soared. In February 2000, following the murder of a homosexual army private at Fort Campbell, Kentucky the Pentagon added a 'Don't Harass' clause in an attempt to curb the rampant harassment of homosexuals on military bases.

In the ten-year period from 1994 (when 'Don't Ask, Don't Tell, Don't Pursue' was introduced) to 2004, the US military discharged over 10,000 personnel because of their homosexuality. The discharge rate was not constant, peaking in 2001 and then dropping off as the Global War on Terror got under way. In March 2003, a spokesperson for the Servicemembers Legal Defense Network (an advocacy group for gays and lesbians in the military) proclaimed: 'When they need lesbian, gay and bisexual Americans most, military leaders keep us close at hand,' and asserted that thousands of gay troops were currently serving with the US military in the Middle East.[75] This statement reinforced a claim by Clinton in his autobiography that the Pentagon had knowingly allowed more than 100 gays to serve in the 1991 Gulf War and then dismissed them after the conflict when their services were no longer needed.[76]

Homosexuality remained a civil criminal offence in Great Britain until the passage of the Sexual Offences Act in 1967. Henceforth homosexual acts between consenting adults aged 21 and over were decriminalised. But homosexuality remained a military offence under the 1955 Army and Air Force Act and the 1957 Naval Discipline Act. In December 1994, the Ministry of Defence removed the provision for service personnel to be charged under military law – and potentially court-martialled – for their homosexuality. The new guidelines stressed, however, that homosexuality was considered incompatible with service in the armed forces and if a service member was discovered to be or admitted to being a homosexual they would be administratively discharged.

The British ban on homosexuals (the term 'homosexuals' in this context includes lesbians, gays and bisexuals) serving in the military remained in force until January 2000. The decision to remove this ban was taken only after a September 1999 ruling by the European Court of Human Rights had effectively made it legally unsupportable. Press reports noted that up to 600 personnel had been dismissed from the British military in the past decade because of their homosexuality and that they were now able to apply to rejoin the services. The removal of the ban brought the British military into accord with the majority of its NATO allies, such as France, Germany and Canada (but not the US), all of which allow professed homosexuals to serve in the armed forces.

Openly homosexual personnel have been permitted to serve in the armed forces of the Netherlands since the 1970s. Canada and Australia both removed bans on homosexuals serving in the military in 1992. Israel removed a similar ban in 1993. Some countries have partial bans on homosexuals serving in the military; for example, Greece bans homosexual officers. Other countries retain complete bans; for example, in the Turkish armed forces, military personnel who admit to being, or are determined to be homosexual are discharged from duty on charges of indecency.

From a historical perspective, homosexuals have always been present in the armed forces. The elite fighting unit of the ancient Greek city-state of Thebes was the Sacred Band, which consisted of 150 male homosexual couples. The Theban general Gorgidas established the Sacred Band in 378 BCE. Plutarch claimed that the rationale for forming this unit was the belief that the intense relationship between lovers would cause them to fight more fiercely at each other's side than soldiers motivated merely by the bonds arising out of membership of a city-state. Furthermore, Plutarch implied that these warriors would rather die in battle than disgrace their lovers by fleeing the battlefield.[77]

Initially the members of the Sacred Band were interspersed throughout the front ranks of the Theban hoplite phalanxes to strengthen the resolve of the other soldiers. Later they fought as a formed unit to make their gallantry more conspicuous, being out in front of the other Thebans and the first to close with the enemy. The Sacred Band fought valiantly and fiercely in battle for nearly 33 years. In particular, they were credited with being instrumental in the Theban victories over Sparta in 375 BCE and Leuctra in 371 BCE. The martial success of the Sacred Band enabled Thebes to be the dominant Greek city-state for a generation. In 338 BCE, at the battle of Chaeronea, the Athenian and allied troops fled, leaving only the Thebans to stand against the massive army of Philip II of Macedon. The Sacred Band was soon surrounded but fought on until all were killed. Philip was so impressed by their bravery that he raised a monument over their collective grave.[78]

Although there has never been another military unit formed entirely out of homosexuals, homosexuals have fought with distinction throughout many conflicts. Homer's *Iliad* told of how Achilles, the greatest warrior of the Trojan War, had a comrade named Patroclus who 'I [Achilles] loved beyond all other companions, as well as my own life'.[79] In *Homosexuality and Civilization*, Louis Crompton observes that Homer suggested 'a lover-like relation [between Achilles and Patroclus] and yet provides no indication that this love took on an explicitly sexual form'.[80] Later Greek authors, though, such as Plato and Aeschylus, did state that Achilles and Patroclus had a sexual relationship.

It has long been suspected that Alexander the Great was bisexual. Although Alexander had several wives, including a Persian princess, Roxana of Bactria who bore him a son, these marriages arose out of political rather than personal considerations. According to the Roman historian Quintus Curtius, when Alexander advanced into the Persian province of Hyrcania a Persian general presented him with a peace offering of a Persian boy named Bagoas. Curtius described Bagoas as 'a eunuch of remarkable beauty and in the very flower of boyhood, who had been loved by Darius, and was afterwards to be loved by Alexander'.[81] Bagoas accompanied Alexander on his campaign into India. Plutarch recounted an incident when Alexander held a festival to celebrate his arrival in the Gedrosian capital:

His favorite, Bagoas, won the prize for song and dance, and then, in all his festal array, passed through the theatre and took his seat by Alexander's side; at sight of which the Macedonians clapped their hands and loudly bade the king kiss the victor, until at last he threw his arms around him and kissed him tenderly.[82]

But Alexander's closest companion was not Bagoas nor his wives but rather his boyhood friend Hephaestion. Plutarch does not specifically state that Alexander and Hephaestion were physical lovers, though they may well have been when both were younger. In his biography of Alexander, Robin

Lane Fox states: 'Alexander was only defeated once, the Cynic philosophers said long after his death, and that was by Hephaestion's thighs.'[83]

Alexander had studied *The Iliad* and saw himself as Achilles incarnated, while Hephaestion was his Patroclus. To dramatise this conviction, when visiting the ruins of Troy, Alexander offered his sacrifice at the shrine of Achilles, while Hephaestion made his offering at Patroclus' shrine. Additionally, when the queen mother of Persia was brought before Alexander as a captive, she mistakenly bowed before Hephaestion, who was somewhat taller than Alexander. Alexander commented: 'Never mind, Mother. For he too is Alexander.' When Hephaestion died of a fever in 324 BCE, the grief of Alexander, like that of Achilles, was all-consuming. Alexander hanged the doctor who attended Hephaestion, refused to eat or drink for three days, and slashed his own hair.[84]

In ancient Greece, pederastic love was socially acceptable between a young man and an older, more powerful man, who would act as the younger man's mentor as well as being his lover. The Romans were not so liberal in their acceptance of homosexual relationships. While they did not hold such relationships to be morally wrong, they felt that to be the passive partner, that is, to submit to penetration, was to be feminised and humiliated. The Romans viewed homosexual relationships primarily as a form of dominance that usually occurred between masters and slaves.

Suetonius in the *Lives of the Twelve Caesars* stated that the young Julius Caesar was suspected of having a homosexual affair with King Nicomedes IV of Bithynia in 79 BCE. Caesar, then serving on the personal staff of a Roman governor, was sent to Bithynia to obtain a fleet of ships to aid the Romans in their war against Mytilene. It was widely known that Nicomedes was homosexual and the fact that Caesar 'dawdled' so long at his court and returned to Bithynia soon afterwards (supposedly to settle a debt of the king) laid Caesar 'open to insults from every quarter'. He was slandered as being the 'queen of Bithynia' and when he returned to Rome following his Gallic triumph the soldiers following behind his chariot sang: 'All the Gauls did Caesar vanquish, Nicomedes vanquished him.' But Caesar was also a noted seducer of women, including among his conquests the wife of his rival Pompey as well as the Egyptian Queen Cleopatra, with whom he fathered a child. The elder Curio referred to Caesar in one of his speeches as 'Every woman's husband and every man's wife.'[85]

Other (suspected) homosexual warriors of note include Prince Eugene of Savoy, William of Orange (who later became William III of England), and Frederick the Great of Prussia, who lived with his male companions at Potsdam in a court devoid of women. Homosexual relationships were also common among the samurai class in pre-Meiji Japan. In Japan during the Shogun period, young apprentice warriors were often taken as lovers by powerful older men, who acted as their mentors and assumed responsibility for their training. The lover of the original shogun, Minamoto Yoritomo, was a young officer in the Imperial Guard. At least half of the 26 shoguns who ruled Japan from 1338 to 1837 had homosexual lovers.[86]

Promiscuity, marriage booms and war brides

Like most men, most soldiers are not homosexual; and for many of these men, war often means separation from the one they love. However, for others it offers an opportunity for licentious behaviour.

The social upheavals of war, whereby men and women are separated from their sweethearts or partners and thrust into foreign environments to grapple with uncertain futures, encourages a breakdown of prudery and a lessening of moral restraint. In the opening months of the First World War, newspapers carried accounts of 'khaki fever' – a troubling female promiscuity – with women flocking to the towns surrounding army training depots. This sordid state of affairs galvanised into action the self-appointed moral arbiters of society. In 1914, the Headmistresses' Association suggested the formation of a female police force to monitor and control the behaviour of young women. Consequently, the National Union of Women Workers formed over 2,000 Women's Patrols from the ranks of middle-aged, middle-class women. These women, working in pairs, patrolled alleys and public parks to flush out fornicators. In London their beat also included cinemas, where they sought to put a stop to any 'inappropriate' behaviour occurring in the darkened rooms. The moralising sermons delivered by these women were seldom appreciated by the couples they accosted. In late

1914, the Women's Police Volunteers were formed. A prime concern of the female police was immoral behaviour by women, so one of their duties was to conduct preventive patrols near army camps.[87]

A similar surge in female promiscuity, particularly among teenage girls, was evident during the Second World War. In America these girls were known as 'Cuddle Bunnies' or 'Victory Girls'. They frequented railway stations and bus terminals in the hope of attracting some attention from transiting soldiers, or hung around in bars to pick up soldiers on leave. Few Victory Girls were prostitutes in the traditional sense, their promiscuity largely arising from the need for social contact and a sense of adventure.

The potential death of soldiers in combat has served as both an excuse for male promiscuity and a means of justifying it. The soldier on his last leave before shipping out, attempting to seduce his date by claiming that soon he may be dead, is a cliché of the military genre. But that does not mean that it does not happen. One particularly anxious group of men in this situation are virgins, who do not want to face combat without having had sex.

Impending separation for the duration of the conflict, and possibly per-manently, brought an intensity and an urgency to romances of the Second World War that was missing from the less hurried courtships of peace-time. William Manchester, who was about to be shipped out to the Pacific theatre, recalled how quickly a wartime romance progressed. 'It seemed remarkable at the time, but of course it wasn't. In peacetime we wouldn't have reached this point until the eighth or ninth date, but the clock and the calendar were moving relentlessly.'[88]

The impending separation of couples brought about by war can gener-ate a marriage boom. Anthony Rhodes was welcomed to his regimental mess by a colleague on the opening day of the Second World War with the comment: 'Three of our officers have just got married. Seems like war, doesn't it.'[89] Moreover, a wartime bride recalled:

My fiancé rang me up and said, 'I've been posted overseas and we could get married tomorrow!'. So we went and got married. You could get a special licence to get married in a day. It was the thing to do. It was rather like saying, 'Let's go out to dinner.' That was the war. After we married, he went off to the Middle East and I didn't see him for five years.[90]

Many women felt it was their patriotic duty to marry a soldier. 'The pressure to marry a soldier was so great that after a while I didn't question it. I have to marry sometime and I might as well marry him. That women married soldiers and sent them overseas happy was hammered at us', reminisced Dellie Hahne about her wartime marriage in the USA. 'We had plays on the radio, short stories in magazines, and the movies, which were a tremendous influence in our lives. The central theme was the girl meets the soldier, and after a weekend of acquaintanceship they get married and overcome all difficulties.'[91]

Between 1940 and 1942, 1,000 US servicemen were married every day, an increase of 20 per cent over the national pre-war marriage rate. Rick Atkinson writes in *In The Company of Soldiers* that the February 2003 announcement of the deployment of the US 101st Airborne (Air Assault) Division to Kuwait resulted in a flurry of hastily arranged marriages. One accommodating local magistrate even came to be known as the 'Love Judge'.[92] Similar surges in the number of marriage licences granted in early 2003 occurred in Onslow County, North Carolina, home to Camp Lejeune (US Marine Corps), Liberty County, Georgia, where Fort Stewart (US Army) is located, and San Diego County, home to the Marine Corps base of Camp Pendleton.[93] Many of the soldiers rushing into marriage most likely felt the pressing need to establish some emotional security, to feel that back home they had something worth fighting for. Others were probably motivated by the need to 'mark out their territory', that is, to assert a claim on a woman before she was taken by someone else. Such men were (perhaps foolishly) relying on the sanctity of marriage to discourage interlopers during their absence overseas.

The urgency and intensity of wartime romances often masks incompatibility. J. Glenn Gray points out that:

The soldier who returned after the war to marry his sweetheart of a night or two often found heartache and disillusion, as the statistics of such marriages reveal, for the attraction of both sides was too obviously a product of the immediate situation and the war. Another soldier or another girl under similar conditions might have satisfied as well the need for affection and physical love.[94]

The probability of wartime marriages ending in divorce is increased when the bride and groom are of different nationalities. A well-known example of wartime marriages was the so-called 'GI Brides'. These were women, mainly aged from eighteen to 23, and mostly from the UK, who married American servicemen stationed in Europe during the Second World War and subsequently migrated to the United States. When the British women first met the Americans, most of the young British men were in the services and deployed away from home. The American servicemen received higher pay – an American private in England had an annual salary of £750 while a British private was paid less than £100 – had sharper looking uniforms than their British counterparts and also had access to goods that were heavily rationed in wartime Britain, such as cigarettes, chocolate and nylons. The large number of relationships that developed between British girls and the 1.5 million American servicemen in Britain prior to D-Day naturally led to some resentment among the British servicemen, who famously considered the US soldiers 'overpaid, oversexed and over here'.

Approximately 115,000 women, children and men entered the United States under the provisions of the War Brides Act of 1945 and the Alien Fiancés Act of 1946. Although the GI Brides received the most press, there were also a large number of Canadian war brides, due to the fact that approximately half a million Canadian servicemen were stationed in Great Britain during the war. From January 1942 to February 1948, the Canadian government paid the transportation and associated costs to move approximately 43,000 war brides and their children (some 21,000) from the UK to Canada. Many of these young women experienced severe culture shock upon arrival in Canada, both in relation to their husbands, now out of uniform and somewhat less glamorous, and as regards the living conditions.

A number of the Canadian war brides had trouble adapting to the life-style of rural areas, such as the absence of electricity and the presence of the ubiquitous outhouse.[95] For some the culture shock proved too great. Approximately 10 per cent of the Canadian war brides returned to Europe.

Australia experienced a similar 'friendly invasion' by American troops during the Second World War. The first American servicemen arrived in Australia in late December 1941. Over the next four years they were joined by more than a million of their compatriots, who were either stationed in Australia or transited through on their way to the front or on leave. As had occurred in Great Britain, the better-paid, better-supplied and better-dressed Americans drew the ire of the Australian servicemen. Anti-Americanism was the cause of numerous brawls, some of which degenerated into full-scale riots. The most famous of these clashes occurred in Brisbane on the evenings of 26 and 27 November 1942 – the so-called 'Battle of Brisbane'. The official report into the riot, which left one Australian soldier dead and several others suffering minor gunshot wounds, recorded that ill-feeling had developed due to 'the spectacle of American troops with Australian girls, particularly the wives of absent soldiers, and the American custom of caressing girls in public'.[96]

The first wartime marriage between an American serviceman and his Australian bride occurred in March 1942. By the end of the war there were 15,000 war brides in Australia, the majority of whom subsequently emigrated to the United States. But a number of these brides had second thoughts. Public concern for the plight of these women was such that a bill was introduced in Parliament enabling those brides who had not yet left Australia, and who wished to do so, to seek a speedy divorce from their GI husbands.[97] The traffic in war brides, however, was not one way. Approximately 15,000 war brides migrated to Australia following the end of the First World War, during which large numbers of Australian men had fought in Europe and the Middle East. These women were joined by 25,000 or so war brides who arrived in Australia from Europe, Canada, the Middle East and Asia during and in the period immediately following the Second World War (though Japanese war brides, married to Australian soldiers who served with the occupying force in Japan, were not permitted to migrate to Australia until 1952).

* * *

The nature of the contract between the soldier and the state is that the needs of the military will always be paramount. For the soldier this usually means existing relationships will be interrupted for the duration of the conflict. Rather than in the marital bed, the soldier will be forced to slake his sexual longings in the brothel, with the attendant risk of contracting a venereal disease. He may also be plagued by insecurity as he worries about the potential infidelity of his wife or girlfriend, knowing all the while that he is powerless to do anything about it. For such soldiers some comfort is gained by commiserating with their companions, many of whom are facing the same insecurities. But for other soldiers their burden must remain hidden. They must cloak their sexuality because to reveal it would risk being ostracised by the very men whose support they require to survive on the battlefield. Others will rush into incompatible marriages and forever wonder what might have been. Most soldiers, to a greater or lesser extent, will come to regret the illicit liaison between Aphrodite and Ares.

Chapter 6

KILL OR BE KILLED: LIVE AND LET LIVE

Although wars are fought by nations, the responsibility of having to seek out and kill the enemy falls upon only a few men. Supposedly, as portrayed in popular fiction and the cinema, soldiers can kill without hesitation or compunction – but the reality of combat is far more complex. Western culture emphasises the sanctity of human life and this instils in many a deep-seated aversion to killing another man. This reluctance to kill must be overcome by soldiers if they are to operate effectively on the battlefield. Not all will be able to do so. A British soldier of the First World War wrote in a letter to his family:

> And there is one thing I am most thankful for – and I think you will be too. I have never knowingly killed or injured a man. I don't think if it had come to the point I could have done it, and most mercifully, although I have been in action many times and 'over the top' twice, God has prevented me from being brought to the test.[1]

Günter Koschorrek, who fought with the German Army on the Eastern Front during the Second World War, noticed that one of his comrades kept missing the enemy when he was firing, despite being one of the best shots during training. The soldier showed a distinct reluctance to engage the enemy, and when forced to do so he closed his eyes so that he did not have to see whom he was firing upon.[2] The soldier in question was not lacking in courage, and bravely saved his comrades by jumping out of his trench and striking a Soviet soldier in the chest with his rifle butt. Tragically, as the Russian fell, his Kalashnikov discharged and the German soldier fell dead. A few hours before, he had confessed to a fellow member of his unit that 'his religion forbade him to shoot people. In front of God we are all brothers.'[3]

The conflict of conscience that arises from placing soldiers in situations where the taking of life is expected can result in severe emotional distress. Studies of combat fatigue by US Army psychiatrists in the Second World War found that the fear of killing, rather than being killed, was the most common cause of battle failure, closely followed by the fear of letting their comrades down in combat.[4] A Soviet sniper of the Second World War recalled: 'When I first got the rifle, I couldn't bring myself to kill a living being: one German was standing there for about four minutes talking, and I let him go … When I first killed, I was shaking all over … I felt scared: I'd killed a person.'[5] The resistance towards killing a fellow man is so strong that it can even overcome the instinct for self-protection. A British chaplain of the First World War recounted one such incident:

> I saw a sergeant this afternoon. He had let a German off. His pistol was at the man's head but he had not the heart to pull the trigger. The result was that he was badly wounded by a bomb, others of the enemy party having time to come up unobserved. The dear man was wounds all over but, in spite of that, said he was very glad he had not killed the German.[6]

Killing at close quarters: trauma and exultation

Most members of the military are spared the responsibility required of the infantry, who must seek out and close with the enemy and either kill or capture him. Gwynne Dyer writes in *War* that gunners, bomber crews and sailors usually display less reticence than the infantry in firing their weapons. He concludes that this is partly because there is intense peer pressure to carry out their role, arising from the fact that they are under the direct observation of their comrades and that, in the case of crew-served weapons, if any man did not carry out his assigned role then the weapon would not fire. The technical requirements of their respective weapon systems are also a contributing factor, allowing the soldiers/sailors/airmen to concentrate on the mechanics of getting their weapon to fire rather than contemplating its effect. 'We've got to get the bombs on target. We've got 10 minutes to do it. We've got to make a lot of things happen to make that happen. So you just fall totally into execute mode and kill the target,' recounted Fred Swam, the pilot of a US Air Force B-1 bomber, when asked about a mission during the 2003 Iraq War.[7] But the most important factor in making the firing of their weapons possible is the intervention of distance and machinery between them and their enemy – usually they cannot see the enemy and so can convince themselves that they are not killing fellow humans. Dyer comments: 'Gunners fire at grid references they cannot see; submarine crews fire torpedoes at "ships" (and not, somehow, at the people in the ships); pilots launch their missiles at "targets".'[8] This sentiment was confirmed by Samuel Hynes, who flew dive bombers for the US Marine Corps in the Pacific campaign of the Second World War. Hynes recalled: 'In an air war you are not very conscious of your enemies as human beings. We attacked targets – a gun emplacement, a supply dump, a radar station – not men.'[9]

In *On Killing*, Dave Grossman concludes that there is a correlation between the physical and empathetic proximity between opponents and the difficulty and associated trauma of killing a fellow man. Physical distance can be measured along a spectrum ranging from long-range killing (bombing and artillery) to physical contact (bayoneting and hand-to-hand combat). As the physical distance lessens, the empathetic distance must correspondingly increase to overcome man's innate resistance to killing.[10]

During the Second World War, Allied aircrew, drawing upon what is sometimes called the 'morality of altitude', firebombed Hamburg (70,000 deaths), Dresden (80,000 deaths) and Tokyo (225,000 deaths). At bombing height the aircrew could not hear the screams, nor could they see the burning bodies of their victims, most of whom were the elderly and women and children, since men of soldiering age were typically at the front. 'This was remote control. All we did was push buttons', recalled John Ciardi, who served as a B-29 gunner on firebombing missions over Japan with the US 20th Air Force during the Second World War. 'I didn't see anybody we killed. I saw the fires we set.'[11]

From a practical perspective the aircrew understood the destruction they were unleashing, but the physical distance separating them from their victims permitted them to deny it emotionally. The aircrew therefore consoled themselves with the belief that they were destroying a legitimate military target rather than condemning thousands of people to horrific deaths. At the mid-range of physical distance, the weight of rifle fire directed at the enemy may enable a soldier to deny that he was personally responsible for the death of any particular enemy soldier, as he is able to rationalise that someone else fired the fatal shot. But such cognitive dissonance does not provide a salve for all soldiers. Alexander Aitken, a member of the New Zealand Army, fired on a Turkish soldier at Gallipoli at exactly the same time as a comrade. Recalling the incident, he wrote that:

> It seemed that the two shots could raise nine possibilities, in three of which at any rate I might have killed or had a part in killing a fellow human being. This, of course, was what I was there for, but it seemed no light matter, and kept me awake for some time. I could come to no conclusion except that individual guilt in an act of this kind is not absolved by collective duty nor lessened when pooled in collective responsibility.[12]

At close range the soldier is confronted with the physical effects of his actions and the undeniable certainty that he is directly responsible for someone's death. In *The Bridge at Remagen*, the historian Ken Hechler wrote of the aversion of the American troops to close-range killing: 'This was the part of war that the men hated most. They had almost grown used to killing the enemy at long range, especially when they had seen their own comrades die. But there was something terrible about shooting a man whose facial expression they could watch close up.'[13]

Some soldiers will experience a brief feeling of elation (survivor euphoria) immediately after a close-range kill on the battlefield. 'Ground combat is personal, not like dropping bombs from thirty thousand feet on impersonal targets. It is a primordial struggle, you and the other guy exchanging rounds at a few-metres distance. Emotions flow with an intensity unimaginable to the nonparticipant: fear, hate, passion, desperation', wrote James McDonough in his memoir of his service as an infantry platoon commander in the Vietnam War. 'And then – triumph! The enemy falls, lies there lifeless, his gaping corpse a mockery of the valiant fight he made. Your own emotions withdraw, replaced by a flow or relief and exhilaration, because he is dead and not you.'[14]

But this elation may be rapidly overwhelmed by a feeling of guilt and revulsion so strong that the soldier may become physically sick. A German infantryman recalled experiencing such a sensation after he killed a French soldier during the First World War. The German was taking part in an attack on the enemy's trenches when he was suddenly confronted by a French corporal with his bayonet at the ready:

> I felt the fear of death in that fraction of a second when I realised that he was after my life, exactly as I was after his. But I was quicker than he was, I pushed his rifle away and ran my bayonet through his chest. He fell, putting his hand on the place where I had hit him, and then I thrust again. Blood came out of his mouth and he died. I nearly vomited. My knees were shaking.[15]

Private Simpson served with the Australian 13th Battalion during the Gallipoli campaign of the First World War. He too killed an enemy soldier so close that he could feel his final breath. 'I got [a Turk] in the neck,' Simpson related in a letter home, '... made me feel sick and squeamish, being the first man I have ever killed ... I often wake up and seem to feel my bayonet going into his neck.'[16] A generation later, Guy Sajer wrote of the emotional torment he felt when he killed a Russian partisan at close range.

> I felt as if my skull enclosed a black void, and that a nightmare enclosed me, like a fever … I stared at the corpse lying face down on the ground in front of me. I couldn't really believe that I had killed him, and waited for the tide of blood which would soon begin to seep from beneath his body. Nothing else mattered to me. The weight of the drama which had just occurred was so overwhelming that I could only stare at the motionless body.[17]

William Manchester was also sickened by his first battlefield kill: a Japanese sniper. The sniper had just shot two Marines in the adjoining company and Manchester realised that the sniper's fire would soon be directed towards his own section. He ran towards the sniper's position and fired. His first shot missed but the second smashed into the man's femoral artery. He continued to fire as the Japanese soldier's life ebbed away. As the threat was now removed, the realisation of the finality of his actions enveloped Manchester:

> A feeling of disgust and self-hatred clotted darkly in my throat, gagging me … Then I began to tremble, and next to shake, all over. I sobbed, in a voice still grainy with fear: 'I'm sorry.' Then I threw up all over myself. … At the same time I noticed another odor; I had urinated in my skivvies [underwear].[18]

Not all soldiers are affected in this way after they have killed a man. Some, such as snipers, are seemingly able to kill men without compunction, men who posed no direct threat to the firer. James Sims, a paratrooper who took part in Operation Market Garden during the Second World War, noted that: 'Snipers were altogether different and seemed to enjoy killing their enemies, and were disliked as much by their own side as their enemies.'[19] Sims recounted an incident where a badly wounded German soldier, who had managed to drag himself across a road towards his own lines, was shot through the back of the head by a sniper moments before he would have reached safety. 'That upset me – not only me, it upset quite a few of us. Nobody likes snipers – and I said, "What the hell did you do that for? He was out of the battle." The Welshman [sniper] said, "Well, he was the enemy, he was a German wasn't he?"'[20] Jack Coughlin, who served as a sniper with the US Marine Corps during the 2003 invasion of Iraq, was coldly professional in relation to killing men, focusing on the technical aspects of the shot, such as the distance to target and the effect of wind on the bullet's flight, rather than his shared humanity with the enemy. 'I did not think for a moment about him being another human being with a family, and in fact didn't really think about him as a person at all', recalled Coughlin of his feelings as he lined up an Iraqi soldier in the crosshairs of his rifle's scope. 'He was a threat who had stepped into my world, and those would be the last steps he ever took.'[21] Dan Mills, a sniper with the 1st Battalion, the Princess of Wales's Royal Regiment, which deployed to Iraq in April 2004, explained that:

> The difference with being a sniper is you can see the man's face when you kill him. You can see everything about him, because you've probably been studying him for minutes, or even hours. So when you pull the trigger you have to be able to separate yourself from the knowledge that you're taking a life. There's no point in putting someone through eight weeks of highly physically and mentally demanding training if when the moment comes, all he's going to do is think about the wife and kids the target might have at home. He may well be a father of eight, and have four grandparents to feed too. But he's the enemy and that's that. Tough shit.[22]

Evan Wright, a journalist who was embedded with a US Marine Corps unit during the 2003 invasion of Iraq, noticed that some Marines experienced a visceral thrill when killing and bragged about it later. Wright commented: 'It's not just bragging. When Marines talk about the violence they wreak, there's an almost giddy shame, an uneasy exultation in having committed society's ultimate taboo and having done it with state sanction.'[23] But the apparent ease with which the young Marines killed belied a hidden psychological cost. One of them later confided to Wright: 'I felt cold-blooded as a motherfucker shooting those guys that popped out of the truck. Dog, whatever last shred of humanity I had before I came here, it's gone.'[24] A similar comment was made by the Soviet sniper quoted earlier: 'I've become a beast of a man: I kill, I hate them as if it is a normal thing in my life. I've killed forty men.'[25] Likewise, Colin Sisson began to question the psychological price of his war service after he killed a man in Vietnam: 'I sat down beside him, remembering the enjoyment of killing him, and felt terrible. What am I becoming? I asked myself. The answer to this question was too disturbing to contemplate.'[26]

The need to dehumanise the enemy

The British soldiers of the 1st Battalion, the Princess of Wales's Royal Regiment displayed little hesitation when firing upon insurgents in the heat of battle while deployed to Iraq in 2004. 'You identify the enemy, place the sight on to him, pull the trigger and he is gone, but you don't think about it because you are still taking incoming fire,' recalled Dave Falconer.[27] But when ordered to collect the bodies of the insurgents and bring them back to camp for identification, the soldiers were struck by the finality of their actions. 'Killing someone, then looking at them afterwards as you place them in your Warrior, is something I could have done without, personally', wrote Chris Broome. 'Even though this is our job, it still does not stop you replaying the dead faces in your mind time and time again and asking the questions why, and what if? The next time I hear a comment like "I want to get stuck in and kill somebody", I will remind him of what you may have to live with for the rest of your life.'[28]

The psychological cost of killing another man may not become apparent until some time after the event, and is a key factor in the onset of post-traumatic stress disorder. 'When we kill another human being, there's a price to pay', recounted a helicopter pilot who served with the US Marine Corps during the Vietnam War. 'We try to put a barrier around our heart and our emotions, but there is a price to pay … we talk about how much we want to kill the enemy, but it's still going to come back and haunt you, because it's an unnatural act.'[29]

Military training seeks to enable soldiers to kill on the battlefield without hesitation. 'You're taught by the army to conceive of them as the faceless enemy, not a real person. In training you spend time shouting "Kill, kill, kill!" to help make it a normal thing to do,' First Lieutenant John Yaros explained to Oliver Poole, a journalist embedded with the US Army's 3rd Infantry Division for the push towards Baghdad.[30] Another officer commented to Poole: 'The simulation of battle training is so realistic that it resembles almost exactly what occurs in the field. They have lifesize tanks that resemble precisely the ones we've been shooting out here. There are dummies of soldiers popping out of doors that resemble the scenario in house-to-house searches.' Indeed, so successful was the battlefield inoculation training provided to the American soldiers that the officer concluded: 'Half these soldiers feel they've been on a field exercise. The problems are going to come when it sinks in that this time it was people they've been shooting at.'[31] Earlier, Poole had spoken to two American soldiers who were still so hyped-up after a skirmish with some Iraqi soldiers that 'they were literally bouncing on the spot, unable to stand still':

'Did you see us? We took those two out. The captain spotted them on a ridge and we turned, I saw them and badabadabada,' one said, mimicking the sound of machine-gun fire, his hand imitating grasping a gun. 'I shot right across the first one, and it caught the grenades strapped to his front and he exploded. This other guy was scrabbling in the ground for an AK-47. Badabadabada. His head was shot straight off.'[32]

But a few days later, when Poole encountered the men again, the soldier who had shot the Iraqis confided to him that: 'I didn't feel any remorse then. I wanted to kill them. But yesterday I was sitting up on the tank thinking, "Damn, I did kill guys".'[33]

At close range it becomes very difficult for a soldier to deny the humanity of his enemy. Dave Grossman states that 'looking in a man's face, seeing his eyes and his fear eliminate denial … Instead of shooting at a uniform and killing a generalised enemy, now the killer must shoot at a person and kill a specific individual.'[34] Al Slater served as a company commander with the US Marine Corps in Vietnam in 1967. While patrolling in the demilitarised zone, Slater's company was fired upon by a North Vietnamese soldier. Taking advantage of the pause when the enemy soldier stopped firing to reload, Slater ran forward and shot him in the face at close range. Slater later recalled the 'look in his face and I can see the fear in his eyes'. He would replay the incident over and over in his mind. 'It's easier to fight from afar. It's hard to realise that our enemy, who we hate so bad, is human. We look at them as an object. They look at us the same way. It didn't keep me from wanting to kill them. I would just rather kill them from afar.'[35]

To overcome man's innate resistance to close-range killing, armies seek to develop an empathetic distance between their soldiers and those of the enemy. An empathetic distance is cultivated by emphasising differences. This action diminishes the feeling of shared humanity with those deemed the enemy. 'Somehow everything German gave one the creeps', felt Hervey Allen, a member of the US 111th Infantry Regiment, 28th Division as he sifted through the refuse left in a dugout by withdrawing German soldiers in the closing stages of the First World War. 'It was connected so intimately with all that was unpleasant, and associated so inevitably with organised fear, that one scarce regarded its owners as men. It seemed *then* as if we were fighting some strange, ruthless, insect-beings from another planet ...'[36]

Racial and ethnic differences are accentuated to establish a cultural distance. This process is materially assisted when soldiers already have a deeply ingrained belief in their own country's cultural superiority.

A common method of developing a cultural distance between soldiers and their foe is the use of epithets for enemy soldiers, such as 'hun', 'boche', 'gyppos', 'gook', 'chink', 'slope', 'zipperhead', 'dink', 'raghead', 'sand nigger' or 'camel jockey'. Paul Fussell, who served with the US Army during the Second World War, pointed out: 'We always called the Germans "Krauts", doubtless to bolster our sense that we were killing creatures very odd and sinister and thus appropriate targets of contempt.'[37] Frederick Downs, who fought with the US Army during the Vietnam War and personally killed a number of 'dinks', wrote:

> In order to kill, the soldier is taught to dehumanize the enemy, to kill targets. Any hesitation, any thought that he is killing someone's father, son, or brother, and the soldier may be slow to pull the trigger. Soldiers kill French Frogs, German Krauts, slanty-eyed Japs, Rebs, Yanks, Pepper Bellies, Dinks, Onion Eaters, Chinks, Ragheads, and so it continues: an endless list throughout time.[38]

Bernard Szapiel, who served with the Australian infantry in Vietnam in 1967, provided an example of the practical effect of dehumanising the enemy when he recalled an incident where he was required to bury some dead Vietnamese soldiers.

> We were in a hurry, so we dug very shallow graves but they wouldn't fit so we started jumping on the arms and bodies to get them to go in. But that didn't work so we thought, 'Oh well, we'll make an easy job of it,' so we just cut the arms off with machetes and threw them in and piled the dirt over the top. It was like cutting up a sheep. Our training was so good that you never saw them as people, they were just an animal, really, nothing else but an animal.[39]

In Vietnam the process of dehumanisation was assisted by the 'body count' mentality, under which American soldiers were encouraged to think of the dead Vietnamese in terms of a gross number, rather than as fellow humans who had families and friends who would mourn their death. The dehumanisation of the Vietnamese was furthered by the practice employed by some American units of awarding in-country R&R to the soldier or Marine credited with killing an enemy combatant.

The practice of dehumanising the enemy continues to the present day. Joshua Key recalled that when he underwent basic training in 2002, he and his fellow recruits were lined up on the bayonet range facing a life-sized dummy and told to imagine that it was a Muslim man. As they repeatedly thrust their bayonets into the dummies, one of their commanders shouted through a microphone: 'Kill!, Kill! Kill the sand niggers.'[40] 'The hajjis, Habibs, rag heads and sand niggers were the enemy, and they were not to be thought of with a shred of humanity,' wrote Key.[41]

Establishing empathetic distance not only frees soldiers from hesitation but also from guilt after killing an enemy. David Hackworth, in his memoir, *About Face*, recounted that the first time he killed enemy soldiers (North Koreans):

> I dropped four guys point-blank with my M-1, each dead with a six-o'clock-sight picture in the chest, just like the good book said. I felt no guilt – few of us did; I'd been trained too well, and besides, the enemy had been utterly dehumanised throughout my training. *They aren't men, they're just gooks.*[42]

Racial differences are more readily accentuated and internalised than cultural differences. The psychologist Samuel Stouffer found that among the American troops he interviewed in the Second World War, around 44 per cent 'would really like to kill a Japanese soldier', but only around 7 per cent felt the same way about killing a German.[43] This is a direct reflection of the shared ethnicity of many American and German people. Additionally, most Germans and Americans are of the same race (Caucasian). In *Citizen Soldiers*, Stephen Ambrose writes: 'There was little racial hatred between the Americans and the Germans. How could there be when cousins were fighting cousins? About one-third of the US Army in ETO [European Theatre of Operations] were German-American in origin.'[44] No such common cultural heritage existed with the Japanese. Additionally, the Japanese were Asians and thus different in appearance to the majority of Americans. The greater publicity given to the wartime atrocities committed by the Japanese forces, coupled with the anger over the Japanese surprise attack on Pearl Harbor, were also contributing factors to the level of hatred American soldiers felt towards their Japanese counterparts.

The racial and cultural differences between American and Japanese soldiers resulted in some US servicemen considering the Japanese as somewhat less than human. This lack of empathy for the Japanese resulted in the desecration of their bodies. Gold teeth were commonly extracted with a K-bar knife, but perhaps the most notorious practice involved American soldiers stripping the flesh from the decapitated heads of Japanese soldiers and sending the bleached skulls home as macabre souvenirs. The situation got so out of hand that in September 1942 the commander-in-chief of the Pacific Fleet felt it necessary to issue an order stating: 'No part of the enemy's body may be used as a souvenir.'[45] A similar lack of recognition of their enemy's common humanity would result in American soldiers during the Vietnam War cutting off the ears of slain Vietnamese and wearing them as a necklace, just as primitive hunters would wear the tusks of slain boars.

Creating moral distance

In modern society, it is generally more politically acceptable, and less likely to lead to wartime atrocities, to emphasise the moral rather than cultural distance between armies and nations. Developing a moral distance entails establishing the enemy's guilt in conjunction with asserting the legality and legitimacy of one's own cause. Hence killing the enemy becomes an act of justice: he is being punished for his crimes.[46]

Of course, each side will seek to claim the moral high ground. The Union forces of the American Civil War convinced themselves that the killing of their southern brethren was necessary to end the secession of the Confederate states and preserve the Union, while the Confederate soldiers spoke of northern aggression and unjustified interference in the internal matters of the southern states. 'So, figuring as a whole, the responsible parties for the great flow of blood should not be laid at the Southerner's door', wrote William Fletcher in his memoir of his service with the Confederate Army, 'for the matter of property rights, as granted by the Constitution, should, at least be proof enough to exonerate; and the cloak of defense of preserving the union was not the kernel in the nut ...'[47] Conversely, the Union soldier Hamlin Alexander Coe wrote of Confederate deserters in morally unambiguous terms: 'They are really disheartened and have given up the idea of establishing a southern confederacy. It is only surprising to see how willing they are to confess their guilt, and I believe they mean to do better in the future.'[48] Moreover, Rice Bull commented on the destruction wrought upon the property of the plantation class of South Carolina (the first state to secede from the Union) as Sherman's army made its way north through the Carolinas in early 1865:

> The grudge held against South Carolina and her people by many soldiers was very intense; many times they ruthlessly destroyed property when they heard it belonged to an active secessionist. They excused their actions by saying that they wished such people to suffer for their responsibility in bringing upon our country the Civil War.[49]

When establishing the moral high ground the basis of the enemy's guilt may be atrocities committed against a third party or simply aggression – that is, they started the war. Such was the case in mid-1914, when the propaganda efforts aimed at the British public in general, and potential soldiers in particular, focused on two German crimes: the violation of Belgian neutrality and the atrocities committed by the Germans in their efforts to crush Belgian civil resistance.

In December 1914, the British government asked Viscount Bryce to chair an independent committee to investigate the alleged German brutality towards civilians in occupied Belgium. The resulting *Report of the Committee on Alleged German Outrages* was released in May 1915 and translated into 30 languages. The report was based upon more than 1,200 depositions, mostly from Belgian refugees, though some British troops were interviewed as well. A number of captured German diaries were also used as sources.

The report detailed the systematic execution of civilians and the looting and burning of Belgian villages by German soldiers, many of whom were apparently drunk when these acts were committed. Specific atrocities included: killing a woman and then cutting off her breasts, cutting off limbs and hands, using civilians as screens when firing upon Belgian troops, murdering priests, raping and then bayoneting women, and murdering a child by nailing it by its hands and feet to the door of a farmhouse. The report provided two graphic accounts of German soldiers bayoneting infants and then leaving the dying child attached to the bayonet while the rifle was hoisted into the air. The report also claimed that German soldiers had bayoneted and cut the throats of wounded British soldiers and had feigned surrender and then opened fire. The report concluded by stating: 'Murder, lust, and pillage prevailed over many parts of Belgium on a scale unparalleled in any war between civilised nations during the last three centuries.'[50]

There is little doubt that the Germans burnt numerous villages and executed a large number of civilians (about 5,500 in total) during their occupation of Belgium. McKinley McClure, serving with the US Army's 12th Field Artillery Regiment, tells us of an incident where an old Belgian lady pointed out to him a bullet-pocked wall where her husband and brother, along with some other men from the village, had been executed by the Germans in 1914.[51] But some of the more gruesome atrocities of which the German soldiers were accused, such as the bayoneting of babies, were most likely the invention of propagandists. Furthermore, from the German perspective, many of these actions were militarily justifiable because of the tactics of the Belgian partisans (which included stringing wire across streets to decapitate German soldiers travelling in staff cars). A German officer recounted in his memoirs that:

Warfare in Belgium soon became a hideous experience because the population took part in the fight. Whenever they had the chance they shot down German soldiers ... There was little defence against that sort of warfare because the streets were full of civilians and so were the houses. Unless they shot first, nobody knew where the enemy was. It was nerve-wracking in the extreme and resulted in savage and merciless slaughter at the slightest provocation.[52]

Phillip Knightley notes in *The First Casualty* that the depositions given to the Bryce inquiry vanished after the war and that a Belgian commission of inquiry in 1922 'failed markedly to corroborate a single major allegation in the Bryce report'.[53] But such issues were now largely irrelevant. The war was over. The Bryce report had achieved its aims: the German crimes justified British participation in what was essentially a continental European war, and the report helped turn American public opinion against the Germans.

Another salient example from the First World War of the process of cultivating moral distance is the propaganda arising from the death of the British nurse Edith Cavell. Cavell had moved to Belgium in 1907 and was appointed matron of the Berkendael Medical Institute (a training school for nurses) in Brussels. When Germany invaded, Cavell, then holidaying in Britain, returned to Belgium and harboured fugitive Allied soldiers in the institute. In conjunction with some Belgian civilians, she also actively assisted the soldiers to escape to neutral Holland. She was arrested by the German authorities on 5 August 1915 and charged with having aided the escape of some 200 British, French and Belgian soldiers. She confessed and was found guilty by a military court. Four Belgians who had assisted Cavell were also sentenced to death. She was executed by firing squad at dawn on 12 October 1915. Her execution became a cause célèbre, notwithstanding the fact that under the accepted rules of war the Germans were justified in executing her as a spy. Newspaper reports condemned German barbarism and Cavell became a martyr. Avenging her execution became a recurring theme in recruiting campaigns. Numerous propaganda posters were produced, generally showing Cavell's body, still dressed in her nurse's uniform, prostrate at the feet of a German soldier, or a photograph of Cavell with the caption: 'Murdered by the Huns'.

A more contemporary example of establishing a moral distance arose after the Iraqi invasion of Kuwait in August 1990. It was widely reported in the Western press that Iraqi soldiers had removed hundreds of premature Kuwaiti babies from incubators. The soldiers allegedly left the infants to die on the hospital floors and sent the incubators back to Iraq. The story originated with a fifteen-year-old Kuwaiti, identified only as 'Nayirah', who tearfully testified to witnessing this atrocity before the US Congressional Human Rights Caucus on 10 October 1990. President George Bush referred to the incident six times in the next five weeks as a prominent example of the evil nature of Saddam Hussein's regime. Furthermore, during the US Senate debate on whether or not to approve military force to evict the Iraqis from Kuwait, seven senators specifically mentioned the death of the 'incubator babies'.[54] Chris Hedges, a journalist who reported on the US-led liberation, discovered when he arrived in Kuwait and checked with the Kuwaiti doctors that the allegations were a fabrication.

> But by then the tale had served its purpose ... Nayirah turned out later to be the daughter of the Kuwaiti ambassador to the United States, Saud Nasir al-Sabah. She did not grant interviews after the war and it was never established whether she was actually in the country when the invasion took place.[55]

Throughout the 20th century, the aggressor nation in a conflict has usually attracted international condemnation. Thus the Allied nations could claim moral justification when responding to German aggression in the First and Second World Wars, General Dwight Eisenhower even categorising the June 1944 landings in occupied France in the morally unambiguous terms of a 'Great Crusade'. Likewise, the North Koreans, Argentines and Iraqis (among others) were all condemned through the mechanism of a United Nations Security Council Resolution for invading a neighbouring country.

The ultimate aim of emphasising the moral distance between combatants is for this moral distance to be internalised by soldiers. Once internalised, soldiers will usually seek out confirmatory evidence and disregard contradictory facts. Thus reports of enemy atrocities, even those based on the flimsiest of evidence, will be readily believed and will serve to reinforce the underlying moral justification for military action.

Mechanical distance and conditioning men to kill

In contrast to a moral distance, a mechanical distance between soldiers and their foe is established when technology masks the humanity of combatants. For example, a soldier equipped with a thermal sight on his rifle can momentarily convince himself that he is firing at a green blob rather than another human.[56] 'On the thermal imaging screen on my TOGS [Thermal Observation and Gunnery Sight] I could see Iraqi infantry advancing towards us. I felt curiously removed from the actual blood and guts of warfare although men were being cut down and dying,'[57] recalled James Hewitt as his Challenger tank engaged Iraqi troops during the opening stages of the ground offensive of the 1991 Gulf War.

The humanity of combatants is also masked when the opposing soldiers are encased in a metal shell, such as in an aeroplane or a tank. 'Air fighting is a very detached sort of warfare', recalled Frank Carey, a pilot with the Royal Air Force during the Battle of Britain, 'being fought, as it were, between machines with the human factor very much submerged in a "tin box". Once in a while, for a few fleeting seconds when someone bales out, one can suddenly be aware that humans are actually involved but, as the parachute descends, machines quickly regain the centre of the stage once more.'[58] A similar mindset enabled Ray Holmes, another pilot with the Royal Air Force, to state: 'I wanted to shoot an aeroplane down, but I didn't want to shoot a German down';[59] as if the two events were not inextricably connected. Some 60 years later, as James Newton lined up an Iraqi T55 tank in the sight of his TOW missile launcher, he felt that: 'I wasn't thinking in human terms: it was helicopter versus tank. The training had kicked in and it was as if I had been pre-programmed to delete any thoughts about fellow souls. All I could see was 60 tons of Russian armour.'[60] The need to develop this mechanical distance is encapsulated in the following comment from a British Army officer who fought in the Falklands War: 'It's all very well shooting a target, but when the target is speaking and you realise it's a human being, things are actually rather different.'[61]

External factors also influence a soldier's decision to kill in combat. S.L.A. Marshall's research during the Second World War highlighted the impetus to fire generated by officers and NCOs moving among their men and ordering them to shoot. This conclusion was confirmed by a 1973 study by Kranss, Kaplan and Kranss, which found that 'being told to fire' was the most critical factor in making soldiers engage the enemy on the battlefield.[62] Another factor enabling men to fire in combat is the anonymity provided by group participation. Getting others to share in the killing process alleviates individual feelings of responsibility.[63] Marshall noted that in almost all the combat actions he reviewed, the unit's machine guns were fired. Significantly, a two-man team operated each machine gun.[64] Realistic combat training also enables soldiers to kill their enemy on the battlefield.

Modern combat training techniques are based on operant condition-ing, the central tenet of which is that all human behaviour is influenced by past rewards and punishments. During the First World War, marksman-ship training for soldiers involved them lying prone in a grassy field and calmly firing at a bull's-eye target. After a series of shots were fired, the target would be checked and the firer's score calculated. However, the chief limitation of training in this manner was its artificiality. Max Plowman recounted how 'having thoroughly enjoyed practice upon a target, I begin to feel squeamish on being told that firing low is a mistake because the head is the most vulnerable part of a man. Bull's-eyes on a target are a pleasure; but when the power to make them becomes applied ... then a gruesome sense of mean inhumanity begins to assert itself ...'[65]

Modern marksmanship training seeks to simulate, in a safe environ-ment, the actual conditions of combat. A soldier will stand in a weapon pit wearing full combat equipment and olive-drab man-shaped targets will pop up briefly in front of him at varying ranges (the conditioned stimulus). The soldier must instantly aim and fire (the target behaviour). If he hits the target it will immediately fall backwards – just as a living target would – providing immediate positive reinforcement of the target behaviour. The latest refinement of marksmanship training replaces the man-shaped targets with video images of actual humans. These images are controlled by an interactive computer program that responds to the actions of the soldier. If the soldier accurately engages the target within a defined timeframe the target will drop. Praising high-scoring firers and awarding them marksmanship badges, to differentiate them from their less capable peers, confers additional positive reinforcement.[66] Shooting is considered a basic but essential skill that is required of every soldier. Those who do not achieve the required standard of firing accuracy will suffer mild punishment in the form of retraining, as well as loss of face among their peers. These factors further reinforce the desired target behaviour.

One aspect of the battlefield that is not simulated on the rifle range is the olive-drab man-shaped targets firing back. Most soldiers therefore find their initial exposure to combat quite perplexing. Someone he has never met and with whom he has no personal conflict is actively trying to kill him. Fritz Nagel, who served with the German Army throughout the First World War, recalled that his reaction to his baptism of fire was: 'Those people on the other side were trying to kill us. It seemed incredible to me.'[67] This sentiment was shared by many other soldiers, including Philip Caputo, whose initial thoughts following a sniper's bullet slamming into a tree a few inches above his head were: 'Why does he want to kill *me*? What did I ever do to *him*?'[68] Soldiers need to make a shift from personal motives being a basis for an individual's actions to these actions being merely a reflection of organisational objectives. Soldiers kill each other simply because it is their job. Their political leaders are in disagreement and young men killing other young men who are wearing a different uniform will supposedly resolve this dispute.

Rehumanising the enemy

As noted, creating an empathetic distance between soldiers and their enemies helps soldiers overcome their innate resistance to killing a fellow man. Maintaining this distance is made easier when the foe poses a real or perceived threat to a soldier or their comrades' survival and the foe is dressed in a different uniform, thereby identifying him as the enemy. When these factors are stripped away the enemy becomes less threatening. Consequently, the empathetic distance is decreased and their common humanity is embraced. During the First World War, Emilio Lussu was preparing to shoot an Austrian soldier from a concealed position, assuring himself that 'it was not a man I was looking at, but an enemy', when suddenly the soldier lit a cigarette. Lussu wrote in his memoirs that:

This cigarette formed an invisible link between us. No sooner did I see its smoke than I wanted a cigarette myself; which reminded me that I had some with me. All this took place in a moment; but the act of aiming, which had been automatic, became rational. I was forced to remember that I was taking aim, and that I was aiming at somebody. The finger that was on the trigger slackened its pressure.[69]

Lussu felt that he could not justify to himself killing the enemy soldier – who posed no danger to him – in such circumstances, and lowered his rifle.

Robert Graves recalled that the only time he refrained from shooting at an enemy soldier was when he spied a man taking a bath in the German support lines through the telescopic sight of a sniper rifle. Graves could not bring himself to shoot a naked man; he passed the rifle to a sergeant who promptly killed the German.[70] A generation later, Eugene Sledge, who had just narrowly avoided being killed by the fire of a Japanese machine gun, an incident that filled him with a 'cold, homicidal rage and a vengeful desire to get even', was suddenly confronted with three Japanese soldiers fleeing from a pillbox. Each of the fleeing men held his rifle in his right hand and his pants up with his left. Sledge stated: 'This action so amazed me that I stared in disbelief and didn't fire my carbine.'[71]

Encountering a dead or wounded enemy soldier can help break down the empathetic distance between foes. 'Looking at the individual German dead, each took on a personality', reminisced Robert Rasmus, who fought with the US Army in Europe during the Second World War. 'These were no longer an abstraction. These were no longer the Germans of the brutish faces and the helmets we saw in the newsreels. They were exactly our age. These were boys like us.'[72] During a fierce firefight in Vietnam in July 1967, an American soldier, James Donahue, came across a wounded Viet Cong soldier lying in a bomb crater. Donahue's initial impulse was to kill the man but he decided instead to capture him. Donahue pointed his rifle at the soldier's head and approached the crater. He then saw the extent of the man's wounds:

[E]verything from his waist down had been blown away. It looked as though someone had dumped a pile of Franco-American spaghetti down the slope behind him. The blast had cut him in half, spilling his insides out on the ground. I'd seen a lot of bad wounds but nothing like this – I found it unbelievable that he was still alive.[73]

Instantly Donahue's attitude to the enemy soldier passed from hostility to pity. The soldier pleaded with Donahue to kill him and end his misery. But Donahue could not. 'A couple of minutes earlier I could have killed him in a heartbeat, but now I couldn't pull the trigger … He was no longer the enemy, but a human being not so different from myself.'[74]

Battlefield truces to collect the wounded and bury the dead also help break down the empathetic distance between combatants. '(B)oth sides were busily employed burying the dead and bringing in the wounded; French and English promiscuously mixed, and assisted each other in that melancholy duty as if they had been intimate friends', wrote Joseph Donaldson of a brief truce during the battle of Fuentes de Oñoro (May 1811) in the Peninsular War. 'So far did this friendship extend, that two of our lads who spoke French, went up that night after dark to the enemy's picket, and having conversed and drank wine with them, returned unmolested to their company.'[75] '(I)t took three days to bury them, and during that time the soldiers of the three countries were mixed up together, picking out and burying their own men', recalled Henry Franks of the Heavy Cavalry Brigade of the truce that followed the battle of Inkerman in the Crimean War. 'Those men, although unable to speak each other's language, fraternized together, exchanging and smoking each other's pipes, and when the grog was served out, in many cases drinking one another's healths …'[76]

Empathetic distance is also lessened when the enemy displays qualities on the battlefield, such as bravery under fire, that elicit admiration from his foes. A German soldier of the First World War recounted an incident during the rapid German advance on 21 March 1918 (the opening day of the German spring offensive) when he had the opportunity to fire on some withdrawing British soldiers. 'I just saw the last one running away back up his trench about 200 metres away. I didn't shoot him though; the last one away was the bravest and I wasn't going to shoot him in the back.'[77] Another incident where respect for the bravery of the enemy stayed his opponent's hand was recounted by Johann Voss in *Black Edelweiss*, his memoir of his service with the Waffen-SS during the Second World War. He wrote of an incident where an American soldier jumped out of his tank in full sight of the Germans and at great personal risk helped some wounded men onto the deck of the tank and drove them to safety. Voss recalled: 'Those of us witnessing the scene, whether nearby or more distant, instinctively felt there was no honour to be won by firing upon this death-defying act of comradeship.'[78] In effect, the more the enemy displays the qualities soldiers look for in themselves and their comrades, the less abstract he becomes and the more his foe comes to think of him as a fellow soldier.

An incident in Vietnam, recounted in Philip Caputo's *A Rumor of War*, serves to further illustrate how the lessening of the empathetic distance between foes can affect combat soldiers. The men of Caputo's platoon had killed a Viet Cong soldier. No photographs or identity papers were found on the dead soldier, a factor for which Caputo was grateful. 'I wanted to think of him, not as a dead human being, with a name, age, and family, but as a dead enemy. That made everything easier.'[79] But this feeling of disassociation was fleeting and was soon overtaken by remorse as Caputo and his men realised they had killed a fellow man; a man whose death would result in his family and friends suffering an irrevocable loss. In contextualising the death of the enemy soldier they changed from focusing on differences to focusing on similarities. One of the soldiers in the platoon expressed what Caputo felt was now a collective emotion: 'They're young men, they're just like us, lieutenant. It's always the young men who die.'[80]

Stripping away the illusions

But young men (late teens to early twenties) are precisely the resource required for the efficient functioning of armies. Young men are generally economically dispensable, as most have had only basic schooling and have yet to acquire the education or vocational experience that would make them economically valuable to a society. Young men usually do not have dependants, they are physically strong and their personalities are malleable. The military offers young men discipline and a rigidly structured environment that helps ease the passage from adolescence to adulthood by delaying, or at least minimising, the acceptance of personal responsibility for the conduct of their lives. Previously their parents looked after them and made all major decisions, such as where they would live and so on. Now the military makes those decisions for them. Young men are searching for role models and peer acceptance and are likely to be responsive to military conditioning, which one historian described as a 'heavy-caloried diet of fight films, fight talks, speed hikes, obstacle courses, firing, grenading, bayoneting, constant appeals to his competitive instincts, flags, bands, marching-in-step'.[81]

Jean-Baptiste Barres tells us that as he marched out of Paris in 1805 to campaign against the Austrians, 'war was the one thing I wanted. I was young, full of health and courage, and I thought one could wish for nothing better than to fight against all possible odds ...'[82] Ernst Jünger wrote in his memoir of the First World War, *Storm of Steel*, that as he and his compatriots made their way to the front, they felt as they had 'grown up in an age of security, we shared a yearning for danger, for the experience of the extraordinary. We were enraptured by war.'[83] Laurie Lee recalled that when he arrived in Spain in December 1937 to fight as a Republican volunteer against the Fascists, 'I was at that flush of youth which never doubts self-survival, that idiot belief in luck and a uniquely charmed life, without which illusion few wars would be possible.'[84] In a similar vein, the former Marine pilot and historian Samuel Hynes concluded:

War brings to any society its electric, exhilarating atmosphere, and young men rush to join in it, however grim the stories of war they have read and accepted as the truth. Every generation, it seems, must learn its own lessons from its own war, because every war is different and is fought by different ignorant young men.[85]

These young men are filled with the vigour of youth and have a poor appreciation of the finality of some of their actions, both on the enemy and also on themselves – that is, until they have been exposed to combat.

Soldiers may be exposed to the after-effects of enemy action before encountering the enemy, for as they move forward they will often cross paths with wounded and killed soldiers moving rearward. Edmund Wheatley recalled that as his unit marched towards Waterloo, 'we marched nearly three hours unknowing our destination, until three or four wounded passing convinced us that every step was bringing us nearer and nearer to the scene of slaughter.'[86] William Fletcher, a Confederate soldier who served with the 5th Texas Infantry during the American Civil War, wrote of his impressions as his unit moved forward to engage the Union forces during the Seven Days Battles (Peninsula campaign) in June 1862: 'You will hear the roar of the battling front; see the wounded going and being carried to the rear; and if advancing, as we were in this instance, passing the dead and dying …'[87] When Max Plowman entered the railway station at Boulogne on his arrival in France in July 1916 he encountered wounded soldiers disembarking from a hospital train: 'These blood-stained heads come as a sharp reminder of our destination. Quite surely we are bound for the places these men have left.'[88] He continued on to Étaples where he encountered a British cemetery: 'Lord! How many have died already! The ground is smothered with wooden crosses.'[89] Jack Short poignantly recalled his arrival at Omaha Beach shortly after D-Day: 'As you looked up on the beach, you could actually see the making of GI cemeteries. That was your first shock, regardless of all the combat training you had.'[90] A few miles to the east, John Foley came ashore on Sword Beach and 'stared in fascination at the inverted rifles stuck into significant mounds of earth, each balancing a tin hat on the butt.'[91] Fay Lewis, who served as a nurse with the Australian Army in the Vietnam War, remembered that: 'We arrived at Tan Son Nhut and landed beside planes taking coffins back to America. I think that was the first thing we saw.'[92] Wounded and killed men soon dispel any notions that soldiers may still have concerning the so-called 'glory of war'. Further exposure to the battlefield will continue the process of stripping away the illusions about war nurtured by inexperienced soldiers.

In *Band of Brothers*, Stephen Ambrose chronicles the experiences of Easy Company of the 506th Parachute Infantry Regiment, 101st Airborne Division during the Second World War. On D-Day, elements of Easy Company, in a fierce close-quarters attack, eliminated a German 105mm artillery battery and its security force of 50 German paratroopers. A gauge of the intensity of this assault and the heroism displayed by the Easy Company soldiers is that of the eighteen men who took part in the attack, one was recommended for the Medal of Honor (but was instead awarded the Distinguished Service Cross), four were awarded the Silver Star, and eight were awarded a Bronze Star. Ambrose comments that one of the key factors contributing to the assault's success was that it was the American soldiers' baptism of fire. The men had taken chances in combat that they would not take again. They had displayed, in Ernst Jünger's phrase, 'the courage of inexperience'.[93] Two of the soldiers involved in the attack later confessed to Ambrose: 'You don't realise, your first time. I'd never, never do again what I did that morning.' While his companion stated: 'I was sure I would not be killed. I felt that if a bullet was headed for me it would be deflected or I would move.'[94]

Ambrose writes of the distinct change in the paratroopers' attitude to combat as they waited in England for their next mission. 'The recruits were excited, tense, eager, nervous. The veterans were worried.'[95] The veterans' initial fear, that they would let their comrades down in combat, no longer troubled them. Having survived one combat operation they were confident in their ability to function under fire, but they had now replaced that generalised fear with more specific horrors. In particular, the paratroopers feared being shot while floating down to the ground and being unable to defend themselves. As the veterans got ready to board the transports, a solemn, gloomy mood permeated the aerodrome. In distinct contrast to the Normandy drop, no one was getting Mohawk haircuts and there were no shouts of 'Look out, Hitler! Here we come.' When the drop was cancelled because Patton's 3rd Army had advanced past the drop zone, rather than disappointment there was laughter, cheering and dancing.[96]

Ernie Pyle felt that the most noticeable characteristic of a unit that had not yet engaged the enemy was their eagerness for combat. He observed:

'After they had a crack at them for a few months, I knew they would be just as eager to let somebody else have a turn at it.'[97] Günter Koschorrek, reflecting upon the attitude of the men of his unit before their battlefield initiation, wrote: 'we were all so full of eagerness and the desire to win! And how impatiently we waited for the opportunity to fight at the front!' But this enthusiasm quickly faded. 'Now, after exactly three weeks in combat, no one talks of heroism or enthusiasm any more. On the contrary, the only wish is to get out of this death trap alive. This is not war as we imagined it would be, and of which we talked.'[98] A British officer who served in the Falklands War recounted that prior to their initial battle at Goose Green, the soldiers of the 2nd Battalion of the Parachute Regiment 'were so eager to get going we had to restrain them on the start line, the platoon commanders telling them to wait.' 'After Wireless Ridge,' he recalled, 'one man told me that he would never have got up from that start line if his platoon commander hadn't moved off and shouted "Follow me". They had grown wiser and knew how unpleasant it would be.'[99]

Once their initial enthusiasm for combat has receded, servicemen may come to feel they have done their part and that others should step up and man the front line. During the Second World War the embittered members of the US Army's 1st Division – which had made assaults in the North African, Sicilian and Normandy campaigns – bitterly complained: 'The Army consists of the 1st Division and eight million replacements.'[100] Philip Neame, one of the company commanders in the 2nd Battalion of the Parachute Regiment, noted that after the battle of Goose Green during the Falklands War, 'Without a doubt there was also a reaction setting in that we all rather felt we had done our bit, now let the rest of the Army do theirs.'[101] The battalion chaplain, David Cooper, felt the same way and told the 5 Brigade commander that 'it was somebody else's turn other than 2 Para to earn medals.'[102] This decrease in bravado can be manifested in the unwillingness of the infantry to close with the enemy. Instead they come to rely on the massed firepower of artillery, armour and the air force to obliterate their objective. This increased reluctance to risk their lives in combat also results in the soldiers beginning to question the forces that keep them exposed to the dangers of combat.

Few soldiers subscribe to the sentiments of the Roman poet Horace, who proclaimed '*Dulce et decorum est pro patria mori*' (It is sweet and proper to die for one's country). Is it patriotism that drives men forward? Robert Graves proclaimed: 'There was no patriotism in the trenches. It was too remote a sentiment, and rejected as fit only for civilians.'[103] George Coppard recalled that among his fellow British soldiers of the First World War, 'To most of us it was not a matter of patriotism any longer – that had burned itself out long ago. What remained was a silent bonding together of men who knew that there was no other way out but to see the thing through.'[104] A survey of over 1,500 American soldiers during the Second World War found that less than 2 per cent listed patriotism as their major incentive to keep fighting.[105] An American soldier observed: 'You're fighting for your skin on the line. When I enlisted I was patriotic as all hell. There's no patriotism on the line. A boy up there 60 days in the line is in danger every minute. He ain't fighting for patriotism.'[106] These sentiments were echoed 50 years later by a Marine Corps lieutenant colonel, who commented to the journalist Chris Hedges as they prepared to cross into Kuwait in 1991: 'none of these boys is fighting for home, for the flag, for all that crap the politicians feed the public. They are fighting for each other, just for each other.'[107] A generation earlier in 1971, another journalist, Max Hastings, asked some American soldiers patrolling through the jungles of Vietnam what were they fighting for. One of them wearily replied: 'Survival. To get out of here.'[108] Quite simply, the only right that soldiers usually fight for is the right to go home.

Loyalty to the group

Is it fear of punishment that keeps men in the line? Over time, even the most draconian punishments begin to lose their effectiveness, for soldiers will resort to self-inflicted wounds and other such measures to escape from combat. Rather, the underlying basis of any effective combat force is small group loyalty. A guide for officers in the US Air Force encapsulates this point in the comment: 'It will be seen that it is not primarily a cause which makes men loyal to each other, but the loyalty of men to each other which makes a cause'.[109]

In peacetime, many soldiers, particularly those of the British Army, will display fierce loyalty to their regiment. The regiment provides them with their 'corporate' identity, differentiating them from other soldiers and linking them to the heroic deeds of their predecessors. US Marines will display a similar affection for the Marine Corps – the Few, the Proud – which is the most successful of the American services in instilling *esprit de corps*, though this depth of feeling is also attained in some specialised army formations, such as the Airborne or the Rangers. But in combat, a soldier's primary focus is the members of his section and even more so his combat 'buddy', with whom he will share his foxhole, his meals, his fears and his triumphs. He will be aware of the larger organisations to which his section belongs, but for most practical purposes these superior formations are irrelevant to the soldier's day-to-day existence.

The organising of an army into small administrative elements has a lengthy historical pedigree. The smallest such element of the Imperial Roman Army was the eight-man mess unit (*conturbernia*). Genghis Khan organised his Mongol Army of the early 13th century into *arbans* of ten men. The corresponding element in the Prussian Army of Frederick the Great was the seven-man *Kameradschaft*.

Contemporary British and American armies employ ten-man sections/squads. These small groups of men (termed by sociologists the primary group) become their own little societies and by necessity must provide for their members recognition, approval, and a feeling of personal value: that each man is significant as an individual rather than just as a means to an end.[110] By living together, sharing discomfort and danger and becoming utterly interdependent on each other for survival, a unique bond of comradeship develops among these small bodies of men. This bond is particularly strong among the crews of armoured vehicles. 'An enduring sense of comradeship regardless of rank grows up between the four occupants of a tank, much heightened in time of war,' wrote James Hewitt, who commanded a squadron of Challenger tanks during the 1991 Gulf War. 'For weeks on end you live together in a small confined space with the unspoken knowledge that if you take a direct hit in the wrong place you will most probably die together.'[111]

Julian Thompson, who commanded a British brigade during the Falklands War, surmised that one of the key differences between the Argentine and British troops in this conflict was the source of their combat motivation. He felt that the Argentine conscripts

lacked this feeling that we had – that we were fighting for each other. They seemed to have this idea that they were fighting for their flag, for their country, which is a very fragile foundation on which to base morale, because in the stress of battle it evaporates. Whereas, if you're fighting for yourself, your comrades, for each other, that sustains you in the moments when you think you might be losing.[112]

These bonds of fraternity are strengthened by the unselfishness that permeates these small groups of comrades on the battlefield. An Australian Army officer serving at Gallipoli during the First World War wrote in a letter home: 'The trench is no place for a selfish natured man where almost everything is common property, just for the asking.'[113] An American soldier of the Second World War remembered such an occasion fondly: 'Men, hungry, days without enough food to meet the expenditures their duty demanded of the body, carefully dividing a small piece of bread equally, among four or five buddies, and each man, though the room was ink dark, knowing he'd get his share.'[114] Mansur Abdulin, who served with the Soviet Army during the Second World War, related how the company's bread, sugar and tobacco ration would be divided up and that everyone would take one of 40 portions in the complete knowledge that all portions were equal.[115] Adrien Bourgogne, slowly starving as he trudged his way west from Moscow in 1812 among the remnants of Napoleon's Grande Armée, fortuitously obtained some potatoes that he neglected to share with his comrades. Such is the strength of the bonds of fraternity among soldiers that many years after this event he still reproached himself for his actions that night, confessing in his memoirs that 'I shall never forgive myself for this selfishness.'[116]

1. *(right)* A Soldier's Journey: After the battle for Hill 200 on Peleliu, a weary US Marine sits down amid the rubble and weeps (26 September 1944).

2. *(below)* A Soldier's Journey: US Marine Master Gunnery Sergeant Frank Cordero inhales the scent of a letter from his wife after receiving mail for the first time since his unit left Kuwait to advance on Iraq three weeks before (7 April 2003).

3. The Cost of War: The body of a German soldier found beside an abandoned dugout during the First World War (circa 1917).

4. The Cost of War: Officers and men of the USS *Intrepid*, killed during the battle for Leyte Gulf in the Philippines, are buried at sea on 26 November 1944.

VD

Hello boy friend, coming MY way?

The 'easy' girl-friend spreads Syphilis and Gonorrhœa, which unless properly treated may result in blindness, insanity, paralysis, premature death

IF YOU HAVE RUN THE RISK, GET SKILLED TREATMENT AT ONCE. TREATMENT IS FREE AND CONFIDENTIAL

5. *(left)* Love, Sex and War: Second World War VD poster by Reginald Mount (1943).

6. *(below)* Love, Sex and War: An official German military brothel at Brest (France) during the Second World War, with its hours of operation prominently displayed (1940).

GEŒFFNET VON 10 uhr BIS 21 uhr

JEDER DEUTSCHE SOLDAT HAT UM 21 uhr DAS HAUS ZU VERLASSEN

DER ORTSKOMMANDANT

7. *(above)* Love, Sex
and War: A Japanese
soldier with some
'Comfort Women'
during the Second
World War, one of
whom is noticeably
pregnant (undated).

BOYS!
REMEMBER

NURSE CAVELL

PUBLISHED BY THE STATE PARLIAMENTARY RECRUITING COMMITTEE

8. *(left)* Kill or be
Killed: Live and Let
Live: Propaganda
poster produced
by the New South
Wales (Australia)
State Parliamentary
Recruiting
Committee
concerning the
execution of Edith
Cavell by the
German Army. The
artist is Virgil Reilly
(1915).

9. Kill or be Killed: Live and Let Live: A grief-stricken American infantryman, whose buddy has been killed in action during the Korean War, is comforted by another soldier. In the background a corpsman methodically fills out casualty tags (28 August 1950).

10. Kill or be Killed: Live and Let Live: A snapshot taken by a British officer showing German and British troops fraternising on the Western Front during the Christmas Truce of 1914 (25 December 1914).

11. Kill or be Killed: Live and Let Live: Bodies of three American soldiers lying near a half-sunken landing craft on Buna Beach, New Guinea (February 1943). This was the first image to show dead American troops to be approved for publication by the US Office of War Information censors during the Second World War.

12. Killing Your Own – The Death Penalty: A Union soldier, Private John Thomas Barnett of the 11th Pennsylvania Cavalry, being executed for desertion and highway robbery on 17 September 1862.

13. Killing Your Own – The Death Penalty: The execution of a German spy, Richard Jarczy, by a US Army firing squad on 28 April 1945. Jarczy admitted that he had been trained for sabotage and espionage behind the Allied lines. (There are no photographs available of British military executions during the First World War.)

14. Killing Your Own – The Death Penalty: (l to r) Lieutenant Harry 'Breaker' Morant, executed by the British Army on 27 February 1902 after being found guilty of murder; Private Thomas Highgate, executed by the British Army on 8 September 1914 after being found guilty of desertion (the first British soldier to be executed during the First World War); Private Edward 'Eddie' Slovik, executed by the US Army on 31 January 1945 after being found guilty of desertion (the only US soldier executed for desertion since the American Civil War).

15. Killing Your Own – Friendly Fire: US Air Force Caribou transport plane hit by friendly artillery fire at Ha Phan, Vietnam on 3 August 1967. The three crew members were killed.

16. Killing Your Own – Friendly Fire: Sergeant Ken Kozakiewicz (left) breaks down in an evacuation helicopter after hearing that his friend, the driver of his Bradley Fighting Vehicle, was killed in a 'friendly fire' incident that he himself survived (28 February 1991). Michael Tsangarakis (centre) suffers severe burns from ammunition rounds that blew up inside the vehicle during the incident. They and the body of the dead man (right) are on their way to a MASH (Mobile Army Surgical Hospital).

When soldiers are in the field, all property is communal; they will even pass around their letters from home, though, as Hugh McManners observed during the Falklands War, these letters usually had the 'sports' page omitted.[117] 'The most striking change that came over the individual after he had entered the active zone, and knew that the Jap was just around the corner, was the way in which *self* disappeared,' recalled an officer of the 1st Battalion of the Royal Scots Regiment who fought in Burma during the Second World War. 'It was share and share alike with everything you had. No thought of getting anything in exchange.'[118]

One of the greatest slurs to a soldier's reputation is to be accused of holding back – of not sharing with his mates (including the contents of care packages sent from home). This spirit of generosity often comes as a surprise to outsiders. Michael Herr admired how the Marines he reported on in Vietnam always went out of their way to look after his comfort and safety, even when doing so entailed personal risk – such as offering up their helmets and flak jackets because Herr had turned up without his own.[119] The journalist Max Hastings was also struck by the generosity of the soldiers he accompanied to the Falklands in 1982:

> As the days went by, I became more and more moved by the simple fact that we lived in a world in which everyone was seeking to help and support each other. It is an extraordinary experience to find oneself among thousands of people united in pursuit of a common purpose. Instead of peacetime Britain, with the competition and pettiness and selfishness which are fundamental to most of our civilian lives, here every hour men were striving to do things for each other – often, I found, to do them even for me. Wherever one went, there were offers of food, transport, a poncho, a sleeping bag, a bunk, a 'wet', a bar of 'nutty'.[120]

But it is not only material objects that a soldier offers up to his immediate comrades, as his mere presence may help them overcome their fears on the battlefield. When a soldier is beginning to succumb to such fears, often the realisation that others around him are still fighting provides enough motivation to fight on. This mutual support, generally not based on any personal quality but rather on a simple physical presence, is common to many battlefields and is evident in the tendency of men to group together when under fire, even though they know this action to be tactically unsound and likely to attract the enemy's fire. An Australian soldier, waiting on the start line to assault the German trenches at Pozières in July 1916, noticed the men bunching up. He would later write in his diary: 'It is strange how men creep together for protection. Soon, instead of four paces interval between the men, we came down to lying alongside each other, and no motioning could make them move apart.'[121]

Such is the strength of the loyalty that binds these small groups together that men who are safely out of combat will feel guilty that they are not sharing the same risks as their mates and will hasten their return to the front. Vasily Grossman recounts an incident where two wounded Soviet soldiers snuck out of a hospital and walked some 30 kilometres to rejoin their unit fighting in Stalingrad. When they were being loaded onto a vehicle to be returned to the hospital, they both pleaded to stay with their battalion.[122] Bernard Fall in *Hell in a Very Small Place*, his account of the battle of Dien Bien Phu, tells how Sergeant Major Abderrahman, an Algerian soldier who had been medically evacuated to Hanoi with grenade splinters in both thighs, received a letter from his comrades informing him that an attack on their position by the Viet Minh was likely. Abderrahman defied the doctors' orders, smuggled himself aboard an aircraft and returned to the besieged fortress on 12 March 1954 – he was captured when the position was overrun three days later.[123] William Manchester received a superficial gunshot wound during the battle for Okinawa and was evacuated back to a field hospital. Yet he went AWOL from hospital to rejoin his section and take part in an amphibious landing behind the Japanese lines. He survived the landing only to be gravely wounded by a shell burst the following day. At the time, his rationale for violating orders and risking death were unclear; it was only after many years of reflection that he understood his real motives:

It was an act of love. Those men on the line were my family, my home. They were closer to me than I can say, closer than any friends had been or ever would be. They had never let me down, and I couldn't do it to them. I had to be with them, rather than let them die and me live with the knowledge that I might have saved them. Men, I now knew, do not fight for flag or country, for the Marine Corps or glory or any other abstraction. They fight for one another. Any man in combat who lacks comrades who will die for him, or for whom he is willing to die, is not a man at all. He is truly damned.[124]

The depth of this feeling of solidarity among combat soldiers is seldom grasped by the men serving in the rear. David Hackworth, hospitalised with a head wound during the Korean War, pleaded with the doctors to release him so that he could return to the front. 'They didn't realise that the guys in the 3d Platoon were my brothers, my family, and I loved them,' Hackworth recalled. 'I'd only been with them three weeks, it was true, but in combat that's a lifetime, and I didn't want to leave them out there alone, if by being there I could help keep them alive, keep them out of a head-wound ward.'[125] His pleading failed to convince the doctors and he was transferred to a general hospital in Japan. More recently, a Marine who had suffered two concussions from explosions while serving in Iraq, and who was told that a third occurrence could lead to severe brain damage, was given the option of returning home. He chose to stay. 'These are my brothers. I feel safer with them than I do anywhere else. I need to be with them.'[126]

In a moving and eloquent passage in his memoir of the Vietnam War, Philip Caputo wrote of the emotional ties that bind together the men in an infantry platoon:

[T]he communion between men is as profound as any between lovers. Actually, it is more so. It does not demand for its sustenance the reciprocity, the pledges of affection, the endless reassurances required by the love of men and women. It is, unlike marriage, a bond that cannot be broken by a word, by boredom or divorce, or by anything other than death. Sometimes even that is not strong enough. Two friends of mine died trying to save the corpses of their men from the battlefield. Such devotion, simple and selfless, the sentiment of belonging to each other, was the one decent thing we found in a conflict otherwise notable for its monstrosities.[127]

It is this passionate comradeship that enables combat effectiveness. Ken Moorefield was a company commander with the US Army in Vietnam in 1969/70. Reflecting upon what kept his men fighting, he surmised:

We had gone beyond the point of saying, 'We're here to defend South Vietnamese freedom' or 'We are here to defend democracy'. The waters had been too politically muddied by then. And I'm not sure that in most wars by the time a man gets to the front lines that kind of appeal works. He'll die for the man on his left and his right, maybe for his platoon commander or company commander. What drives him on day to day, moment to moment, is the strength of the personal bonds within the unit itself.[128]

Among Israeli soldiers this comradeship is known as *achavatt lochameem* (literally combatant's brotherhood) and its most frequent manifestation on the battlefield is the persistence shown by the Israeli army, regardless of additional casualties incurred, to recover their wounded and retrieve the bodies of their dead. This concept of a combatant brotherhood is not confined to the Israeli army. For example, the fifth stanza of the US Army's Ranger Creed states: 'I will never leave a fallen comrade to fall into the hands of the enemy.' It was such a sentiment that made Michael Durant, a pilot with the US Army who was held as a prisoner of war in Mogadishu for eleven days in mid-1993, never doubt that his fellow soldiers would attempt to rescue him.

It was simply not acceptable in our special ops culture to leave a man behind, dead or alive, under any circumstances. That was part of what made it possible to do the job. You might be wounded, cut off from your comrades, surrounded by the enemy, but someone would be coming to get you. You might be killed in action, but you were going to be buried at home.[129]

Rage and revenge

In practical terms, this depth of feeling between a soldier and his immediate comrades means that a direct threat to the members of his primary group, even more so the death of a member of the group, provides the necessary impetus to overcome the inherent resistance among soldiers to killing the enemy. Joseph Donaldson, serving with an artillery battery during the siege of Cadiz in the Peninsular War, looked about and saw that 'the ramparts were strewed with the dead and wounded; and blood, brains, and mangled limbs.' He pointed out that 'When any of our comrades fell, it excited no visible feeling but revenge. "Now for a retaliatory shot!" was the word; every nerve was strained to lay the gun with precision: and if it took effect, it was considered that full justice was done to their memory.'[130]

Siegfried Sassoon was largely indifferent to the Germans until his closest companion in the trenches (a fellow officer) was killed. He tells us that 'I went up to the trenches with the intention of trying to kill someone. It was my idea of getting a bit of my own back.'[131] This rage returned when a comrade fighting alongside Sassoon was shot through the head by a sniper: 'all feelings tightened and contracted to a single intention – to "settle that sniper" on the other side of the valley.' He charged the German position, lobbing Mills bombs into the trench as he ran forward. The Germans hastily retreated without suffering any casualties in the face of Sassoon's attack. Sassoon then disregarded an order to retire from the front lines because 'I definitely wanted to kill someone at close quarters.'[132]

On the other side of no man's land, the German soldiers were driven by the same base emotions. Ernst Jünger remembered that when one of his soldiers was shot and killed, 'His comrades stayed a long time at their shooting-slits afterwards, hoping to exact revenge. They seemed to feel personal enmity for the Britisher who had fired the mortal shot.'[133] Following the death of his companion, who was crouching beside him in a foxhole, during the German campaign in Russia in the Second World War,

Günter Koschorrek became enraged: 'That bloody sniper! If only I could catch him, it would give me the greatest pleasure to wipe him out – without compunction, even if he were on his knees in front of me and begging for mercy.'[134] 'Seeing Gale [a member of his section] killed shocked me as our first casualties had done, and I think enraged me,' recounted George MacDonald Fraser, a British soldier who fought the Japanese in Burma during the Second World War. 'I wanted a Jap then, mostly for my own animal pride, no doubt, but seeing Gale go down sparked something which I felt in the instant when I hung on my aim at the Jap with the sword, because I wanted to be sure. The joy of hitting him was the strongest emotion I felt that day.'[135]

Philip Caputo was initially sympathetic towards the Viet Cong soldiers killed by his platoon and observed that his men felt a sense of guilt and pity for the slain enemy. 'We retained a capacity for remorse and had not yet reached the stage of moral and emotional numbness.'[136] But as the war progressed and his unit began to suffer casualties – among them some of Caputo's fellow officers as well as members of his former platoon – darker feelings took hold of his psyche.

> I burned with a hatred for the Viet Cong and with an emotion that dwells in most of us, one closer to the surface than we care to admit: a desire for retribution. I did not hate the enemy for their politics, but for murdering Simpson, for executing that boy whose body had been found in the river, for blasting the life out of Walter Levy. Revenge was one of the reasons I volunteered for a line company. I wanted a chance to kill somebody.[137]

The Vietnam veteran Gary McKay bitterly recalled an incident where he helped gather up the remains of seven of his fellow Australian soldiers following a rocket-propelled grenade striking their armoured personnel carrier and igniting the onboard ammunition. He wrote: 'I felt a quiet rage inside, waiting to exact revenge for what had happened.'[138] Ted Cowell served as a 'tunnel rat' with the Australian field engineers in Vietnam in

1966. He later recounted how he reacted when his best mate was killed while clearing a Viet Cong tunnel complex:

> That was the end of me. I went back in and I was down there for seventeen hours going through the complex, unhooking little booby traps and in general killing any bastard I got my hands on … In actual hand to hand combat I killed three in the complex, and then the next morning I got a bloke by the river. That was a bit sadistic, but I had great joy in doing it … I could have shot him in the back of the head I suppose, but I didn't. I dropped the scope and blew his family jewels off and he ran himself out of blood.[139]

A similar rage filled an Argentine conscript during the Falklands War: 'the only thing I wanted to do, my obsession, was to avenge my fallen comrades. Whenever I saw one of my friends hit it was worse, it just made me want to continue fighting, it didn't matter for how long or at what cost.'[140]

Thus an escalating, self-perpetuating cycle of violence takes hold on the battlefield, and wider motives are subsumed by a localised desire to 'settle the score'. Günter Koschorrek was initially shocked by the violence of the battlefield until a comrade took him aside. 'This is the way of war, with its constantly increasing avalanches of hate,' he explained to Koschorrek. 'It begins with an attack and then combat. The two enemies fight for their lives, and it develops into grim determination and over-reaction from both sides. That leads to revenge and retaliation, in accordance with the motto "As you do unto me, so I do unto you!"'[141]

Gottlob Bidermann detected a change in the attitude of his comrades in the German infantry as they advanced into Russia. They 'became consumed with a remorseless ever-increasing rage; and the fevered minds could concentrate only on revenging fallen comrades, to kill the enemy, and to destroy.'[142] Tim Collins, who commanded a British infantry battalion during the 2003 invasion of Iraq, observed that: 'Both the insurgents in the streets of Iraq and many of the GIs fought out of a need for vengeance, the cycle of retaliation gathering pace with every fresh killing.'[143]

In *Achilles in Vietnam*, the psychologist Jonathan Shay writes that a motivational technique widely employed by the US military in Vietnam on soldiers who had recently lost a comrade was to tell them: 'Don't get sad. Get Even!'[144] 'I really loved fucking killing, couldn't get enough,' one of the Vietnam veterans interviewed by Shay stated brusquely. 'For every one that I killed I felt better. Made some of the hurt went away [sic]. Every time you lost a friend it seemed like a part of you was gone. Get one of them to compensate what they had done to me. I got very hard, cold, and merciless.'[145] Tim O'Brien provides an account of a revenge attack on a Vietnamese village where women were beaten, babies and children killed and the huts levelled by napalm. The attack followed the deaths of two popular soldiers in O'Brien's company from a booby-trapped artillery round concealed near the village. O'Brien bitterly recalled: 'But Chip and Tom were on their way to Graves Registration in Chu Lai, and they were dead, and it was hard to be filled with pity.'[146] Colin Sisson, serving with the New Zealand Army in Vietnam, tells us of an incident where one of his comrades was killed by some NVA (North Vietnamese Army) soldiers:

> Every man in Five Platoon stood up, feeling shocked and angry. Sensing the mood of his men their Platoon Commander made a snap decision. To his section commanders gathered around him he uttered only three words: 'Go get 'em.' The three sections of 5 platoon fanned out at the run through the bush with only one intent – to kill. The Communists had scattered, but the one who had killed Des became trapped between two converging sections … [and] died in a hail of bullets from twenty incensed men whose one thought was revenge. They continued firing at the fallen figure until their anger and their magazines were exhausted, and left an unrecognizable, bloody mess.[147]

Similar motivational techniques were employed in Iraq. David Bellavia writes that just prior to his unit (2nd Battalion, 2nd Infantry Regiment, 1st Infantry Division) assaulting Fallujah in November 2004, the battalion commander assembled the men and told them: 'We're gonna go out there and kick their asses. They killed our own. Twenty-seven of our brothers are dead and these are the assholes who are responsible. This is personal for me, and it should be for you.'[148]

The philosophy of live and let live

Ultimately this cycle of violence will begin to exhaust itself. Edward Costello, who fought in the Peninsular War against the French, noted that in the early stages of the campaign both the English and French picquets frequently fired upon each other, producing a limited but steady stream of casualties. 'We had not yet established that understanding with the enemy which avoided unnecessary bloodshed at the outposts, which afterwards tended much to humanise the war,' wrote Costello.[149] As the campaign progressed, Costello tells us that 'the French soldiers and ourselves began to establish a very amicable feeling, apart from duty in the field. It was a common thing for us to meet each other daily at the houses between our lines, when perhaps both parties would be in search of wine and food.'[150] On one memorable occasion Costello had to retrieve a colleague who had got drunk in the company of two French soldiers. Hamlin Alexander Coe detected a lessening of the animosity between the Union and Confederate picquets during the American Civil War: 'Both parties seem to have mutually agreed not to fire upon each other, and while the Rebs are upon the south bank and our boys upon the north bank of the river, they have changed papers, traded coffee for tobacco etc.'[151] The local armistice lasted for several days until the Confederates' picquets were moved back from the river bank. 'It is supposed they were getting too intimate with our boys,' thought Coe. 'And, indeed, they were becoming friendly, proving to me clearly that if the privates of both armies were turned loose they would settle this war in a hurry.'[152] Maurie Pears wrote of a localised truce between the Australian and Chinese troops during the Korean War, whereby 'They left us alone and we left them alone, except when battalion patrols passed through us and the wire.' Pears tells us that only once was the truce broken by the Chinese without cause: 'The platoon went berserk at this breach of the truce and we poured small arms and fire support onto the Chinese position, some standing and hurling abuse as well as shots ... The pure frenzy of our response must have made them think, as they did not snipe again.'[153]

The localised truce written of by Pears may also occur on a much more generalised scale. Following the widespread slaughter of the opening campaigns of the First World War, war weariness permeated the British trenches. The Western Front had lapsed into stalemate, whereby neither side could achieve a decisive breakthrough and each combatant would respond in kind to any offensive action, that is, sniping would be met with sniping, artillery fire with artillery fire, and so on. A British officer, Charles Sorley, wrote in a letter home that the troops realised that 'to provide discomfort for the other [the Germans] is but a roundabout way of providing it for themselves.'[154] Many of the troops therefore adopted a 'live and let live' philosophy. The front-line soldiers had realised that attacking their enemy, rather than conferring a tactical advantage, would only provide an incentive for retribution and thus increase their own discomfort. Another example of this sentiment comes from the final days of that war, when Horace Baker, a member of the American Expeditionary Force, had an opportunity to fire upon some German soldiers.

> I raised my rifle and took deliberate aim. Thinks I, 'I can get one now and know I got him.' But just as I was ready to pull the trigger I thought of something else: 'I can get two at least before they can tell where I am.' So I aimed again but failed to shoot again for a pesky thought almost knocked me down. It was, 'What if I do kill two or a half dozen and they swarm over the field and kill me, how much will that profit me?' So I decided to let them live and risk chances of living myself.[155]

Paul Fussell tell us that during the Second World War 'both we [the US Army] and the Germans found that in the absence of orders to the contrary, the best policy was to leave well enough alone.'[156] Half a century later, Anthony Swofford pondered why his small patrol of Marines had not been subjected to a follow-up assault following an Iraqi rocket attack during the opening stages of the 1991 Gulf War:

I will never know why those men didn't attack us over the rise. Perhaps we shared an aura of mutual assured existence, allowing us to slowly approach one another and prepare to engage, but finally when the numbers were crunched, the numbers were bad for both sides, and the engagement thus sensibly aborted. If wars were fought only by the men on the ground, the men facing one another in real battle, most wars would end quickly and sensibly. Men are smart and men are animals, in that they don't want to die so simply for so little.[157]

Probably the best-known manifestation of the live and let live philosophy among soldiers is the Christmas Truce of December 1914. By early December, this point of view had begun to spread throughout the trenches of the Western Front. The onset of wintry conditions precluded large-scale offensives, so many of the front-line troops saw no point in making life in the trenches any more difficult than it had to be. Neither antagonist fired at the other during mealtimes and, where the trench-lines were close, soldiers would throw newspapers, cigars and rations between the trenches. In many sectors, troops on both sides accepted the status quo and friendly banter echoed along the lines. This banter was mainly conducted in English, as many of the German soldiers had previously worked in England. The lessening of the enmity towards the Germans alarmed the British High Command. In early December 1914, II Corps HQ issued an order specifically forbidding fraternisation with the enemy, stating that it destroyed the desired offensive spirit.[158]

In *Silent Night: The Story of the World War I Christmas Truce*, Stanley Weintraub notes that while the British High Command might have been prepared to overlook localised truces, a widespread Christmas truce was another matter.[159] To counter this possibility, the High Command issued a message on Christmas Eve warning of a possible German attack over the Christmas–New Year period. On Christmas Eve, instead of attacking, the Germans placed small Christmas trees on the parapet of their trenches and that evening the sounds of Noël singing drifted over no man's land.

The following morning signboards appeared along the front proclaiming, 'You No Fight, We No Fight' and 'Merry Christmas'. Cautiously the

British and German soldiers left the safety of their trenches and advanced into the middle of no man's land, where localised truces were agreed upon. The soldiers exchanged food, tobacco, alcohol, and various souvenirs such as uniform buttons and insignia. They also took advantage of the truce to bury the dead who had fallen in no man's land, each side assisting the other. Impromptu games of football were even played between the British and German soldiers, a ration tin or a sandbag stuffed with rags substituting if an actual football was not forthcoming. Scattered localised truces also took place in the sectors held by the French and the Belgians. The truce largely ended that night, either by random acts of violence or by a pre-arranged signal, such as shooting into the air, though little in the way of firing occurred on Boxing Day in many sectors. The British General HQ issued a statement announcing: 'After a comparative lull, on account of the stormy weather, the Allies and the Germans are again actively engaged in Northern France and Belgium.'[160] Weintraub points out that:

> [M]any troops had discovered through the truce that the enemy, despite the best efforts of the propagandists, were not monsters. Each side had encountered men much like themselves, drawn from the same walks of life – and led, alas, by professionals who saw the world through different lenses.[161]

The following year, to ensure that the events of Christmas 1914 were not repeated, orders were issued by both sides strictly forbidding all fraternisation with the enemy. Intermittent shellfire was maintained throughout Christmas Eve and Christmas Day 1915 as a further incentive for the troops to remain in their trenches. The efforts of the staff were largely successful, with attempts at fraternisation invariably being met with a response from a machine gun or shellfire. Ernst Jünger recalled:

> We spent Christmas Eve in the line, and, standing in the mud, sang hymns, to which the British responded with machine-gun fire. On Christmas Day, we lost one man to a ricochet in the head. Immediately afterwards, the British attempted a friendly gesture by hauling a

Christmas tree up on their traverse, but our angry troops quickly shot it down again, to which Tommy replied with rifle grenades. It was all in all a less than merry Christmas.[162]

Furthermore, the events of 1915, such as the German use of poison gas at the Second Battle of Ypres in April and the sinking of the *Lusitania*, a British cargo and passenger liner, by a German submarine in May, solidified a deepening enmity between the combatants. A British Army officer, recalling the spirit of Christmas 1915 in the trenches, commented: 'things have got past the stage when one can fraternise with the enemy, there is too much hatred flying about.'[163]

Although the Christmas Truce of 1914 is the most famous example of the live and let live philosophy during the First World War, it was not an isolated incident. In *Trench Warfare 1914–1918: The Live and Let Live System*, Tony Ashworth concludes that this system persisted and evolved throughout the war, despite the best efforts of the HQ staff to eradicate it. Initially battlefield truces occurred around mealtimes, particularly breakfast. 'The command "stand-down" was almost invariably followed by a lull along the whole front,' Sidney Rogerson fondly recalled. 'Hostilities were temporarily suspended by mutual, if mute, consent … For anything from an hour to two hours the most vicious noise to be heard in the trench was the sizzling of frying bacon.'[164] Temporary truces also arose so that both sides could collect their wounded or bury their dead, or when environmental factors, such as the flooding of the trenches, produced such general misery that neither side saw much advantage in adding to their misfortune by firing at the enemy and inviting a similar response. Jünger observed such an incident in December 1915 when heavy rain had collapsed the trenches. He wrote: 'The occupants of both trenches had emerged from the morass of their trenches on to the top, and already a lively exchange of schnapps, cigarettes, uniform buttons and other items had commenced between the two barbed-wire lines.'[165]

In some cases, overt fraternisation would occur during these battlefield truces which, when detected by superior officers, resulted in prohibitions and threats of punitive action against future transgressors. As a result, subsequent truces would usually be more discreet. Winston Churchill witnessed such a truce when he visited the French trenches on Vimy Ridge in December 1915. He was struck by the difference between this sector and that held by his battalion (2nd Battalion, Grenadier Guards):

> It was pretty quiet. The lines are in places only a few yards apart, but a much less spiteful temper prevails than on the part of the Guards. There you cannot show a whisker without grave risk of death. Here the sentries look at each other over the top of the parapet: & while we were in the trench the Germans passed the word to the French to take cover as their officer was going to order some shelling. This duly arrived ... [166]

A less obvious manifestation of the live and let live system was inertia, whereby soldiers refrained from firing at the enemy in the expectation that this action would be reciprocated. Although both sides would meet aggression with equal or greater aggression, neither wished to be the first to break the peace. 'The military situation at le Touquet was curious,' wrote George Coppard, 'for it seemed as if both sides, the Germans and ourselves, had tacitly agreed that this part of the line should be labelled "Quiet", it being understood that if one side started up any bloody nonsense, then the other would follow suit.'[167]

The HQ staff also sought to stamp out battlefield inertia because it did not accord with the desired objective of maintaining an offensive spirit. The staff countered inertia by requiring frequent reports detailing the amount of ammunition expended. These reports, when compared to those of other units, indicated the sectors where inertia predominated. To counter the efforts of the staff, soldiers responded by ritualising their firing: weapons would be fired high, or into unoccupied areas, or at precisely the same time each day. This ritualisation of firing would be interpreted by the enemy as a desire to maintain the spirit of peaceful coexistence, despite the insistence of their respective staffs on expending a pre-determined quantity of ammunition. Small arms fired high wounded or killed by chance rather than design, while soldiers could easily take precautions when faced with the predictable timing and impact of artillery fire (despite specific instructions being issued to the gunners that shelling was to occur at irregular intervals). The staff responded to the practice of ritualised firing by ordering that raids be carried out. Raids were short-duration attacks with limited objectives such as the capture of a prisoner, to identify an opposing unit, or to kill as many of the enemy as quickly as you could and return to your own trenches before the inevitable counter-attack could be mounted. Tony Ashworth concludes that:

> [B]efore a raid mutual trust, confirmed by ritualised violence, prevailed among antagonists; after a raid, a background expectancy of mutual caution was general … Such a situation inhibited the re-emergence of peace; for where mutual distrust had replaced trust, the cycle of aggression and escalated counter-aggression was likely to recommence at any time.[168]

Unsurprisingly, the tactical advantages of raids were appreciated more by staff officers than the men who had to carry them out. Max Plowman wrote: 'one sees a raid as a foul, mean, bloody, murderous orgy which no human being who retains a grain of moral sense can take part in without the atrophy of every human instinct.'[169]

The mounting of raids would invariably produce retaliation in kind. Carl von Clausewitz noted: 'Even where there is no national hatred and no animosity to start with, the fighting itself will stir up hostile feelings: violence committed on superior orders will stir up the desire for revenge and retaliation against the perpetrator rather than against the powers that ordered the action.'[170] Throughout history, commanders have exploited the psychological desire to 'get even' to maintain the necessary feeling of animosity towards the enemy. Stanley Weintraub points out that in mid-December 1914 the British High Command initiated a series of small but costly raids on the enemy's trenches for the sole purpose of provoking an aggressive response from the Germans. At the time, as we have seen, the British generals were increasingly concerned about the growing number of incidents of battlefield fraternisation between the British and German troops (rightly so, as the Christmas Truce occurred a few weeks later).[171]

The staff were justified in their concerns about the persistence of the live and let live system, as the efficient functioning of an army is very much dependent on its ability to prevent introspective questioning by its soldiers about why and for whom they are fighting and dying. A British officer of the First World War wrote after the war: 'I still wonder what might have happened if the friendly meetings between German and English in No-Man's-Land, where they exchanged cigars for cigarettes and chattered together, had not been nipped in the bud.'[172]

The need for consensus and support

While abstract ideals such as duty and patriotism may inspire those on the 'home front' – for whom dying in war is also largely an abstraction – they soon lose their appeal once the bullets become real. Soldiers then search for a reason to justify the continued exposure of themselves and their comrades to danger.

As noted earlier, if the war is being fought to ensure national survival and their family and friends are directly threatened, this factor will provide justification for soldiers to continue to risk their lives for the 'cause'. But in the absence of a direct threat, soldiers may experience internal unease. They may come to feel that they are fighting for an unrepresentative, uncaring government, or that their sacrifices serve only to enrich certain elements of society. The combination of the slaughter of their comrades and the apparent futility of various battles will cause many soldiers to question the value system that exposes them to the squalor of war.

This questioning by soldiers about why they are fighting reinforces the importance of developing a national consensus supporting the war. The US Army's *Textbook of Military Medicine: War Psychiatry* states that it is a 'national consensus about the war that legitimates [a soldier's] behaviour in combat and validates the suffering, deprivation, guilt and fear that he experienced'. The textbook goes on to note the difference in national consensus, and the resultant effect on returning soldiers, between a conflict of national survival and campaigns that 'required only fractional commitment of national resources and did not put national survival in jeopardy'.[173]

In relation to campaigns where national survival was not at stake, such as Korea, Vietnam, Afghanistan and Iraq, the textbook points out that:

> Governmental assertions about the national interests that were at stake were not particularly credible, and the longer the wars went on the more threadbare became public support. Civilian inconvenience and involvement were modest, and because casualties were few and affected but few families, there was little pressure in the society as a whole for validation of the war effort …[174]

The key point being made here is that soldiers have a pronounced need to believe that their efforts and the war in general are supported by their society. This belief in the support of the home front becomes eroded when soldiers are seemingly abandoned to their fate, for example when they are required to remain in combat until wounded or killed. As noted in Chapter 3, the policy of the US Army and the US Marine Corps during the Second World War was to keep their rifle companies in the field for extended periods and to make up losses with individual replacements. William Manchester commented that the practical effect of this policy on the fighting man was that: 'Unless he became a casualty, lost his mind, or shot himself in the foot – a court-martial offence – he stayed on the line until he was relieved, which usually happened only when his outfit had lost too many men to jump off in a new attack.'[175]

An example of the debilitating psychological consequences of this policy is illustrated by an encounter between J. Glenn Gray and an American deserter during the Second World War. The soldier had been deployed for over two years and had fought in the African, Italian and French campaigns. His aggregate combat experience eventually overwhelmed him and he deserted. When Gray found him he was living in a crude hut in the forest and was surviving on battlefield refuse. The soldier, close to tears, professed that:

> All the men I knew and trained with have been killed or transferred … I'm lonely. They promised me I would be relieved or rotated, but

nothing ever happens. I can't stand the infantry any longer. Why won't they transfer me to some other outfit? The shells seem to come closer all the time and I can't stand them.[176]

The feeling of social abandonment is reinforced when soldiers return home on leave and encounter a civilian population largely oblivious and indifferent to the adverse conditions the soldiers are facing in combat. A British soldier of the First World War wrote with a touch of plaintiveness: 'One thing I found when I eventually got home was that my father and mother didn't seem in the least interested in what had happened. They hadn't any conception of what it was like … they had no idea of what kind of danger we were in.'[177]

Soldiers may also develop a feeling of revulsion for the bloodthirsty attitude exhibited by the civilian population towards the enemy. George Orwell, who fought on the Republican side during the Spanish Civil War, noted that: 'One of the most horrible features of war is that all the war-propaganda, all the screaming and lies and hatred, comes invariably from people who are not fighting.'[178] The building of the necessary national consensus supporting the war requires that whole societies are indoctrinated, with the aim of developing an abstract hatred of the enemy. This abstract hatred is fed by government propaganda, long-range bombing, and ever-increasing casualty lists. Most civilians will have no direct contact with the enemy and so are unable to mitigate this abstract hatred through personal observation of the enemy's shared humanity. Gray surmises that:

A civilian far removed from the battle area is nearly certain to be more bloodthirsty than the front-line soldier whose hatred has to be responsible, meaning that he has to respond to it, to answer it with action. Many a combat soldier in World War II was appalled to receive letters from his girlfriend or wife, safe at home, demanding to know how many of the enemy he had personally accounted for and often requesting the death of several more as a personal favour for her![179]

Hugh McManners recalled that the members of the Falklands taskforce steaming south did not share the jubilation of the British press (and presumably elements of British society) following the sinking of the Argentine cruiser *General Belgrano* on 2 May 1982, resulting in 323 deaths (and the infamous 'Gotcha' headline from the *Sun* newspaper). He wrote angrily in his diary when informed of the sinking that:

> My feelings are not of the glee reported by the idiot press on board *Canberra*, but more akin to the Royal Marine Colonel who when told of the sinking by a press man, used a four letter expletive and returned to his office to carry on with his work. Of course there are clowns who are pleased. They seem to be the ones least likely to be personally involved in future fighting.[180]

From enemy to friend and back again

The abstract hatred felt by civilians towards the enemy means that most civilians find it difficult to comprehend that soldiers will often empathise with enemy combatants. Joseph Donaldson, who served with the 94th Regiment of Foot during the Peninsular War, wrote:

> How different were our feelings in this respect from many of our countrymen at home, whose ideas of the French character were drawn from servile newspapers and pamphlets ... but I myself must confess, in common with many others, that I was astonished when I came in contact with French soldiers, to find them, instead of pigmy, spider-shanked wretches, who fed on nothing but frogs and beef tea, stout, handsome-looking fellows, who understood the principles of good living, as well as any Englishman amongst us; and whatever may be said to the contrary, remarkably brave soldiers.[181]

Edmund Wheatley, a British officer serving with the 5th Line Battalion of the King's German Legion at Waterloo, confessed in his diary to admiring his French foes more than his German allies.

On the opposite ascent stand hundreds of young men like myself whose feelings are probably more acute, whose principles are more upright, whose acquaintance would delight and conversation improve me … When I looked at my own comrades I could not conceive why my animosity was diverted from them in preference to the French who are, by far, more commendable characters than these heavy, selfish Germans.[182]

The Union soldier Rice Bull, lying among the wounded following the battle of Chancellorsville during the American Civil War, discerned a difference in attitude between the soldiery of the Confederacy and the civilians who visited the battlefield after the fighting had concluded.

They were mostly old men and women and children. These elderly people did not have the kindly feeling for us shown by the soldiers that had fought a few days before. The battle had, it would seem, created a feeling of respect on the part of the soldiers. Usually when these people came they were on horseback. They would ride up close to the wounded seemingly filled with hatred.[183]

This lessening of the enmity between combatants may paradoxically arise out of an attack. At dawn on 19 May 1915 at Gallipoli, over 40,000 Turkish soldiers tried to push the men of the Australian and New Zealand Army Corps (Anzac) invaders back into the sea. The gallant assault had little chance of success against the entrenched Anzacs but the Turks bravely pushed on, so that by around noon 10,000 Turks lay dead or wounded in no man's land. The Anzacs, who above all else respected courage, largely discarded the abstract hatred for the Turk that had characterised the first month of so of the campaign and instead of being simply the 'enemy' he became 'Abdul' or 'Johnny Turk'. An armistice was arranged five days later to bury the dead. The opposing forces met in no man's land and exchanged photographs and cigarettes. One of the Australian soldiers who took part in this armistice wrote in his diary: 'The time was taken up by making friends with the Turks, who do not seem to be a very bad chap after all. After today most of our opinions on the Turks were changed, they certainly play the game better than the Germans do.'[184] The historian Bill Gammage characterised the subsequent stages of the Gallipoli campaign as a 'friendly but determined rivalry'.[185] The solidarity between the Turkish and Australian forces was encapsulated in an eloquent tribute written for the Anzacs killed at Gallipoli by one of the Turkish commanders, Colonel Mustafa Kemal (later known as Atatürk), after the war.

> Those heroes that shed their blood and lost their lives … You are now lying in the soil of a friendly country. Therefore rest in peace. There is no difference between the Johnnies and Mehmets to us where they lie side by side here in this country of ours … you, the mothers, who sent their sons from faraway countries wipe away your tears; your sons are now lying in our bosom and are in peace. After having lost their lives on this land they have become our sons as well.[186]

Moreover, as an indication of the enduring respect between the combatants at Gallipoli, the Australian, New Zealand and Turkish forces were looking forward to holding a combined Anzac Day ceremony in Korea in 1951, the first time such a ceremony would have occurred, with all three nations providing contingents to the United Nations force fighting the Korean War. Anzac Day (25 April) commemorates the initial landing of Australian and New Zealand troops at Gallipoli, and apart from the formal remembrance ceremonies this particular occasion would have included a soccer match between the Turks and the Australians.[187] Unfortunately, a major attack by the Chinese forces that began on the night of 22 April 1951, and which saw the Australians deployed to the valley of the Kapyong River to delay the Chinese advance, put paid to any commemorative activities.

A feeling of fraternity among combatants based on common suffering was also evident during the Second World War. An American soldier of that war noted: 'most of us felt a good deal of respect and sympathy for the average German soldier. Many of them were there whether they wanted to be or not, and suffered the same hardships we did.'[188] While digging into the chalky earth of the Russian steppe, the German soldier Henry Metelmann reflected that: 'Being well aware that our opponents were suffering the same hardships in their own foxholes, the personal feeling of hatred for them completely evaporated and was to a large extent transferred to our officers …'[189] Stanley Fennell, a sapper with the Royal Engineers, was captured by German troops during the Italian campaign. He was treated 'very reasonably' by the Germans, whose simple explanation for doing so encapsulates the comradely spirit that can develop between soldiers of opposing nations: 'We are the front swine. You are the front swine.'[190]

Some soldiers may come to realise that it is only political happenstance that nations fight against rather than beside each other, for history is punctuated with examples of former allies that have become enemies. Ernst Jünger's German regiment (the 73rd), which fought against the British and French during the First World War, was known as 'Les Gibraltars'. The 73rd wore the battle honour 'Gibraltar', in the form of a ribbon sewn on the left sleeve of their uniforms, in remembrance of their predecessors in the Hanoverian Guards who had defended Gibraltar alongside some British soldiers, during the 'Great Siege' of 1779–83 by the armies of France and Spain.[191] Furthermore, it was British and Prussian troops who defeated the French under Napoleon at the battle of Waterloo in 1815. Slightly less than a century later, British and French troops now faced the Germans across the fields of Belgium. Niall Ferguson makes the same point in *The Pity of War*, noting the imperial rivalry between Britain and Russia, and Britain and France. 'Few contemporaries in 1895 would have predicted that they would end up fighting a war on the same side within twenty years. After all, the collective diplomatic memory of the previous century was of recurrent friction between Britain, France and Russia.'[192]

The Second World War would witness several rapid transformations from ally to enemy. In May and June 1940, British and French troops fought alongside each other in an unsuccessful attempt to stem the German blitzkrieg through the Netherlands, Belgium and France. The campaign ended with the evacuation of British, French and Belgian troops from the beaches of Dunkirk. The French government signed an armistice with Germany on 22 June 1940 and the Vichy regime, under Marshal Henri Philippe Pétain, became the de facto government of France. Winston Churchill feared that the French fleet could be used by the Germans to threaten Britain's maritime supply lines. He ordered the ships of the French Navy to either fight alongside the Royal Navy or be neutralised, with the option of being impounded in a British-controlled port. On 3 July 1940, less than two weeks after the signing of the armistice between France and Germany, a French naval squadron at Mers-el-Kébir harbour in French North Africa (now Algeria) was attacked by the Royal Navy after the French admiral had refused to comply with a British ultimatum. An old French battleship, the *Bretagne*, was sunk and several other French ships were damaged and run aground. The British attack concluded on 5 July, by which time almost 1,300 French sailors had been killed by the forces of their former ally. The French responded by dive-bombing the British fleet on its way back to Gibraltar.

One of the most rapid transformations from ally to enemy during the Second World War occurred following the signing of an armistice between Finland and the Soviet Union on 19 September 1944 (the actual fighting had ceased two weeks earlier). At this time, the German 20th Mountain Army (around 200,000 men) was deployed to Finland and had been fighting alongside the Finnish soldiers against the Soviets in the Karelia region. A condition of the armistice was that the Germans were to be driven off Finnish soil. The German forces began to withdraw through Norway but when this did not occur quickly enough to satisfy the Soviets, skirmishes arose between the Finnish and German troops. These skirmishes escalated into what became known as the Lapland War, which lasted from September 1944 to April 1945, when the last German troops were driven out of Finland. The Finns suffered around 4,000 casualties in this conflict (of which roughly one quarter were killed), the Germans approximately 3,000 casualties (one third of whom were killed). In addition, around 1,300 German soldiers became prisoners of war. The German soldiers based in Italy would face a similar situation following the armistice between the Italian and Allied forces that came into effect on 8 September 1943. The German troops immediately sought to disarm their former allies to prevent the Italians' weapons being used against them. The Germans would fight on in Italy until the end of the war.

Censoring the reality of war

One of the crucial enablers of combat effectiveness is to place soldiers in a position where they feel that they have to fight. Some soldiers will take out their frustration with being placed in this position on the enemy, as by resisting and not capitulating immediately, enemy combatants expose soldiers to further risks and privations and prolong their separation from their families. The enemy combatants therefore become an obstacle preventing soldiers from realising their personal goals, an obstacle that will respond only to direct force. A British soldier of the First World War bitterly recalled the enmity that he and his comrades felt towards the Germans: 'We were bent on his destruction at each and every opportunity for all the miseries and privations which were our lot.'[193] Few soldiers will stop to consider that the enemy soldiers have probably arrived at the same realisation. So the opposing soldiers blame each other for their predicament, rather than acknowledging that they are kept in the field at the behest of the civilian population. One soldier who did blame the civilian population for the hardships endured by the troops at the front was Siegfried Sassoon. He sent a letter of protest to the newspapers while recuperating from wounds received in France during the First World War.

> I am not protesting against the conduct of the War, but against the political errors and insincerities for which the fighting men are being sacrificed ... I believe that I may help to destroy the callous complacence with which the majority of those at home regard the continuance of agonies which they do not share, and which they have not sufficient imagination to realise.[194]

But Sassoon was perhaps being a touch unfair, as the public's ignorance of the suffering of their nation's soldiers is often an intended result of government policy. For example, during the First World War British officers were required to censor all outgoing mail. Additionally, many soldiers practised self-censorship in their letters as they did not wish to unnecessarily alarm their families. When combined with the rigid press censorship that was maintained in Britain throughout the war, it is hardly surprising that the British public remained largely ignorant of the conditions on the front line.

Subsequent conflicts also saw self-censorship exercised by soldiers. Kevin Mervin, who served with the British Army in the 2003 invasion of Iraq, felt that 'Writing letters back home was easier said than done ... We felt we had to tone things down a bit and make a joke of the weather, or write about how bored we were ... What we didn't do was write about the fire-fights, the Scud attacks, and the artillery and mortar bombardments.'[195] And as James Newton made his way to the Persian Gulf with the Royal Navy, he too felt the need to shield his family and friends from the reality of the war:

> What's the point in making a stressful situation even worse? Naturally, family and friends back home often go into denial about their loved ones, trying not to imagine the worst that might happen to them. So, as a general rule, servicemen on the phone to those back home try and affect a blasé attitude to what is happening, keeping them in the dark about the harsher realities and not letting on about anything that might be upsetting.[196]

But to a certain extent, the public, at least those who have not been caught up in the actual fighting (such as civilians living in bombed cities), have always been shielded from the harsh realities of war. Government perception has usually been that 'home front' morale would be adversely affected by the releasing of images of the nation's dead soldiers. Such a policy can result in complacency among those distant from the conflict and enlarge the gulf between those who must fight and those for whom they are fighting.

During the Second World War, the Pentagon did not release to the press any photographs of dead American soldiers until the battle for Tarawa Atoll, part of the Gilbert Islands, which took place in late November 1943. Tarawa Atoll is less than three square miles in area, yet it took a three-day assault to wrest it from its Japanese defenders. A thousand Marines were killed, 2,000 wounded, and all but 146 of the Japanese garrison of 4,800 were killed in this battle. The military hierarchy decided to release these photographs because they felt it was time the home front understood the true nature of the war. The photographs showed lines of dead Marines lying face down on the beach among the wreckage of landing craft. Their release generated a public uproar. The public's reaction was perplexing to the Marines on Tarawa. William Manchester wrote:

> The men on Tarawa were puzzled. The photographers had been discreet. No dismembered corpses were shown, no faces with chunks missing, no flies crawling on eyeballs; virtually all the pictures were of bodies in Marine uniforms face down on the beach. Except for those who had known the dead, the pictures were quite ordinary to men who had scraped the remains of buddies off bunker walls or who, while digging foxholes, found their entrenching tools caught in the mouths of dead friends who had been buried in sand by exploding shells.[197]

The public's reaction to these images was largely due to the shattering of the blissful ignorance of those staying at home; ignorance deliberately cultivated by the American leadership and which suggested that victory could somehow be achieved without Americans being killed (of course many American soldiers had already been killed, but the effect of these deaths was localised). The public could no longer maintain this fallacy when presented with the incontrovertible evidence of the dead Marines lying prostrate on the beach.

The Pentagon continued to enact policies to shield the American public from the ugly reality of combat after the Second World War. During the Vietnam War, the mutilated bodies of American soldiers were shipped home in lead-lined, sealed coffins, accompanied by a notice forbidding the opening of the coffins. A soldier escorted the coffin to the funeral to ensure that if the remains were marked non-viewable the coffin remained sealed.

From January 1991 (the start of the Gulf War) until April 2009 (when the ban was removed by the recently elected Obama administration), the Pentagon banned the media from taking pictures of the caskets of service personnel being returned to the United States. The Pentagon's sensitivity to this issue was particularly evident in April 2004. That month a photograph of a cargo plane filled with 21 flag-draped coffins of American servicemen appeared in the country's major newspapers. The servicemen had been killed in Iraq and were being returned to America for burial. The photograph was taken by a civilian employee of a defence contractor, who was subsequently sacked for violating US government and company regulations. The Pentagon, when announcing the photographic ban in 1991, cited the need to protect the privacy of soldiers' families. Yet the soldiers' families, along with reporters, were also banned from being present at Dover Air Force Base (AFB), Delaware, when the bodies arrived back in the US.[198]

Dover AFB houses America's largest military mortuary. The flights carrying the remains arrive at night and are met by a military honour guard from the US Army's 3rd Infantry Regiment – the 'Old Guard' – accompanied by a chaplain and a senior representative from the service of the deceased servicemen and women. In 2004, a pilot, responsible for flying elements of the honour guard to Dover AFB, and a frequent witness to the 'receiving the remains' ceremony, remarked bitterly: 'A democracy's lifeblood, after all, is an informed citizenry, and this image is nowhere in the public mind. The men and women arriving in flag-draped caskets do not deserve the disrespect of arriving in the dark confines of secrecy.'[199] An article in the *Sydney Morning Herald* in November 2003 drove home this point: 'So the American dead and injured from Iraq pass through a politically imposed void, until their coffin – or stretcher in the case of the wounded – arrives in the back blocks of Idaho or Texas, by which time they have long ceased to be a prime-time or national story.'[200]

The British government has also tried to maintain the public's ignorance of the traumatic effect of modern munitions striking the bodies of soldiers. When the critically wounded Robert Lawrence was returned to the Royal Air Force base at Brize Norton from the Falkland Islands, he was informed, to his surprise and dismay, that his parents would not be able to meet him.

All family and all press were banned from meeting us at Brize Norton, I learned only later, because it appears that they didn't want the severely injured people, the really badly burned and maimed young men, to be seen. The press had, however, been allowed to take pictures earlier at Brize Norton of a group of the walking wounded, who had returned home with a scar on the cheek, perhaps, or an arm in a sling or a slight limp or a couple of crutches. These, apparently, represented the agreeable image of wounded heroes …[201]

Lawrence bitterly recounted in his memoir of the war, *When the Fighting is Over*, that the severely wounded Falklands veterans were not permitted to participate in the Lord Mayor's victory parade in London. Additionally, although allowed to attend the thanksgiving ceremony held at St Paul's cathedral for those wounded and killed in the Falklands conflict, he was ordered not to wear his uniform, even though he was still an officer in the British Army.[202]

Estrangement between soldiers and civilians

But whatever the reason or circumstance, a source of irritation for soldiers is the difference in attitude towards the war effort between those on the front line and those at home. The soldiers in the trenches of the First World War were appalled by the actions of striking war workers. A British officer angrily recorded in his diary:

[S]trikes in the U.K., bitterly resented by the troops, caused trouble … It seems monstrous that, in the same desperate struggle, men volunteering as soldiers, undergoing great risks and hardships for about two shillings a day, should be liable to be shot for disobeying orders in the field, while munition workers, miners and other civilians, well paid, safe at home and engaged on work vital to the needs of the Army, should be at liberty to 'down tools' when they like without incurring any penalty. The strikers were, unwittingly no doubt, committing manslaughter, just as artillerymen would be doing by quitting their guns in action, their fire being required to cover an attack.[203]

The estrangement between soldiers and civilians continued into the Second World War. An American paratrooper bitterly recounted in a letter to his parents following the death of a comrade that:

> He wasn't twenty years old. He hadn't begun to live. Shrieking and moaning, he gave up his life on a stretcher. Back in America the standard of living continued to rise. Back in America the race tracks were booming, the night clubs were making their greatest profits in history, Miami Beach was so crowded you couldn't get a room anywhere. Few people seemed to care. Hell, this was a boom, this was prosperity, this was the way to fight a war. We read of black-market restaurants, of a manufacturer's plea for gradual reconversion to peacetime goods, beginning immediately, and we wondered if the people would ever know what it cost the soldiers in terror, bloodshed, and hideous, agonising deaths to win the war.[204]

Farley Mowat, who served with the Hastings and Prince Edward Regiment of the Canadian Army in the Italian campaign of the Second World War, harboured similar feelings:

> Although we were very short of reinforcement, the news from home told of a continuing evasion of overseas conscription by MacKenzie King's Liberal government; of anti-war riots led by Fascist sympathizers; of strikes by war workers for higher pay; and of sacrifices being less than stoically endured by the civilian population which was having to submit to the horrors of sugar rationing.[205]

Soldiers above all resent the elements of civil society who seek to profit from the misery of war. The nature and duration of a conflict was initially defined by the limitations of human muscular endurance using readily attainable elements, such as simple clubs or spears. But since the first sword was forged and the first chariot built, a merchant class has arisen to manufacture the arms needed for waging war. Although initially operating on a cottage industry basis, such as the ubiquitous village blacksmith of pre-industrialised Europe, increased complexity in design coupled with economies of scale has resulted in the manufacture of armaments and munitions becoming the almost exclusive domain of large organisations. The two world wars were fought between industrial colossi, with victory or defeat ultimately determined by the industrial and technological capabilities of the protagonists.

The industrial organisations that produce the equipment and munitions required by the military are not established or maintained for altruistic purposes. Rather they exist to generate financial returns for their shareholders and remuneration for their management and workers. The elements of society financially interconnected with these organisations noticeably profit from the increased consumption of equipment and munitions during a war. In *Citizen Soldiers*, Stephen Ambrose cites the fact that, on the outbreak of the German Ardennes offensive in December 1944 (the Battle of the Bulge), 'In New York the stock market, which had tumbled after the German retreat from France in anticipation of an early peace, became bullish again.'[206]

Dwight Eisenhower, in his farewell presidential address of January 1961, warned of the growing political influence of what he termed the 'military-industrial complex'.

This conjunction of an immense military establishment and a large arms industry is new in the American experience. The total influence – economic, political, even spiritual – is felt in every city, every State house, every office of the Federal government … In the councils of government, we must guard against the acquisition of unwarranted influence, whether sought or unsought, by the military-industrial complex. The potential for the disastrous rise of misplaced power exists and will persist.[207]

The prophetic nature of Eisenhower's warning became increasingly clear over the next decade as America propelled itself into the morass of the Vietnam War. The direct cost of this war to the United States was approximately US$150 billion, resulting in the enrichment of the US corporations that supplied arms and armaments to the US Department of Defense. The profitability of the Vietnam War for these corporations is best illustrated by examining the consumption rate of one of the key items of military equipment used in Vietnam, the ubiquitous helicopter.

During the main period of hostilities (from August 1964 to March 1973), 4,865 US helicopters were lost to enemy ground fire – as Michael Herr memorably commented in *Dispatches*, 'choppers fell out of the sky like fat poisoned birds a hundred times a day'.[208] These losses also entailed a large cost in human life, with deaths among helicopter crew accounting for approximately 10 per cent of all US servicemen killed during the Vietnam War. At an average cost per helicopter of US$250,000, these losses constituted the destruction of equipment worth around $1.2 billion.

During the Vietnam War, the US Army had a finite number of helicopter squadrons, each of which required a set number of helicopters. It is unlikely that any squadron deployed to South Vietnam deficient in any of its helicopters; so every helicopter destroyed meant the purchase of a replacement, representing a 'war bonus' for, among others, the Bell Helicopter Corporation (3,305 Bell UH-1 or 'Huey' helicopters were lost in Vietnam).[209] That is, without the equipment losses arising from the Vietnam War, the US Army would not have been required to purchase an additional 5,000 helicopters ($1.2 billion in sales) from US corporations. Undoubtedly, a similar time period during peacetime would have resulted in some helicopters being lost because of mechanical failure or pilot error. But these losses would not in any way approximate to those incurred because of enemy fire during the Vietnam War.

Assuming that corporations are not making their products for the military as a public service and operating at a loss, the greater the demand for their products, the greater their profitability and the greater the yield to shareholders on their investment. The demand for armaments and munitions will obviously be greater during a period of conflict than during a period of peace. So from a purely financial perspective, these corporations and their shareholders would prefer a state of war to a state of peace.

A commonly held belief among British soldiers during the First World War was that businesses, particularly food producers, foundries, chemical firms and shipping companies, were engaged in widespread profiteering. For example, Spillers (flour manufacturers) increased their profit from £89,000 in 1913 to £368,000 in 1914, and shipping companies were estimated to have made profits of £262 million in the first two years of the war.[210] In 1919, the *Economist* published the war dividend figures for British industry. Motor manufacturer dividends had tripled, while those of the shipping and iron and steel industries had more than quadrupled.

The British government was aware of the animosity that the perception of wartime profiteering generated among its soldiery. It responded by imposing an Excess Profits Duty in October 1915. Excess profit was calculated by comparing a firm's wartime profit to their most recent peacetime profit, or on the excess return on capital over a normal (peacetime) rate. The initial duty on excess profits was 50 per cent, raised to 80 per cent by 1918. The Excess Profits Duty was repealed in 1922. The British government enacted a similar taxation measure during the Second World War. Initially the rate for Excess Profits Tax was 60 per cent, rising to 100 per cent by the end of the war. The tax was repealed in 1946.[211]

War is still viewed as an opportunity to increase profits by many companies, and by some individuals as a chance to increase the return on their investments. As the United States readied itself for war in the winter of 2002/2003, not all segments of American society were concerned with whether a pre-emptive strike was justified by the threat Iraq posed to the US. Rather, CBS MarketWatch was puzzling over the following dilemma: 'Uncertainty about a war with Iraq has left many investment strategists and money managers on Wall Street wondering how best to play the conflict.' Fortunately the Standard and Poor's organisation was also concerning itself with such issues and had analysed its database to identify mutual funds with portfolios weighted toward key defence stocks.[212] The opening week of Operation Iraqi Freedom saw New York's Dow Jones Industrial Index experience its greatest rise in two decades.

'Life goes on'

Although the realisation that others have profited from their suffering embitters returning soldiers, their overall sense of disillusionment is also due to an unrealistic expectation of public interest in the war and sympathy for the soldiers' plight. The military, to a certain extent, is a self-contained society. This factor, when combined with the intense emotional investment that naturally arises when soldiers are asked to risk life and limb for a cause, somewhat blinds them to the fact that, quite simply, for civil society 'life goes on'. Charles Delvert, a French veteran of Verdun on leave in Paris, noted with disdain the crowds of well-dressed women promenading in the parks on the arms of their escorts, and the affluence of restaurant patrons. Delvert sarcastically remarked: 'Life is good … one can understand these people behind the lines resigning themselves to the war … What is consoling is that one may be perfectly sure that if one perishes in the barbed wire, they will not be too much affected by the loss.'[213] Upon returning to Germany after the epic retreat from the Volga, the indifference of his countrymen to the safe return of himself and a few remaining comrades galled Henry Metelmann: 'Smiling wryly, we reminded each other that Hitler himself had promised the soldiers that the gratitude of the Fatherland to them would be ensured forever. But we realised that these had merely been words, and the cold reality was quite different.'[214]

A generation later, Bob Gibson returned to Australia after completing a tour of duty in Vietnam and was shocked by the fact that 'Everyone was walking on the street like there was no war going on in Vietnam. I mean, it was business as usual, everyone was getting on with life.'[215] Similar feelings were held by Lewis B. Puller Jr. while lying in a hospital bed after being severely wounded in Vietnam, as he bitterly recalled:

I was also quite disturbed by the boisterous activity that had been going on all day on the street twelve floors below my window. A constant flow of traffic had shuttled spectators to and from the John F. Kennedy Memorial Stadium for the Army–Navy game, and I could not reconcile the differing circumstances of the mob outside with the suffering of my platoon back in Vietnam. It seemed insane that men I knew should be grovelling in the mud or being blown to bits while most of America was concentrating on point spreads.[216]

The same disenchantment at the apparent indifference of civilians to the suffering endured by members of the military was felt by the soldiers and sailors returning from the Falklands. 'I asked to go down to Southsea and drive along the sea-front before we finally returned home. As I drove along the front there were people having their summer holidays; they were eating icecreams, sitting in their deckchairs and passing the time of day, enjoying themselves,' pointed out a naval officer who had been sent home after his ship was sunk during the Falklands War. 'I felt as if I wanted to pick them up and shake them and say: "Look, there is a war going on, people are getting killed".'[217] An Argentine conscript, who had been severely wounded during the Falklands War, recounted that when he was released from hospital:

We went to a bar. I expected everyone to be miserable because of what had happened, but everyone was having a lot of fun in the bar as if we had just won the World Cup. There was no trace of worry or sorrow on people's faces. In Buenos Aires it was as though nothing had happened. There had been a war down there and it went wrong and it was all over now, but no one was really interested in what had actually happened, in how many had died. People seemed even happier because the war was over. It was a party atmosphere. I had just seen what human beings can do to each other and all the sacrifices that our troops had made, but in Buenos Aires all that was worthless. No one was interested.[218]

The media's lack of interest in the plight of the soldier confounded servicemen and women who had fought in Iraq. A British soldier, recently returned to the UK from operational service in Iraq, complained to a comrade: 'You read the paper, you watch the TV. Iraq hardly gets a mention.

It's like nobody even knew we were there. But we were fightin' a war, you know? A full-blown bloody war.'[219] The commanding officer of the 1st Battalion, the Princess of Wales's Royal Regiment, Lieutenant Colonel Matt Maer, also bemoaned the dearth of press coverage of the fighting in Iraq:

> I was depressed as the summer progressed at the lack of focus on Iraq compared to other events, particularly *Big Brother* and Euro 2004. Both of these saw the media seeking heroes ... the former in a false 'reality' situation. It jarred in its contrast to the very real 'reality' the soldiers were facing, really heroically across Iraq, every day.[220]

A similar sentiment was expressed by Andrew Exum, who wrote of reading the 'Year in Review' section of the *New York Times* of 29 December 2002.

> I skipped ahead to March to see what they had to say about Operation Anaconda. I had lived a part of history and was anxious to see what others had to say about what we had done in Central Asia. But there was no mention of the battles in the *Times* calendar. The death of comedian Milton Berle made the March listing, as did Kmart's announcement that the retailer was closing 284 stores ... Reading this and all the other end-of-the-year summaries in other periodicals, I became incensed. They hadn't even mentioned – at the very least – any of the American soldiers who had died over there.[221]

Kayla Williams was also incensed by the apathy of elements of American society towards the ongoing conflict in Iraq and Afghanistan when she returned home after completing her tour of duty with the US Army. 'I came home, and the only things people were interested in were things just beyond my comprehension. Who cared about Jennifer Lopez?', recounted Williams in her autobiography *Love My Rifle More Than You*. 'How was it that I was watching CNN one morning and there was a story about freaking ducklings being fished out of a damn sewer drain – while the story of soldiers getting killed in Iraq got relegated to this little banner across the bottom of the screen?'[222] Another US soldier who felt the same way was

Teddy Spain, who was the commander of the US military police forces in Baghdad in 2003–2004. He bitterly recalled watching Fox News on the evening of the day one of his soldiers was killed. 'It talked about Michael Jackson, and about Martha Stewart, and so on, and about fifteen minutes into it, they said "Oh, and yeah, we lost a soldier in Baghdad today".'[223] Joshua Key, returning home on a mid-tour break, was struck by the incongruity between his day-to-day existence in Iraq and the attitude to the war of the people back home:

> America felt like a dreamland. It seemed that not a soul in the country had the faintest clue about what I had been living every day in Iraq. My buddies were in danger, and I couldn't stop thinking about them. But outside the military base, people in Colorado Springs carried on as usual – going to work, sporting events, malls and movies. Walking about the city, a person visiting from another country would have had no idea that the United States was at war.[224]

Such feelings of 'us and them' may make soldiers, following their reintegration into society, experience a sense of being used, and they may come to believe that what they had endured was not appreciated by the nation they served. Hal Moore lamented that in relation to the campaign in the Ia Drang Valley in South Vietnam, where 305 American soldiers were killed in October–November 1965, 'our countrymen knew little and cared less about our sacrifices'.[225] Reflecting upon his service during the 1991 Gulf War, Anthony Swofford wrote: 'I know that none of the rewards of victory will come my way, because there are no rewards, not on the field of battle, not for the man who fights the battle.' Swofford stated bitterly that in his opinion, 'the rewards accrue in places like Washington, DC, and Riyadh and Houston and Manhattan, south of 125th street, and Kuwait City. The fighting man receives tokens – medals, ribbons, badges, promotions, combat pay, abrogation of taxes, a billet to Airborne School – worthless bits of nothing, as valuable as smoke.'[226]

* * *

One of the key tenets of Western society is the sanctity of human life. Yet soldiers are expected to disregard their upbringing and kill in combat someone they most likely have never met and with whom they have no personal conflict. Militaries have had hundreds of years of trial and error to refine their processes to overcome soldiers' resistance to killing, but even after the indoctrination and the intense conditioning, soldiers will still struggle against their conscience and may feel guilt and remorse when they first kill in combat. Sent out to kill or be killed on the battlefield, soldiers may come to feel abandoned by their society but are sustained by the intense fraternity they develop with their immediate comrades. These men become a soldier's 'family' and he will generally do whatever is required to protect them. One of the key paradoxes of warfare is that it is not hate but rather love that motivates soldiers to kill.

Chapter 7

KILLING YOUR OWN – THE DEATH PENALTY

S oldiers expect the enemy to attempt to kill them on the battlefield – after all, that is the nature of their profession. Death in combat is commonly glorified and romanticised, but the thought of being killed by your comrades is abhorrent to all soldiers. The death penalty has therefore been the ultimate punishment with which to enforce military discipline. Many commanders have felt it was only the threat of the death penalty that overcame soldiers' natural instinct of self-preservation and kept them on the battlefield. In effect, the knowledge that they might be killed if they stayed and fought was counterbalanced by the realisation that they would surely be killed if they fled.

Up to the beginning of the 20th century, wars were generally fought by men grouped together in tight formations to maximise the application of force, whether that be through muscle power or gunpowder. The area of greatest personal risk was invariably the leading edge of these forma-tions, as it was here where contact with the enemy was made. The men in the leading ranks would usually be physically prevented from fleeing the battlefield by the ranks of men arrayed behind them, who also pro-vided immediate encouragement when the resolve of the men in the front rank began to wane. Benjamin Harris recounted that during the siege of Copenhagen by the British Army in 1806, a front rank man seemed inclined to hang back, and once or twice turned round in my face. I was a rear-rank man, and porting my piece, in the excitement of the moment I swore that if he did not keep his ground, I would shoot him dead on the spot; so that he found it would be quite as dangerous for him to return as to go on.[1]

The urge to remain on the battlefield for the soldiers placed in the rear ranks was inculcated by officers or NCOs stationed immediately behind them and under orders to summarily execute any man who broke contact. Prussian Army regulations of the mid-18th century stated: 'If a soldier during an action looks as if about to flee, or so much as sets foot outside the line, the noncommissioned officer standing behind him will run him through with his bayonet and kill him on the spot.'[2] In Britain, a similar attitude was expressed in a regimental order issued by the then Lieutenant Colonel James Wolfe while the commanding officer of the 20th Regiment of Foot at Canterbury in 1755. Wolfe wrote: 'A soldier who quits his rank, or offers to flag, is instantly to be put to death by the officer who commands that platoon, or by the officer or sergeant in rear of that platoon; a soldier does not deserve to live who won't fight for his king and country.'[3]

The Mutiny Act and the Articles of War

Although summary executions on the battlefield usually instilled in soldiers the necessary fear to remain in the line, governments felt it necessary to provide a statutory basis for the military's use of capital punishment. In England in 1689, the 1st Foot (Royal Scots) mutinied at Ipswich and declared their allegiance to the deposed Catholic monarch, James II, against the Protestant co-regents of Mary II (James's eldest daughter) and William III (formerly William of Orange). Three weeks later, Parliament passed the Mutiny Act, which stated:

[E]very person being in their majesties' service in the army and being mustered and in pay as an officer or soldier, who shall ... excite, cause, or join in any mutiny or sedition in the army, or shall desert their majesties' service in the army, shall suffer death or such other punishment as by a court-martial shall be inflicted.[4]

The Mutiny Act was renewed annually thereafter. The original legislation applied only to military forces serving in England and Wales, though this was later extended to incorporate first Ireland, then Scotland, then the Colonies and by 1803 the rest of the world. The Mutiny Act, along with the various Articles of War which governed the actions of forces sent abroad in time of war, constituted the statutory framework for the application of military discipline. These separate laws were consolidated into the Army Discipline and Regulation Act of 1879 and subsequently the Army Act of 1881. Correspondingly, the various regulations that governed naval discipline and courts martial were codified into the Articles of War of 1749 (and later 1866). Less than a decade later, these Articles of War would provide the legal basis for the execution of an admiral.

On 14 March 1757, the English admiral John Byng was executed by a firing squad on the quarterdeck of HMS *Monarch* in Portsmouth harbour. His death sentence was due to his failure to defeat the French fleet off Minorca on 20 May 1756, during the Seven Years War (1756–63) between England and France. The defeat of the British Mediterranean Fleet and the loss of the naval base at Minorca, an island off the eastern coast of Spain, was a sharp blow to Britain's national prestige, which at the time was closely linked to the assurance of British naval supremacy. The subsequent public outcry, along with the precarious position in which the loss of Minorca had placed the British government, resulted in the prime minister ordering the arrest of Byng and his return to England to face court martial. Byng was charged with violating the Twelfth Article of War, which read:

Every person in the fleet, who through cowardice, negligence or disaffection, shall in time of action withdraw, or keep back, or not come into the fight or engagement, or shall not do his utmost to take or destroy every ship which it shall be his duty to engage … every person so offending and being convicted thereof by the sentence of a court martial, shall suffer death.[5]

The court martial's finding was that Byng had failed to do his utmost to relieve the garrison on Minorca or to destroy the French warships then besieging it, and was therefore negligent and had not fulfilled his duty. The court, having no other option, sentenced Byng to death, though they added a recommendation for a royal pardon because they did not believe that his failure had arisen due to cowardice or disaffection. King George II refused to pardon Byng and the execution was duly carried out. Byng's execution inspired Voltaire's famous quip: 'In this country [Britain] it is thought proper to kill an admiral from time to time, to encourage the others.'[6]

Byng was not the only senior officer to be executed during the Seven Years War. The loss of the French enclave in India brought about the beheading of General Thomas Arthur Lally on 9 May 1766. In 1756, at the outbreak of the Seven Years War, Lally was given command of the French military expedition to India. Following some initial successes, the French Army suffered a series of defeats, including a failed siege of Madras. The French retreated to the city of Pondicherry on the east coast of India, where they were besieged by the British. The surrender of the city in 1761 largely ended the French presence in India. Lally was shipped back to England as a prisoner of war. He was charged *in absentia* by the French with treason. Lally demanded to be paroled so that he could return to France and defend himself; the British acquiesced. However, unfortunately for Lally, a scapegoat for the French defeat in India was needed to appease public sentiment. Having been imprisoned in France for over two years, he was found guilty of treason on 6 May 1766 and sentenced to death.

Execution as a means of control

The execution of senior officers, however, was the exception rather than the rule. But for the rank and file, executions were a regular factor in military life. The suspect allegiance of many of the foreign troops who fought under Wellington's command in the Peninsular War, coupled with the fact that many of the British soldiers had been given the choice of prison or enlistment, required the frequent use of the death penalty during this campaign to maintain discipline among the rank and file. The troops of the condemned man's formation would be formed up along three sides of a square and the condemned was positioned with his back to the opening of the square, so that he was facing his comrades. The firing party stood a few paces in front of him. The muzzles of the muskets were so close to the condemned man that it was virtually impossible for any of the firing party to miss. The sentence of the court martial was read out, a priest provided spiritual assistance and then the condemned man was blindfolded and made to kneel. Once the execution had been carried out, the assembled troops were marched past the corpse, a soldier recalling after one such execution that 'his mangled corpse presented a sad spectacle, his head was literally blown to pieces'.[7]

Executions were also undertaken when the army was in barracks to maintain discipline among the men. Benjamin Harris recounted the execution in 1802 at Winchester of a serial deserter who would leave one regiment only to join another and claim another enlistment bounty. Harris was one of sixteen soldiers who formed the firing party and confessed that 'I would have given a good round sum (had I possessed it) to have been in any situation rather than the one in which I now found myself; and when I looked into the faces of my companions I saw, by the pallor and anxiety depicted in each countenance, the reflection of my own feelings.'[8] Harris estimated that around 15,000 men had been assembled to witness the execution. The firing party was assembled and the condemned man was brought out, blindfolded and made to kneel down behind a coffin. The officer-in-charge gave the agreed signal (a flourish of his cane) and the men fired. Harris recalled that although the soldier was pierced by several musket balls and fell to the ground, 'his hands wavered for a few moments, like the fins of a fish when in the agonies of death'. Four men from the firing party were sent forward. They placed the muzzles of their muskets to the condemned man's head and fired. The regiments then formed into companies and marched past the body in slow time.[9]

The opposing French forces also had need of the death penalty to maintain discipline. Jean-Roch Coignet tells us of an incident that occurred during the advance of the French Army towards Moscow in mid-1812 when 133 Spanish deserters were recaptured. They were paraded before the colonel of the regiment and made to draw lots. Half were placed on one side, half on the other. The colonel then turned to face them: 'You have run away, you have acted as incendiaries, you have fired upon your officer; the law condemns you to death, and you must submit to your punishment. I could have you all shot, but I will spare half of you. Let that serve as an example to you.' Coignet poignantly recalled: 'They shot sixty-two of them. My God! What a scene it was. I left the spot immediately with a bursting heart ...'[10]

The military forces on the other side of the Atlantic also found it necessary to resort to executions to control their soldiery. During the American Civil War, Daniel Crotty wrote of the execution of a deserter from the Union Army:

> Of course, it is a hard thing to see one of our comrades shot in such a way, but military discipline must have its course. The soldier who deserts his comrades in the hour of danger, deserves all the punishment due him, which is shooting to death by musketry. If he had stood his chances with all the rest, then there would be no need of his coming to such an ignominious death.[11]

By 1864, as the Civil War entered its third year, stern disciplinary measures were needed to counter widespread combat fatigue among the soldiers and to control the growing number of less-than-willing conscripts, substitutes and bounty soldiers in the ranks. Some officers ordered their regiments to fire on other regiments who broke and fled during battle, while others appointed soldiers to trail the assault lines and shoot any member of their own regiment who fell back. Bounty jumpers were particularly at risk of being summarily executed if caught deserting. A bounty jumper would join the army, receive his enlistment bounty payment and then desert at the first opportunity, only to re-enlist in another district or state and receive another payment. In *Embattled Courage: The Experience of Combat in the American Civil War*, Gerald Linderman quotes an account describing the summary execution of several deserters.

Just as the contingent started for Virginia, three bounty men tried to escape and were shot down and killed. In New York City four others attempted to run; sentries shot three and an officer overtook the fourth. 'He made no attempt to arrest the deserter, but placed his pistol to the back of the runaway's head and blew his brains out as he ran … That ended all attempts to escape.' The contingent remained in the harbor for two days while the officer was tried and acquitted.[12]

Linderman also recounts how soldiers who fled the battlefield were forced to stand on top of the Union fortifications at Petersburg wearing a placard proclaiming themselves to be a coward, 'the decision whether theirs was a capital offence apparently left to enemy sharpshooters'.[13] Desertion was also a major problem in the Confederate Army, with over 100,000 soldiers deserting over the course of the war. Less than one in five Confederate deserters was apprehended.

Not all Civil War deserters were summarily executed. Many were formally court-martialled and, if sentenced to death, the execution was often carried out in front of their comrades to deter others of a similar mindset. The Union Army recorded 267 formal military executions during the Civil War, of which 147 were for desertion. The imposition of the death penalty for desertion was not universal. Of an estimated 200,000 Union deserters, 80,000 were apprehended, of which approximately one in 500 was executed.[14]

Most Confederate records did not survive the war, but the number of executions carried out by the Confederate Army may well have exceeded that of the Union Army. For example, on 28 February 1864, 22 Confederate deserters were hanged at Kinston, North Carolina.[15] Nor was this the only incidence of a mass execution. John Casler recounts in his memoirs of the American Civil War, *Four Years in the Stonewall Brigade*, the aftermath of the desertion of approximately 30 soldiers from the 1st and 3rd North Carolina Regiments. The deserters were unable to cross the James River, as every ford and ferry was guarded, and in a bitter fight some were wounded or killed, a few escaped and ten were captured. They were sent to Richmond, court-martialled and sentenced to death; the execution to be carried out in front of the whole division. A large grave was prepared, big enough to hold all ten coffins, and ten posts were planted in the ground, about 50 feet apart. The firing party consisted of 150 men, ten of whom were arrayed in front of each prisoner, with five men as a reserve to finish the execution if the condemned man was not killed in the initial volley. Casler was part of the burial party and wrote: 'It cast a gloom over the entire army, for we had never seen so many executed at one time before.'[16]

The case of Morant and Handcock

Some 40 years later, on 27 February 1902, during the Boer War, the British Army executed two Australians: Lieutenants Harry 'Breaker' Morant and Peter Handcock. Morant, Handcock and a third Australian, Lieutenant George Witton, were charged with various offences committed while serving with an irregular unit of auxiliaries, the Bushveldt Carbineers (BVC). The principal role of the BVC was pacification, in particular suppressing resistance to the British occupation of northern Transvaal, rather than undertaking an active fighting role against the Boer commandos. All three men had previously been members of the volunteer contingents sent by the Australian colonies to the Boer War but had remained in South Africa following the withdrawal of these contingents. Such men were usually attracted to the irregular regiments by the promise of promotion and an increase in pay.

The trial of the Australians had begun in mid-January 1902. Morant accepted full responsibility for the execution of a number of Boer prisoners. He claimed that this action was justified by the conditions in which the BVC operated, and that he was ordered by a superior officer (Captain Percy Hunt) to take no prisoners. But a key issue was Morant's attitude towards the 'no prisoners' order. He revealed that he had not shot any prisoners until Hunt (a close friend of Morant) was killed and his body mutilated by the Boers. The crucial question to be decided by the court martial was whether Morant's subsequent actions were motivated by a loss of compassion for the Boers or by a quest for vengeance.

Fellow BVC officers confirmed Morant's account, in particular Hunt's order that no prisoners were to be taken, though Colonel Hubert Hamilton, military secretary to Lord Kitchener (the British military commander in South Africa), denied giving such an order to Hunt. Morant also told the court that on one occasion he had been reprimanded by Hunt for returning to the BVC's home base with prisoners. Witnesses confirmed this incident. The defence argued that Boers dressing in British uniforms was against the customs of war and so justified them being shot. Members of other irregular units gave evidence that they had also dealt summarily with Boers caught wearing khaki and that they too had been reprimanded for bringing in prisoners. The prosecution countered by stating that even if a superior officer had ordered members of the BVC to shoot prisoners, this order did not constitute a lawful command and so was not a valid defence. Furthermore, the fact that other irregular units had shot Boer prisoners was irrelevant because two wrongs did not make a right. The prosecution also stated that the shooting of Boers wearing khaki was justified only if it could be proved that the Boers had intended to deceive by this action.[17]

On the charge concerning the death of Josef Visser, a captured Boer fighter, Morant was found guilty of murder, while Handcock and Witton were found guilty of manslaughter. Visser, when captured in early August 1901, was allegedly wearing part of the late Captain Hunt's uniform. Morant held a 'drum-head' court martial, which decided that Visser was to be shot. The execution occurred six weeks before Kitchener issued an order empowering senior officers to court martial and execute Boer prisoners of war who were caught wearing khaki for the purpose of deception (such as staging a ruse).[18]

Of the other charges, Morant, Handcock and Witton were found guilty of murdering eight Boers on 23 August 1901 and Morant and Handcock were found guilty of murdering three Boers on 7 September 1901. Significantly, the murdered Boers were not wearing khaki, were not bearing arms and when captured were travelling towards the BVC's base camp to surrender. A verdict of not guilty was delivered on the final charge, that of murdering Daniel Hesse, a Cape-born, German-trained missionary. Hesse had spoken to the eight Boers shortly before they were killed by the BVC troopers and had probably seen their bodies shortly afterwards. Hesse's body, along with that of his African servant, was discovered a day or two later on a remote section of the road to Pietersburg. Both men had been shot, most likely by Handcock on the orders of Morant, but the absence of a witness resulted in the not guilty finding. The court submitted a recommendation for mercy for all three convicted men based on their previous good service and relative ignorance of military rules, and in the case of Morant, due to the provocation of Hunt's death and subsequent mutilation.

On the morning of 26 February 1902, the three officers were individually called into the office of the jail's governor. Morant and Handcock were informed that they were to be shot the following dawn. Witton was told he had been sentenced to death but that his sentence had been commuted to imprisonment for life, Kitchener having considered that Witton had been unduly influenced by Morant and Handcock. The Australian officers were executed at dawn on 27 February (the British Army executed only four men during the Boer War). Morant refused to be blindfolded. His last words were directed to the firing squad of eighteen British soldiers: 'Shoot straight, you bastards! Don't make a mess of it.'

Elements of the case have generated considerable controversy (a number of books have been written about it, and it was also the subject of a feature film); in particular, the belief that Morant and Handcock were made scapegoats to shield senior British Army officers from blame for practices that had become widespread. In February 1901, a year before Morant and Handcock were executed, Colonel Howard, commander of the Canadian Scouts (another irregular unit), was shot going forward to take the surrender of some Boers who had raised the white flag. Howard's successor Captain Charlie Ross, an Australian, made the Scouts stand around Howard's body and swear an oath never to take prisoners again. A Canadian Scout later boasted that 300 to 400 Boers had lost their lives as a direct result of this vow.[19] An officer from the Canadian Scouts testified at Morant's court martial that the Scouts had summarily executed Boer prisoners caught train-wrecking, wearing khaki or murdering British soldiers. Ross was never court-martialled and ended the war as a major general. Craig Wilcox, in his account of the Australian involvement in the Boer War, comments:

> [I]t was easy to ignore or explain away murders performed by angry soldiers in a black rage, even vendettas like that of the Canadian Scouts where some kind of rationale was applied, some kind of ritual directed the crimes to combatants and thus towards an end that headquarters could tolerate. But headquarters could not fail to notice, or fail to act, when a detachment of Bushveldt Carbineers began killing non-combatants openly, repeatedly, and without reason.[20]

The British did not officially inform the Australian government that they had executed Morant and Handcock until some weeks after the event, and then only after an official telegraph had been sent by the government to Lord Kitchener requesting confirmation of the executions. Various petitions to secure the release of Witton, who was then imprisoned in England, helped keep the issue alive. Witton was released in August 1904, two years after the end of the Boer War. He later wrote a book entitled *Scapegoats of the Empire* that served to further inflame public opinion about the alleged injustice of the execution of Morant and Handcock.

The British had executed two Australians serving as soldiers with the British forces (albeit as irregulars) without the concurrence or even the knowledge of the Australian government. This factor, along with the public outrage – though belated and largely ill-informed – over the alleged injustice of the executions, meant the Australian government was no longer prepared to have its servicemen subject to the arbitrary justice of foreign powers. This decision was to have significant repercussions in the First World War.

British use of the death penalty in the First World War

The British War Office records reveal that between 4 August 1914 and 31 March 1920, approximately 20,000 military personnel were convicted of offences for which the death penalty could be imposed. Of these 20,000, the British Army sentenced 3,080 soldiers to death, 346 (11 per cent) of whom were actually executed. With the exception of one man who was executed by hanging, the rest of the condemned men were shot by firing squad.[21] The vast majority of these executions occurred in France, with a small number in the Middle East and Mediterranean theatres of war.

For a British soldier to be executed, two actions needed to occur. First, the soldier had to be sentenced to death by a court martial; and second, the sentence had to be confirmed by the commander-in-chief. The board of a court martial was usually made up of regimental officers, who most likely viewed this duty as a distasteful adjunct to commanding in the field. Service on a court martial was a supplementary duty and, particularly in the latter stages of the war when most regular officers had been replaced by New Army officers, an unfamiliar experience. Consequently, the officers on the board relied heavily on the guidance contained in the *Manual of Military Law*.

The board of a court martial could deliver three verdicts: not guilty; guilty of a lesser or substitute charge (for example, being found guilty of absence from duty without leave (a non-capital offence) rather than desertion (a capital offence)); or guilty. If the board found the defendant guilty they could submit a recommendation for mercy, and often did so. The finding of the court martial was then reviewed by various officers as it passed up the chain of command until it reached the commander-in-chief. As noted, all death sentences had to be personally confirmed by the British commander-in-chief, who on the Western Front was initially Field Marshal Sir John French and then, from 19 December 1915 until the end of the war, Field Marshal Sir Douglas Haig. There is some evidence that courts martial readily meted out the death penalty because they knew that this sentence could be commuted when the case was reviewed by senior officers. The need for the sentence to be confirmed by the commander-in-chief to some extent removed the moral responsibility for the sentencing from the board's hands.

The commander-in-chief's decision to confirm or suspend a death sentence was based on two factors: the seriousness of the crime and the current state of discipline in the offender's battalion, brigade or division.

Some offences, such as the incident when a deserter murdered a military policeman sent to arrest him, would have resulted in the offender being executed regardless of the state of discipline in their unit. But for most offences the commander-in-chief had a degree of latitude as to whether the death penalty would be confirmed. This latitude accounts for the fact that almost 90 per cent of soldiers sentenced to death by a court martial received a reprieve from the commander-in-chief. The primary responsibility of the commander-in-chief in such matters was to maintain the discipline of the army rather than serve as a judicial authority, and it was this responsibility that was probably uppermost in his mind when considering a death sentence.

A further indicator that the aim of capital punishment was exerting control over soldiers rather than dispensing justice was that on 11 November 1918 all outstanding British Army death sentences for purely military offences (such as desertion, as opposed to murder, which was also a civil offence) were commuted to penal servitude. As the requirement to exert rigid control over the forces in the field largely ceased following the signing of the Armistice, the need for men to be executed as an example or deterrent also ceased.

The British High Command during the First World War believed that maintaining the discipline of the army was a crucial enabler of military success, particularly when engaged in an attritional conflict. Gerald Oram, a historian who has extensively studied military executions during the First World War, concludes that the harsh discipline of the British Army existed because 'this was believed to be the only effective means of ordering men drawn from the very bottom strata of society'.[22] Furthermore, Oram postulates that the large number of executions by the British Army was due to the absence of corporal punishment (branding was abolished in the British Army in 1871 – the letter 'D' for deserter and 'BC' for bad character – and flogging ceased in 1881).[23]

During the First World War, the most severe punishment that could be imposed by a battalion commander was 'field punishment', which was intended to be a substitute for corporal punishment. A court martial could impose up to three months' field punishment; a battalion commander was limited to 28 days. Soldiers sentenced to undergo Field Punishment No. 1 could be bound to a fixed object (often a wagon wheel was used for this purpose) for not more than two hours in any one day, not more than three out of any four consecutive days, and not more than 21 days in total (Field Punishment No. 2 did not entail the offender being bound to a fixed object).[24] Field Punishment No. 1 was widely unpopular among the citizen soldiers of the British Army of the First World War, who felt that it was unnecessarily degrading. A British soldier, George Coppard, maintained that: 'Lashing men to a wheel in public in a foreign country was one of the most disgraceful things in the war.'[25] Because of the absence of corporal punishment, coupled with the general ineffectiveness of field punishment or threats of imprisonment in deterring soldiers from committing offences such as desertion, the British High Command saw no viable alternative to capital punishment if they were to maintain discipline.

The British High Command felt that control of the army was most readily achieved through the use of capital punishment as a deterrent (*pour encourager les autres*, to encourage the others, in Voltaire's phrase) and the need for this deterrent was greatest on active service. Consequently, although the British Army Act of 1914 listed 27 offences punishable by death, fifteen of these offences attracted the death penalty only if committed on active service. For example, if a sentry was asleep or drunk at his post, or left his post before being properly relieved, he was liable for the death penalty if he committed the offence while on active service. But if the offence was committed during peacetime service, the soldier, at most, would be imprisoned.

Soldiers were usually charged with objective offences (an exception is cowardice) that permitted definitive proof – that is, they were told to carry out an order and they did not, they were meant to be in the trenches and they were not, and so on. Soldiers were not usually charged unless they were observed to have committed an offence. Therefore, although from a legal perspective there was the presumption of innocence in a court martial, in practice this was often not the case. The issue to be decided at the court martial was usually not one of determining if the soldier had committed the offence (innocence or guilt) but rather whether there were any mitigating factors to account for the soldier's actions. In *Blindfold and Alone: British Military Executions in the Great War*, Cathryn Corns and John Hughes-Wilson write: 'It is significant that in not one of the surviving sets of court martial papers is there any argument that the wrong man had been convicted, or that a completely innocent man had been executed.'[26]

On 19 September 1914, Field Marshal Sir John French, the commander-in-chief of the British Expeditionary Force (BEF), set out his views on discipline in an order to the troops.

> The Commander-in-Chief wishes to impress upon all officers serving upon Courts-Martial that it is their duty to give weight to consideration of good character, inexperience and all other extenuating circumstances, but that, at the same time, they are seriously to consider the effect which the offence in question may have upon the discipline of the Army, upon which its safety and success depend, and if they come to the conclusion that a sentence, however severe, is necessary in the interests of discipline, no feeling of commiseration for the individual must deter them from carrying out their duty.[27]

The value of executions as a means of maintaining discipline and hence control over the soldiers of the BEF was reiterated in the comments made by various reviewing officers as court martial records made their way up the chain of command to the commander-in-chief:

> The evidence of previous convictions for the same offence in this battalion does not support the contention of GOC Division that an example is not necessary. On the contrary, an example is very necessary. This is a very serious military offence and if it is passed over the state of courage of the British soldier is likely to be lowered.[28] (General Hubert Gough, General Officer Commanding Reserve Army, recommending the death sentence be upheld for Private Arthur Earp for the offence of leaving his post without orders from his superior officer)

Cowards of this sort are a serious danger to the Army. The death penalty is instituted to make men fear running away more than they fear the enemy. In the interests of the Service I recommend that the Death Sentence be carried into execution.[29] (Lieutenant General Hunter-Weston, General Officer Commanding VIII Corps, recommending the death sentence be upheld for Private John Bennett for the offence of cowardice)

Conversely, a satisfactory state of discipline in a battalion might be offered up as justification for commuting a death sentence. When asked to confirm the death sentence for Lance Corporal William Moon for the offence of desertion, Moon's battalion commander, Lieutenant Colonel Evans (11th Battalion, Cheshire Regiment), stated: 'I do not consider that the existing state of discipline in this unit makes necessary the carrying out of the extreme penalty in this case.'[30] Unfortunately for Moon, Evans's recommendation was overruled and Moon was shot on 18 November 1916.

Many of the soldiers whose death sentences were confirmed were recidivists, having previously been convicted by court martial, often for the same offence. Previous disciplinary infractions and the battalion commander's opinion of the value of the soldier to the army were decisive factors when the reviewing officers were considering whether to recommend that the death sentence be carried out. For example, Lieutenant Colonel Harold Fargus, the commander of the 95th Infantry Brigade, wrote when recommending that the death sentence be upheld for Private William Bowerman for the offence of desertion:

I consider the extreme penalty should be inflicted because:
(a) the man has already deserted once on active service
(b) he has no intention of fighting for his country
(c) is quite worthless, as a soldier or in any other capacity and is better removed from this world.[31]

The first British soldier to be executed during the war was nineteen-year-old Private Thomas Highgate of the 1st Battalion, Royal West Kent

Regiment. Highgate deserted from his unit during the retreat from Mons and was apprehended, hiding in civilian clothes, at 8:15am on 6 September 1914. Later that day, he appeared before a hastily convened Field General Court Martial charged with desertion. He was found guilty and sentenced to death. The sentence was confirmed and Highgate was executed at 7:07am on 8 September 1914 – less than 48 hours after his arrest.[32] Britain had been at war for just over a month. The last British soldier executed during the First World War, Private Louis Harris (also a deserter), aged 23 from the 10th Battalion, West Yorkshire Regiment, was executed at dawn on 7 November 1918 – four days before the Armistice. Of those executed, the overwhelming majority were privates. Only three junior officers were executed (one of whom, a deserter, was found guilty of murdering a military policeman sent to apprehend him).

Despite the execution of the three junior officers, a blatant double standard was often evident in the disciplinary treatment accorded to officers when compared to the punishments given to junior soldiers. Brigadier General Frank Crozier, then a lieutenant colonel commanding the 9th Battalion, Royal Irish Rifles, recalled an incident involving a young officer, Lieutenant Rochdale. On the night in question, Rochdale had been sent out, by order of the corps headquarters, to hang a sign on the German barbed wire informing the German soldiers that their families were starving at home. A subsequent trench mortar bombardment proved too much for Rochdale's already frayed nerves. He leapt out of his dugout, ran past his men and was discovered the next morning asleep in a disused French dugout behind the lines. Rochdale was tried by court martial for the offences of desertion and cowardice and found guilty. At the same time, another member of the 9th Battalion, a private named Crozier (no relation to the commanding officer) also fled the trenches. Private Crozier was apprehended by the military police about 25 miles behind the front line. He too was found guilty by a court martial. Subsequently, a message arrived from a higher HQ stating that Rochdale was to be 'released from arrest and all consequences' and reinstated to duty with the 9th Battalion. Lieutenant Colonel Crozier refused to reinstate Rochdale, who was transferred to another battalion. Crozier was then asked his opinion about whether the death sentence should be carried out on Private Crozier. He wrote: 'From a fighting point of view, this soldier is of no value. His behaviour has been that of a "shirker" for the past three months,' and recommended that the execution proceed.[33] Field Marshal Haig confirmed the death sentence. Private James Crozier was executed on 27 February 1916. He was eighteen years old.

The majority of the executed British soldiers were tried at a Field General Court Martial, usually presided over by a major (though the president of the court martial could be a captain), with two junior officers making up the rest of the board. A Field General Court Martial was designed to provide a practical expedient when on active service, rather than holding a full General Court Martial. For example, a Field General Court Martial could consist of only two officers (though normally three), while a General Court Martial had a board of five to nine officers. The decision to award the death penalty had to be unanimous. From June 1917, the courts martial were advised by a legally qualified Courts Martial Officer, but throughout the war most of the accused were not provided with a 'prisoner's friend' (defending officer) and so had to prepare and deliver their own defence. In only 50 of the courts martial that resulted in executions did an officer assist the defendant. Julian Putkowski, who has extensively studied the executions carried out by the British Army during the First World War, notes that the courts martial typically took only twenty minutes to find the defendants guilty as charged. Putkowski tells us that: 'The defendants were often poorly educated and sometimes illiterate, inexperienced and inarticulate and all were on trial for their lives.'[34]

The president of the court martial did not inform the condemned man that he had been sentenced to death. Rather the soldier was informed of the death sentence only a few hours before the execution was scheduled to take place. This was in accordance with Section 52 of the Army Act, which detailed the oath to be taken by the members of the court. This oath included the clause: 'You will not divulge the sentence of the court until it is duly confirmed.'[35] As noted, all death sentences had to be confirmed by the commander-in-chief and those upheld were to be carried out as soon as possible. The practical effect of this clause was that the accused would have known that they had been found guilty, as those acquitted were told straight away. But they would not have known that they had been sentenced to death, though they may have suspected so based on the nature of the offence, until no more than a day before the execution (which was always scheduled for dawn).

The transcript of the court martial, with a recommendation for mercy if the court deemed it warranted, was passed up the chain of command. The transcript passed from the convicted man's battalion commander to the brigade commander, then the divisional, corps and army commanders and finally to General HQ, where the commander-in-chief would review it. At each stage the reviewing officer added his recommendation to the transcript with regard to the appropriateness of the punishment, taking into account the circumstances of the offence, the conduct record of the convicted soldier (which was attached to the transcript) and the overall discipline of the man's unit and consequent need for an example to be made. The final decision as to whether the execution would occur resided with the commander-in-chief. He would be guided but not bound by the recommendations of the reviewing officers and frequently overruled them.

The execution party usually consisted of an officer, a sergeant and ten soldiers. Six soldiers formed the firing party (though firing parties of twelve or even twenty men were also used); the other four acted as stretcher-bearers and were tasked with removing the body. The firing party was usually provided from the condemned man's unit and the execution would often be carried out in front of his company or battalion. To help ease the conscience of the soldiers of the firing party, one of the rifles contained a blank round; or alternatively the men of the firing party would be told to ground arms, ordered to turn around, and one or more of their rifles would be unloaded. In theory, the men would never really be sure if they had fired a fatal shot – though the recoil, or rather lack of it, would indicate to a soldier whether he had fired a blank round. In practice, the uncertainty of whether they would actually be firing on the condemned man was most crucial in the moments immediately before the execution. This uncertainty generated an ambiguity that made the firing of their weapon easier – once the volley had been fired, as far as the military authorities were concerned, the consciences of the men in the firing party were incidental.

The condemned man was brought forward under armed escort, blindfolded and tied to a wooden pole or heavy wooden chair. A piece of white cloth or paper was pinned to his chest (over his heart) as an aiming mark. The firing party stood approximately fifteen paces away. The officer would give the signal for the firing squad to come to the present (adopt a firing position) and then the command 'fire'. Once the soldiers had fired, the assisting doctor would move forward, examine the condemned man and, if he was still alive, the officer in charge of the firing party would deliver the *coup de grâce* with his pistol. The regimental details of the soldier, the nature of the offence and the date of execution were incorporated into the General Routine Orders issued by the General HQ. These orders were promulgated to all units and read out to all soldiers on parade.

The value of the death penalty in upholding military discipline was a divisive issue, and how someone felt about it was very much a reflection of where they sat in the army's hierarchy. The view of many professional army officers was that the death penalty was a distasteful but necessary measure to maintain discipline. Brigadier General James Jack wrote in his diary about the execution of two soldiers, which he described as 'a sickeningly terrible end to those poor fellows', but conceded: 'if discipline is not strictly upheld on active service an army may become a rabble.'[36] Many senior NCOs probably held a similar opinion to the officers as to the unpleasant necessity of the death penalty. Conversely, military executions often produced feelings of disgust and resentment among the rank and file, and to be called upon to form part of a firing squad was one of the most distasteful tasks that could be asked of a British soldier. George Coppard recalled with a sense of dread: 'Personally, I was horrified at this terrible military law, and I was scared stiff that one day I would be picked for a firing squad.'[37] A British Army officer observed that news of an impending execution produced 'silence and melancholy' among the soldiers of the condemned man's unit whence the firing squad would be drawn. 'Nobody enjoyed the thought of one [of] our own men being shot by our own troops', the officer affirmed. 'I have seen many men shot, but never one of our own men by our men, and it made me feel very sick.'[38]

An account by Private John McCauley, who witnessed the execution of Private James Briggs of the 2nd Battalion of the Border Regiment in March 1915, provides an insight into the distaste with which many soldiers viewed executions. Briggs had been found guilty of the offence of desertion. The entire battalion was paraded to witness the sentence being carried out. McCauley was one of fifteen soldiers detailed to form the firing squad but not one of the twelve who actually carried out the task. He wrote that:

> Many of those who stood on this strange parade on this cold morning in March were men who had seen death in a hundred hideous forms since the days of Mons. They had faced death a thousand times too, laughing, cheering, shouting and cursing as they leaped to meet it. But this was something different. They just stood in solemn silence … It was a mournful body of men that tramped silently back to billets … We were all unnerved and disconsolate. The firing party filed back to the barn where we were quartered, and several flung themselves down and cried openly and bitterly.[39]

Shell shock and battle fatigue

Of the 346 men executed by the British Army during the First World War, 37 were executed for murder and three were put to death for mutiny. Of the remaining 306 executions, 266 (77 per cent) were for the offence of desertion and eighteen were for cowardice. The remainder were for quitting a post (seven), striking a superior officer (six), disobeying an order (five), sleeping on duty, and casting away arms (two each).[40] A number of the men convicted of desertion and cowardice were most likely suffering from what was then termed 'shell shock' but would later be known as 'battle fatigue'.

In 1914 there was limited knowledge of the psychological effect of war on soldiers. During the first winter of the war, the medical services of the British Army were bewildered by the large number of soldiers who bore no external signs of wounding yet had lost the use of some of their senses. Others were unable to speak, urinate or defecate; some had memory loss; a number suffered physical paralysis and many shook uncontrollably. The Cambridge University psychologist Charles S. Myers, then working at a base hospital at Le Touquet, France examined some of these soldiers.

Myers described their symptoms in an article published in the British medical journal *The Lancet* in February 1915. As nearly all the men referred to Myers first began to display these symptoms following a shell detonating close by, he categorised them as suffering from 'shell-shock'. The initial belief was that they were suffering from the physical effects of the explosion of projectiles. Myers theorised that the vibrations of the exploding shell produced a kind of 'molecular commotion' in the brain, while other doctors proposed that the explosion of the shell drove microscopic particles into the brain or released carbon monoxide.

In June 1916, a follow-up article appeared in *The Lancet* by Harold Wiltshire, an experienced physician who had spent a year at a base hospital in France. Wiltshire observed that the symptoms attributed to 'shell shock' were not displayed by soldiers physically wounded by shell-fire, even though they had greater exposure to the physical and chemical effects of an exploding shell than the unwounded. He commented that the cheerfulness of the physically wounded men was in stark contrast to the morose gloom of those suffering from shell shock. Secondly, he noted that rarely was there any evidence of physical concussion in the shell shock patients, many of whom were not even near an exploding shell when they broke down. Wiltshire concluded that the basis of shell shock was psychological rather than physiological, postulating that the prolonged strain of trench warfare wore down the men's resistance until an emotional shock triggered a psychological collapse.[41]

The High Command tended to view shell shock as a disciplinary rather than a medical matter. Men who shirked their duty in the trenches and sought evacuation to the rear by claiming they were 'shell shocked' (now that 'shell shock' had became a recognisable and hence legitimate condition) reinforced this attitude. The difficulty for the regimental medical officers became sorting out malingerers from those soldiers suffering a genuinely debilitating psychological condition. Myers stated that he 'had seen too many men at Base Hospitals and Casualty Clearing Stations boasting that they were "suffering from shell shock, Sir" when there was nothing appreciably amiss with them save "funk" ...'[42] Lord Moran also detected the tendency of men to self-diagnose themselves with shell shock:

> When the name shell-shock was coined the number of men leaving the trenches with no bodily wound leapt up. The pressure of opinion in the battalion – the idea stronger than fear – was eased by giving fear a respectable name. When the social slur was removed and the military risks were abolished the weaklings may have decided in cold blood to malinger, or perhaps when an alternative was held out the suggestion of safety was too much for their feeble will. The resolve to stay with the battalion had been weakened, the conscience was relaxed, the path out of danger was made easy. The hospitals at the base were said to be choked with these people though the doctors could find nothing wrong with them.[43]

At the end of 1915, the High Command, in an attempt to mitigate what it saw as a means of legitimising malingering, ordered that shell-shocked soldiers who reported for medical aid be classified as 'Shell-shock, W' (for wounded) and 'Shell-shock, S' (for sickness). 'Shell-shock, W' had to be attributable to enemy action, such as being buried by a shell explosion, and a soldier so classified was entitled to a wound stripe (worn on a soldier's uniform to indicate that he was recuperating from wounds) and a pension. 'Shell-shock, S' was considered a mental breakdown – with no associated wound stripe or pension.[44]

A more lenient attitude was displayed towards officers who exhibited the symptoms of a psychological breakdown. Officers who displayed 'nerves' were more likely to be sent back to the rear for rest and/or medical treatment than soldiers in the same condition. 'If an officer crumpled up, [the battalion commander] sent him home as useless, with a confidential report. Several such officers were usually drifting about at the Depot, and most of them ended up with safe jobs in England', wrote Siegfried Sassoon.

'But if a man became a dud in the ranks, he just remained where he was until he was killed or wounded.'[45]

Another example of officers receiving more sympathetic treatment than their men is the circumstances concerning the arrest of Second Lieutenant Eric Poole, 11th Battalion, West Yorkshire Regiment. Poole deserted from his company on 5 October 1916 as it moved up to the front line. Two days later he was apprehended by the military police and returned to his unit. The battalion commander ordered a regimental court of inquiry be held to investigate the incident. The court of inquiry recommended that Poole be charged with desertion. Poole's brigade commander, Brigadier General Lambert, concluded:

> I do not think it will be satisfactory to try this Officer by Court Martial. He was once before, on 7th July after the action at Horseshoe Trench on 5th July, admitted to Hospital suffering from shell shock and I doubt if he is really accountable for his actions. He is of nervous temperament, useless in action, and dangerous as an example to the men. The Battalion is, in my opinion, in an excellent state of discipline and has never failed to carry out in action any duty that had been given it ... I recommend that 2nd Lieut. E.S. Poole be sent home away from the firing line as soon as possible. Before the War he was employed on engineering work in Canada. He could be usefully employed at home in instructional duties or in any minor administrative work, not involving a severe strain on the nerves.[46]

If a soldier, rather than an officer, had committed the same offence as Poole, it is highly unlikely that the solder's brigade commander would recommend that he should be returned to England to avoid further strain to his 'nerves'.

Unfortunately for Poole, General Herbert Plumer, the commander of the 2nd Army, was not as tolerant of officers failing in their duty as Brigadier General Lambert and ordered that Poole be charged with desertion and tried by court martial. He was found guilty and was shot on 10 December 1916 – the first British officer to be executed during the war.

While these incidents may seem to be blatant examples of officers being treated more sympathetically by the system than the rank and file, this leniency was based on practical considerations. A British officer, who had recently sent two of his junior officers with 'nerves' out of the trenches, commented:

> [I]t was essential that they should not be near the men while the sort of ague, which is the outward and visible sign of the disease, was upon them. Some of the men had it too, but I allowed none of them to go back. An officer is a different thing, because on him depends so largely the nerves of the men.[47]

As always, the key consideration in such matters was maintaining the discipline of the unit. An officer displaying 'nerves' could spread fear to the men, while the reverse was rarely true.

Unofficial executions

As noted, officially, 346 men were executed by the British Army following their conviction by court martial; but the number of British soldiers executed by their comrades during the First World War is potentially much higher. Douglas Haig wrote in his diary in early December 1914 that 'we have to take special precautions during a battle to post police, to prevent more men than necessary from accompanying a wounded man back from the firing line.'[48] Consequently, several lines of 'straggler posts' were deployed on the battlefield to collect men drifting or fleeing rearwards, to rearm them if necessary and send them back into the fight. The straggler posts were manned by military police but other elements, in particular the battalion regimental police (infantry soldiers chosen to perform policing duties for their unit), were also tasked with ensuring that no unwounded man drifted to the rear during an attack. The foremost elements of the military police/regimental police were positioned immediately behind the front line and were collectively known as 'battle police'. Gary Sheffield, who has extensively studied the role of the British Military Police, argues that:

'On rare occasions they [battle police] were sometimes used to shoot stragglers. However, most decisions to begin shooting at one's own men seem to have been taken entirely unofficially, at a local level by harassed officers or NCOs faced with routing troops.'[49]

Several incidents of this nature were described by Frank Crozier in *A Brass Hat in No-Man's Land*. He recalled that on the opening day of the Somme offensive the retreat of a large party of stragglers in his sector was turned back only when a junior officer drew his pistol and shot one of the fleeing soldiers. Similar means were used to avert a retreat by elements of the 40th Division at the battle of Bourlon Wood in November 1917. Crozier wrote: 'A revolver emptied "into the brown" accounts for five; a Lewis gun fired into the panic stricken mass puts many on the grass and undergrowth.'[50] Robert Graves recalled a captain from a battalion belonging to one of the Surrey regiments telling him: 'In both the last two attacks that we made I had to shoot a man of my company to get the rest out of the trench.'[51]

The above accounts reflect the perspective of officers as to the need for summary executions on the battlefield; soldiers had a somewhat different viewpoint. A British soldier of the First World War bitterly recounted:

After I had been wounded, I came back to our trench and saw two very young soldiers. They were terrified and had been too scared to follow their mates over the top. Soon after I had left them I met two 'Red Caps' [military policemen] with revolvers who wanted to know where I was going. I showed them my wound and they let me pass. I had only gone a few yards when I heard two pistol shots from close by. I feel *sure* that these two unfortunate boys had been 'executed for cowardice'.[52]

Soldiers were shot by their comrades for attempting to flee the battlefield, for not leaving their trenches, and even for surrendering. On 13 April 1918, Lieutenant Colonel Graham Hutchinson, the commanding officer of the 33rd Battalion, Machine Gun Corps, ordered his men to fire on a party of surrendering British soldiers. These men were part of an improvised unit of cooks, batmen and HQ orderlies that had been cobbled together and sent into battle. Of the 40 men who attempted to surrender, the machine gunners of Hutchinson's battalion shot 38.[53] German soldiers also fired upon their surrendering comrades during the First World War. James Jack was informed of several incidents where British troops had left their trenches to accept the surrender of Germans approaching under the protection of a white flag and both parties were fired upon by other Germans in the vicinity who had no intention of surrendering.[54] Furthermore, Emilio Lussu described an incident in *Sardinian Brigade*, his memoir of his service with the Italian Army during the First World War, when his men shot dead a comrade who was attempting to desert to the Austrian trenches. 'Just as he reached the wire, he fell and lay motionless with his legs buried in the snow, his body bent forward and his arms stretched out. The fire of our whole line was still concentrated upon this now inanimate target. It was some time before I was able to stop the firing in our sector.'[55]

There are no official records of summary executions conducted by members of the British Army during the First World War, which is not surprising as such executions were illegal. These executions were usually spontaneous decisions brought about by extreme circumstances and intended to remedy a localised situation. Summary executions differed from those arising out of a court martial, as the latter were intended to provide a deterrent that modified the behaviour of the entire army. For this reason, details of court martial-sanctioned executions were widely promulgated via official means.

The campaign for retrospective pardons

Although the details of the execution of their loved ones were read out to every soldier in the BEF, the families of the executed men were unable to view the court martial's records because of a legal catch-22 contained in Section 124 of the Army Act. This section stated that the proceedings of a court martial could not be released without the consent of the person who had been tried. As the men had been executed, consent for others to view the proceedings could not be given. Indeed, the records of the courts martial pertaining to the executed soldiers were not publicly released until

1993 – 75 years after the end of the First World War. The records were originally sealed for 100 years, principally to protect the relatives of the executed men from any public shame that could arise from the disclosure of their contents.

The release of these records intensified the campaign to gain a retrospective pardon for the 306 servicemen executed for offences other than murder and mutiny. The justification for many of these requested pardons was that the executed men were suffering from shell shock and were therefore not responsible for their actions. In mid-1993, the British Prime Minister, John Major, in rejecting the call for retrospective pardons, stated:

> Shellshock did become recognised as a medical condition during World War One. And where medical evidence was available to the court, it was taken into account in sentencing and the recommendations on the final sentence made to the Commander-in-Chief. Most death sentences were commuted on the basis of medical evidence.[56]

In mid-1918, in a letter to the British Under-Secretary for War, Haig commented that a medical board, a member of which always had neurological experience, examined all soldiers sentenced to death who were suspected, or claimed, to be suffering from shell shock. Haig wrote: 'The sentence of death is not carried out in the case of such a man unless the Medical Board expresses the positive opinion that he is to be held responsible for his actions.'[57] Despite Haig's statement to the contrary, the opening of the courts martial records in the early 1990s raised disturbing questions about whether some of the executed men were indeed suffering from shell shock when their offences were committed. A case in point is that of Private Harry Farr.

Harry Farr was a pre-war regular soldier who had enlisted in May 1908, though when recalled to the colours on the outbreak of war he was serving as a reservist. Farr arrived in France in November 1914. He initially served with the 2nd Battalion, West Yorkshire Regiment, being transferred to the regiment's 1st Battalion in October 1915. On three occasions during 1915–16 he had reported sick, claiming psychological distress. On the first occasion, in May 1915, after serving six months in the trenches, he was hospitalised for five months with shell shock. In April 1916, he again reported sick because of his 'nerves' and spent a fortnight at a dressing station. In July 1916 he once more reported sick with the same complaint and stayed overnight at the dressing station. The medical officer who had examined Farr on these occasions had been wounded and was unavailable to present evidence at Farr's court martial. Farr's company commander wrote:

> I cannot say what has destroyed this man's nerves, but he has proved himself on many occasions incapable of keeping his head in action & likely to cause a panic. Apart from his behaviour under fire, his conduct & character are very good.[58]

Further evidence of Farr's fragile mental state was provided by his platoon commander, who commented:

> I have known the accused for the last 6 weeks. On working parties he has three times asked for leave to fall out & return to camp as he could not stand the noise of the artillery. He was trembling & did not appear in a fit state.[59]

On the morning of 17 September 1916, Farr reported to his regimental sergeant major (RSM) at the battalion's transport section. He told the RSM he was sick and had fallen out from his company the night before when the company was returning to the trenches. The RSM told him to report to the dressing station. The medics refused to admit or treat Farr as he was not wounded, so Farr returned to the RSM. The RSM then ordered Farr to accompany the ration party that was going up to the front line that evening and rejoin his company. Farr left with the ration party but fell out and returned to the rear. When confronted by the RSM, Farr pleaded that he 'cannot stand it'. According to Farr's statement at his trial, the RSM retorted: 'You are a fucking coward & you will go to the trenches. I give fuck all for my life & I give fuck all for yours & I'll get you fucking well shot.' An escort was detailed to return Farr to the trenches. He began to scream and struggle after walking about 500 yards. According to the RSM's statement, Farr was again warned that if he didn't return to the front he would be tried for cowardice. Farr continued to scream and struggle so the RSM ordered the escort to stop restraining him. Farr then jumped up and returned to the transport section, where he was placed under guard. On 2 October 1916, Farr appeared before a Field General Court Martial.[60]

Farr was not provided with a defending officer and conducted his own defence. At his trial he stated:

> The Sgt. Maj. then told L. Cp. Form to fall out two men and take me up to the Trenches. They commenced to shove me – I told them not to as I was sick enough as it was. The Sgt. Maj. then grabbed my rifle & said, 'I'll blow your fucking brains out if you don't go'. I then called out for an officer but there was none there. I was then tripped up & commenced to struggle. After this I do not know what happened until I found myself back in the 1st Line Transport under a guard.[61]

The court asked Farr why he had not reported sick since being arrested. He replied: 'Because being away from the shell fire, I felt better.' Tellingly, when examined by a medical officer on the day of his court martial, Farr's general physical and mental condition were deemed to be 'satisfactory'.[62]

Farr's court martial lasted approximately twenty minutes. He was found guilty under section 4(7) of the Army Act: 'Misbehaving before the enemy in such a manner as to show cowardice'. Despite his relatively clean conduct record (he had been convicted of overstaying his leave pass by one day immediately prior to his battalion shipping overseas), he was sentenced to death. The commander-in-chief confirmed the sentence and Farr was executed at 6am on 18 October 1916. He left behind a widow and a baby daughter.[63]

The circumstances leading up to the execution of Private Farr were to become a cause célèbre for those seeking a posthumous pardon for the soldiers executed by the British Army during the First World War. In August 2006, the UK Defence Secretary, Des Browne, announced that the government would seek parliamentary approval to pardon more than 300 soldiers executed during that war under the provisions of the British Army Act and the Indian Army Act. The government's intention was to seek a statutory group pardon for all the executed soldiers, with the exception of those convicted of murder. Browne stated:

> I believe a group pardon, approved by Parliament, is the best way to deal with this. After 90 years, the evidence just doesn't exist to assess all the cases individually. I do not want to second guess the decisions made by commanders in the field, who were doing their best to apply the rules and standards of the time. But the circumstances were terrible, and I believe it is better to acknowledge that injustices were clearly done in some cases, even if we cannot say which – and to acknowledge that all these men were victims of war.[64]

The statutory pardon was bought into effect by an amendment to the Armed Forces Bill. This bill received Royal Assent in November 2006. The pardon did not quash the convictions, but sought to acknowledge that execution was not a fate the servicemen deserved; rather, it arose due to wartime contingencies. The pardon was the culmination of a long campaign by relatives of the executed men (among others) to remove the stigma of the executions. In particular, the family of Harry Farr had appealed to the Defence Secretary in 2004 to review Farr's case and when their request for a posthumous pardon was refused, they petitioned the British High Court to review the case.

The death penalty among other nations of the First World War

In addition to British soldiers, the British Army also executed troops from the Dominions. Twenty-eight New Zealanders were sentenced to death; five were executed. Of the 216 Canadians sentenced to death by the British Army, 25 were executed. Additionally, one South African and four members of the British West Indies Regiment were executed. Among the other protagonists of the First World War, an estimated 750 Italian soldiers were executed (out of 4,028 death sentences), as were thirteen Belgians. The

United States executed eleven soldiers out of a total of 44 death sentences handed down (24 of these death sentences were for desertion), though the executions carried out were all for the non-military offences of murder and rape (three Americans were executed while serving with the British and Canadian forces). Official statistics published in Germany in 1929 state that 150 death sentences were confirmed and 48 German soldiers executed during the First World War.

The French executed approximately 600 soldiers during the war under a system of military justice that could be even harsher than that of the British. A case in point is the execution of Second Lieutenants Herduin and Millaud, who were ordered to be shot without trial for the offence of cowardice. Herduin's crime was that during the battle of Verdun, seeing his devastated company about to be encircled by the Germans, he gave the order to withdraw, thereby contravening a mandate by the French commander, General Robert Nivelle, that they were to hold their positions at all costs. The withdrawal of Herduin's troops set off a chain reaction and elements of his regiment (the 347th) fled the battlefield leaving a dangerous gap in the line. An example needed to be made and the two lieutenants fitted the bill (Millaud had also taken part in the withdrawal) – they were executed by men drawn from their own platoons. Both officers had previously been commended for their bravery under fire.[65]

During the First World War the French Army retained, though seldom used, the disciplinary option of decimation. This practice originated among the legions of Imperial Rome, whereby one man in ten would be chosen by lot to be executed after a mutiny or a retreat. Allegedly, such an action was carried out on the 10th Company of the 8th Battalion of the Régiment Mixte de Tirailleurs Algériens. The company was composed of French-African soldiers and the decimation was carried out following their refusal to obey an order to attack. The soldiers were allegedly executed on 15 December 1914 at Zillebeke in Flanders (near Ypres). Decimation was also a disciplinary option available to the officers of the Italian Army during the First World War. Lussu described an attempted sentence of decimation that was foiled only when the firing party declined to execute their comrades and fired high.[66]

In April 1917, after the disastrous Aisne offensive, the French Army suffered widespread mutinies. Five hundred and fifty-four soldiers were sentenced to death out of an estimated 40,000 mutineers (54 divisions were
affected); 49 men were actually executed. In addition to the death penalty,
in mid-1917 the French commander-in-chief, Marshal Henri Philippe
Pétain, instigated the use of *compagnies de discipline* (penal companies).
These companies were formed on the basis of one per division and were
specifically designed to counter the tendency of soldiers to seek imprisonment in order to avoid combat duty. The penal companies were assigned
the most dangerous missions and were filled with men who would otherwise have been condemned to death or sentenced to long periods of
imprisonment. The German Army also made use of penal battalions during the First and Second World Wars.

The variation in the number of executions among the combatant
nations during the First World War reflected differences in their respective
codes of military law. For example, under both the German and French
codes, the offence of sleeping on post did not merit the death penalty, as it
did under the British code. Furthermore, under the German code, capital
punishment could be imposed only for repeated desertions, as a single act
of desertion, even if committed on active service, attracted a maximum of
ten years' imprisonment. The French Army also had a less severe view of
desertion than the British Army, recognising two types of desertion with
the lesser offence punishable by five years' penal servitude. Gerald Oram
concludes that the more lenient view towards desertion by the German and
French armies was a reflection of their conscript basis. He writes: 'A certain
degree of desertion was tolerated, expected even, among soldiers who had
been compelled to serve ... For the British, however, no such tolerance was
considered necessary for men who had enlisted of their own volition.'[67]

The British Army did not execute any Australian soldiers during the
First World War (though two Australians were executed while serving with
the New Zealand Expeditionary Force, which operated in the field under
the British Army Act). Section 98 of the Commonwealth Defence Act of
1903 governed the use of capital punishment for Australian soldiers. This
section stated that only four military offences were punishable by death:

mutiny; desertion to the enemy; traitorously delivering up to the enemy any garrison, fortress, post, guard, ship, vessel or boat; and traitorous correspondence with the enemy. Desertion to the rear was not considered a capital offence. Significantly, a death sentence would need to be confirmed, not by the commander in the field – who was responsible for the discipline of the forces – but rather by the Australian governor-general, who was influenced by domestic politics and public sentiment and so less likely to approve such a sentence.

Unlike the other Dominions, such as Canada and New Zealand, the Australian government, perhaps mindful of the execution of Morant and Handcock during the Boer War, did not allow its soldiers to be tried and punished under the British Army Act. The Australian government insisted on the primacy of Australian law. Initially, Australian troops convicted of serious military offences were sent home in disgrace; however, this action had less deterrent value when the Australian troops were subjected to the horrors of trench warfare on the Western Front.

One hundred and twenty-one Australian soldiers were sentenced to death during the First World War, though none was actually executed. Throughout the war, the British High Command, and later a few Australian generals, including General Sir John Monash, commander of the Australian Corps and the senior Australian Army officer at the end of the war, petitioned the War Council to pressure the Australian government into agreeing to Australian troops serving under the same legal provisions as the British soldiers. Such an action would remove the inequity of troops fighting alongside each other being beholden to different laws. In particular, the generals wanted to be able to impose the death penalty for the offence of desertion. The War Council concurred with the generals and in July 1916 cabled the Australian government asking that it place its troops serving overseas under the provisions of the British Army Act without limitations. The Australian government was about to embark on a campaign leading up to the first of two conscription referendums and was concerned about the adverse effect the introduction of the death penalty would have on recruitment (the Australian force serving overseas was entirely voluntary), as well as the knock-on effects on the referendum.

The Australian official history of the First World War noted: 'To impose the system of the death penalty upon men, who had gone out voluntarily to fight at the other end of the world in a cause not primarily their own, was not compatible with its [the Australian people's] sense of justice.'[68] The Australian government delayed their answer for over eight months, eventually refusing the War Council's request. From January 1918, in an attempt to deter potential deserters by public shaming, the Australian defence minister agreed to a request from the British High Command that the name and town of enlistment of men sentenced to death for desertion would be published in every Australian newspaper. The first such list was published in March and contained only two names.[69]

The absence of the death penalty undoubtedly contributed to some well-documented discipline problems among the Australian troops who fought the First World War. Of the 677 British and Dominion troops convicted of desertion during the first six months of 1917 (an average of almost nine per division), 171 were Australians (34 per division).[70] The Australian official history of the war noted that in March 1918 nearly nine Australians per thousand men serving in France and Flanders were in field imprisonment. The corresponding ratio was one per thousand men for the British troops and less than two per thousand men for the forces provided by Canada, South Africa and New Zealand (though these forces made greater use of suspended sentences, whereby convicted troops were returned to the front line).[71] Yet such was the fighting reputation of the Australian 'Digger' that the British High Command regularly used the Australians as shock troops during crucial battles. The absence of the death penalty was seemingly not a significant factor in determining the overall performance of Australian troops on the battlefield.

A change of heart between the wars

A debate raged in Britain for more than a decade after the end of the First World War over the need to retain the death penalty for military offences. In part, this debate arose because of the changing composition of the British Army. The soldiers of the pre-war army had largely been drawn from the margins of society and it was felt that for such men draconian penalties were needed to maintain discipline. Whether these same disciplinary measures were needed to control citizen soldiers, who had volunteered their services to overcome a national crisis, was another matter.

Many former British Army officers argued for the retention of the death penalty, one commenting in 1927, during the debate over the Army Act in Parliament, that when he was serving with a brigade of the 3rd Division in late 1915, 'They had cracked to a man. You could not send them back to base, yet they were in such a state that they would have willingly taken ten years' penal servitude to stay out of the line. In these circumstances, it was only fear of death that kept them at their posts.'[72]

Yet this particular issue had been addressed during the war. The realisation by the British High Command that for many soldiers a long prison sentence was preferable to service in the trenches resulted in the generals petitioning Parliament to pass the Army (Suspension of Sentence) Act. The provisions of the Act were that even though a soldier had been given a jail sentence for a military offence, he would remain with his unit and continue to serve in the trenches. The jail sentence could be imposed at any time, including after the war had ended. The Act also contained a provision allowing for the remittance of any prison sentence for a period of good conduct or a deed of gallantry.

The Army Act was eventually amended, largely due to the efforts of Ernest Thurtle, a Member of Parliament who had volunteered to join the army during the First World War and had served in the trenches in France. In 1925 the death penalty was abolished for all military offences in peacetime, with the exception of mutiny. In 1928 the death penalty was abolished for eight active service offences, including striking or offering violence to a superior officer, disobeying a lawful order in such a manner as to show a wilful defiance of authority, and sleeping or being drunk while on post as a sentinel. But the offences of desertion, cowardice, quitting a post without orders, mutiny, and treachery continued to attract the death penalty if committed while on active service. Finally, in April 1930, after some intransigence by the House of Lords following speeches from various senior generals, including Field Marshals Viscount Plumer and

Viscount Allenby – who both argued for the retention of the death penalty for offences of desertion and cowardice – an amendment to the Army Act was passed. This amendment abolished the death penalty for all military offences except mutiny and treachery.

A number of British generals during the Second World War would come to lament the watering down of the prescribed penalties of the British Army Act. Desertion was a prevalent offence throughout the British Army during the war. Offences of absence without leave (soldiers absent from their unit for 21 days or less) and desertion (a soldier was declared a deserter from his 22nd day of absence) accounted for 27 per cent of all court martial convictions in 1939–40, rising to 58 per cent in 1944–45. Approximately 31,000 British other ranks were convicted of desertion during the Second World War, while 75,000 were convicted of being absent without leave. Twenty-two thousand deserters were still on the run at the end of the war. Long prison sentences had little impact on reducing the desertion rate. Alex Bowlby, who served with the King's Royal Rifle Corps (Green Jackets) of the British Army during the Italian campaign, detailed in his memoirs of the war the frustrations felt by the men in relation to the lenient treatment of deserters.

'Got a court martial!' he [the deserter] said, as if it was a decoration.
'What did you say to the C.O., Cokey?' asked Humphreys curiously.
'I told 'im I 'ad a feeling I shouldn't go up no more hills.'
We tittered. Baker was furious.
'You'll get three years!' he roared.
Coke grinned.
'And I'll be 'ere when you're pushing up the daisies.'
Baker grabbed him by the throat.
'Say that again and I'll shoot you me fucking self!'[73]

Bowlby noted that Baker's offer to personally drive Coke back to the Corps HQ and hand him over to the military police was not accepted.

In May 1942, General Sir Claude Auchinleck, Commander-in-Chief Middle East, with the unanimous agreement of his army commanders, formally requested that the War Office reinstate the death penalty for the offence of desertion. He was alarmed by the number of 8th Army soldiers deserting as Rommel advanced across North Africa and argued that the deterrent of the death penalty was required for men 'to whom the alternative of prison to the hardships of battle conveyed neither fear nor stigma'.[74] General Harold Alexander also pressed for the reintroduction of the death penalty while commanding Allied forces in Italy during the Second World War. A factor mitigating the disciplinary powers of the military was that many soldiers believed there would probably be an amnesty for deserters after the war, with a pardon being granted to those who had not completed their sentence. Requests for the reinstatement of the death penalty for desertion were not successful. The 1930 provisions of the Army Act, whereby the death penalty could be applied only for the crimes of mutiny and treason, remained in place throughout the Second World War.

In stark contrast to the First World War, only a couple of men were executed by the British Army during the Second World War. In August 1943, three Indian soldiers were hanged after being found guilty of mutiny, and in September 1945 a British soldier, Private Theodore Schurch, was found guilty by a General Court Martial of committing acts of treason during the war. Schurch was hanged on 4 January 1946. Technically, he was executed under the provisions of the Treachery Act of 1940, even though a military court martial had tried him.[75]

Russia and Germany: 'Not One Step Back'

The Soviet forces did not show the same restraint as the British in the application of the death penalty after the First World War. The disciplinary foundations of the Red Army were laid down by Leon Trotsky during the Russian Civil War. Trotsky wrote in his memoirs that:

> An army cannot be built without [repression]. Masses of men cannot be led to death unless the army command has the death penalty in its arsenal ... the command will always be obliged to place the soldiers

between the possible death at the front and the inevitable one in the rear.[76]

Thousands of Red Army soldiers were executed following trial by field tribunal during the Russian Civil War. The sentences of the tribunals could not be appealed against, and were to be carried out within 24 hours. Additionally, to prevent units surrendering or retreating without a fight, Trotsky placed blocking units behind unreliable troops, the first such units being deployed in August 1918. The blocking units were ordered to shoot troops that retreated without permission.

The extensive use of the death penalty by the Soviet Army continued into the Second World War. Units of NKVD (People's Commissariat of Internal Affairs) Special Departments personnel called *Zagradotryady* were formed to stand behind the lines of attacking troops and shoot any man who tried to desert. Furthermore, officers could be brought before NKVD tribunals and executed for cowardice or failure in battle. The rapid German advance in June 1941 at the start of Operation Barbarossa resulted in the arrest, court martial and execution of the commander of the Western Special Military District, General of the Army D.G. Pavlov; his chief of staff, Major General V.Y. Klimovskikh and his signals chief, Major General A.T. Grigoriev; the 4th Army Commander, Major General A.A. Korobkov; and the commanders and commissars of the 30th and 60th Rifle Divisions.[77] The Soviet State Defence Committee recorded that the executed men were considered guilty of a 'lack of resolve, panic mongering, disgraceful cowardice ... and fleeing in the face of an impudent enemy.'[78]

On 16 August 1941, the *Stavka* (Supreme High Command of the Soviet Armed Forces) issued Order No. 270, which declared that all Soviet soldiers who surrendered or were captured were considered traitors. The order stated that any military or political officer who retreated or surrendered would be considered a malicious deserter and that officers who tried to desert in the field could be summarily shot by their superiors. The order also declared that the families of malicious deserters would be liable to be arrested. Fyodor Sverdlov commanded an infantry company during the battle for Moscow and later recounted an incident when, in accordance with Soviet policy, he killed one of his own men. 'There was one soldier, I don't know what his name was, but because of his cowardice and because the combat was very severe he broke down, and he began to run, and I killed him without thinking twice. And that was a good lesson to all the rest.'[79] The Soviet general A.Z. Akimenko, then commander of the 2nd Guards Rifle Division, recalled an incident that occurred on 24 September 1941 when the Soviet forces were reeling from the assault of General Guderian's Second Panzer Army:

> [A] large number of replacement troops from Kursk, numbering about 900 men, committed treachery to the interests of the homeland. As if by command, this group rose up, threw away their rifles, and, with raised hands, they proceeded to the side of the enemy tanks ... I lacked the forces and means required to remedy the situation and to take control of the traitors for future punishment before our Soviet organs. But a traitor is a traitor, and he deserved immediate punishment on the spot. I gave an order for two artillery battalions to open fire on the traitors and the enemy tanks. As a result, a considerable number of the traitors were killed and wounded ...[80]

Soviet officers also resorted to the practice of decimation to restore discipline. In the opening stages of the battle for Stalingrad, some of the troops of the Soviet 64th Division had fled the battlefield rather than face the German onslaught. The colonel commanding the division called a general assembly of regiments and berated the men for not fulfilling their duty. He declared that the assembled men shared in the guilt of those who had run and that they were going to be punished for their cowardice. The colonel then moved to the end of the first row of soldiers, and with pistol in hand, began to move along the row counting the men off in a loud voice. When he reached the tenth man he turned and shot him in the head. He then began the count again. No man broke ranks, each man mentally calculating where he stood in the line and hoping he would be spared. Five shots later, the colonel's revolver was empty and the men were dismissed. Six of their colleagues were left lying dead on the field.[81]

On 28 July 1942, Stalin, in his capacity as the People's Commissar of Defence, issued Order No. 227, following the Germans taking the city of Rostov (the gateway to the Caucasus) and the crossing of the Don river by three German armies. Known as the 'Not One Step Back' order, it stated: 'Each position, each metre of Soviet territory must be stubbornly defended, to the last drop of blood. We must cling to every inch of Soviet soil and defend it to the end.'[82] Anyone who retreated without being ordered to do so was to be considered a 'traitor to the Motherland' and treated as such.

Order No. 227 also directed the formation of penal battalions (*shtraf-baty*) out of men who had 'broken discipline due to cowardice or instability'. These battalions were to serve in the most dangerous sections of the front and were assigned the most hazardous tasks, such as advancing across minefields or undertaking frontal assaults on entrenched German positions. By doing so the disgraced men would be given an opportunity to redeem 'their crimes against the Motherland by blood'. The men in the penal battalions were known as *smertniks* (dead men) because very few were expected to survive.

Vladimir Kantovski was sent to a penal battalion in 1942 and soon realised that his only chance of surviving was to be wounded in combat. His company was ordered to undertake a 'reconnaissance through combat' whereby they would attack an enemy position, causing the Germans to reveal the location of their heavy weapons when they fired upon the advancing Soviet troops. As soon as the men of the penal unit came out into the open the German machine guns began to cut a swathe through their ranks. Kantovski was shot in the shoulder. 'I was wounded and I began to bleed. You had to be heavily wounded to be pardoned, but how can you know whether you are badly wounded or not badly wounded?', Kantovski recalled. 'Until I became convinced that I was heavily enough wounded I didn't dare set off towards the first aid centre. It was very hard to move – my arm was not working, so I had to crawl lying on my back.'[83] At the end of the attack, only nine out of the 240 men in Kantovski's unit had escaped being wounded or killed.

Order No. 227 also directed the formation of Red Army blocking units to assist the Zagradotryady by being stationed behind unstable formations and summarily executing cowards and panic-mongers who tried to flee the battle. A Soviet veteran recalled that the NKVD established numerous roadblocks where they inspected the papers and asked curt questions of the soldiers passing through. Those whose answers were unsatisfactory were shot, and mounds of bodies began to accrue at these checkpoints.[84] Few officers, however, were keen to release their best men to the blocking units and instead chose to fill them with invalids, cowards and other men not wanted in the front-line units. In October 1942 the idea of Red Army blocking units was dropped and this duty returned to be the preserve of the Zagradotryady.[85]

Over the course of the war, 442,000 Red Army soldiers and officers were forced to serve in the penal battalions and a further 436,000 were sentenced to periods of imprisonment. Although precise numbers will most likely never be known, it is estimated that as many as 158,000 Soviet soldiers were formally executed during the Second World War (some 13,500 were executed during the five-month battle for Stalingrad alone). Thousands more were summarily executed with no record of the event being kept.[86]

The German Army also made use of penal units during the Second World War. These units were attached to regular regiments and, as with their Soviet counterparts, were given the most dangerous missions. The men in the penal platoon had committed some infraction of military discipline, such as drunkenness, insubordination, or some offence in the rear (for example, the misappropriation of government property), and were given the opportunity to redeem themselves in combat.

The German Army of the First World War had shown a greater reluctance than the armies of the other European powers to impose the death penalty on its soldiers; unfortunately for the members of the Wehrmacht, this was not to be the case during the Second World War. In 1935 a new capital offence was added to the German code of military law, that of *Zersetzung der Wehrkraft* (undermining fighting power). Though surviving records are fragmentary, the offences of Zersetzung der Wehrkraft (which included self-inflicted wounds) and desertion accounted for approximately 70 per cent of all capital sentences handed down by the German military during the Second World War. Around 12,000 German soldiers were executed during the war, many of whom were killed during the final campaigns as Hitler and his supporters resorted to desperate measures to try to control the disintegrating German Army.[87]

Hitler's paranoia and distrust of his field commanders increased following the assassination attempt on 20 July 1944. He now viewed tactical withdrawals as evidence of cowardice and treason and ordered the court martial and execution of officers who withdrew without written permission. Scapegoats were sought out and executed to atone for military reversals and to generate the desired fighting spirit in the German troops. Following the capture of the bridge over the Rhine at Remagen on 7 March 1945 by elements of the American 9th Armoured Division, Hitler convened a special court martial board of three army officers. The board was not bound by the regulations of German military law and was unique in that it travelled with its own execution squad. The court found five German Army officers guilty of acting in an irresponsible and cowardly manner and sentenced them to death. Four of the officers (three majors and a lieutenant) were executed by firing squad, while the fifth (a captain), who did not appear before the court and had been sentenced to death in absentia, survived only because he was then a prisoner of war of the Americans.[88] But as the Third Reich collapsed, even the pretence of judicial proceedings was abandoned. Squads of *Feldgendarmerie* (military police) were empowered to hang offenders on the spot, their bodies strung from trees and lamp-posts with their crimes proclaimed on placards hung around their necks. Willi Heilmann, a pilot in the Luftwaffe, recalled: 'A chain of hanged men, with "For Cowardice" signs on their chests, was to be seen along all the retreat roads.'[89] Guy Sajer wrote of the execution of two of his comrades who were caught by the military police with some food and alcohol they had taken from a crashed supply truck.

[T]here was a tree, a majestic tree, whose branches seemed to be supporting the sky. Two sacks were dangling from those branches, two empty scarecrows swinging in the wind, suspended by two short lengths of rope. We walked under them, and saw the gray, bloodless faces of hanged men, and recognised our wretched friend Frösch and his companion ... I managed with difficulty to read the message scribbled on the sign tied to Frösch's broken neck. 'I am a thief and a traitor to my country.'[90]

The case of Eddie Slovik

In the European theatre of operations (the only theatre for which detailed statistics are available), 443 US soldiers were sentenced to death during the Second World War. Of these men, 70 were put to death, all but one for the offences of murder and/or rape. Furthermore, while 188 US soldiers were sentenced to death for military-specific offences, only one, Private Edward 'Eddie' Slovik, was actually executed.[91] When Slovik stood before a firing squad on the morning of 31 January 1945, near the village of Saint-Marie-aux-Mines in France, he became the first American soldier to be executed for desertion in the 80 years since the end of the American Civil War.

Slovik had been a petty criminal and was jailed several times. He was paroled from prison in April 1942. He was initially classified 4-F (unfit for duty) because of his police record. His draft classification proved a bonus for Slovik, as many good jobs were available because of the large number of men who had volunteered for or been drafted into the services. The absence of most of the young men also meant that there was less competition for the young single women. Slovik made the most of these opportunities. He secured a job with a plumbing company and seven months after being paroled he was married. Slovik's plans for the future were dealt a serious blow on 7 November 1943 – his first wedding anniversary – when he received a letter from his local draft board stating that they were considering changing his draft classification from 4-F to 1-A (available immediately for military service). The letter directed Slovik to undergo a physical.[92]

The casualties from the Pacific, North African and Italian campaigns, and the resulting need for replacements, generated a manpower shortage that required the lowering of draft standards. In November 1943 Slovik was classified 1-A. He was drafted into the army in January 1944 and began basic training. From the outset, Slovik was a reluctant draftee, rationalising that if he was a poor soldier then they would not send him overseas;

though he did appear to have genuine difficulty with firing his rifle, which made him excessively 'nervous' (in a letter to his wife he confessed: 'I hate guns'). He also suffered from a persecution complex, believing that an undefined 'they' were out to get him. 'I can't understand why they did this to us,' Slovik complained in a letter to his wife.[93]

Slovik arrived in France on 20 August 1944. Five days later he was assigned as a combat replacement to G Company, 109th Infantry Regiment, 28th Infantry Division. Around 11pm on 25 August, while en route to the front, Slovik's party was fired upon. They stopped and the non-commissioned officer in charge of the replacements ordered them to dig in. Slovik, along with a friend, became separated from the rest of their party, who had moved on during the night. The two Americans met up with some soldiers of the Canadian Provost Corps. The Canadians were part of a non-combatant unit who followed the combat troops into towns and posted notices advising the citizens of the provisions of martial law. As the two Americans did not know the location of G Company, the Canadians invited Slovik and his friend to stay with them, though the Americans did send a letter to the 109th informing them they had got lost and advising their current whereabouts. Slovik and his friend remained with the Canadians until 5 October, a period of six weeks. Their situation was not extraordinary, as many other soldiers had also become separated from their units in the push across France. No charges were laid against Slovik and his companion when, on the afternoon of 8 October, they finally joined G Company.[94]

Slovik informed his company commander during his initial interview that he was 'too scared, too nervous' to serve with a rifle company and that he would desert unless transferred to a rear area. Slovik was told that such a transfer could not be approved and was assigned to an infantry platoon. Slovik was handed over to his platoon commander and told not to leave the company area without the permission of the company commander. Slovik soon returned to the company commander and asked whether he could be tried for being absent without leave in relation to the period he spent with the Canadians. Slovik was told that the matter would be investigated and was escorted back to his platoon area, where he was ordered to remain. Despite this order, about an hour later Slovik again sought out his company commander and asked: 'If I leave now will it be desertion?' He was told that it would be. Slovik then deserted from G Company, having served in it for less than two hours.[95]

The following morning Slovik encountered a detachment of the 112th Infantry Regiment and handed a cook a piece of paper, which Slovik told him was a confession. Slovik's regiment was contacted and a military policeman returned him to the 109th's orderly room. His confession was written on a US Army Post Exchange flower order form. In it he confessed to deserting on 25 August while moving forward with the replacement draft and deserting again when he finally arrived at G Company on 8 October. Slovik stated that he had told his company commander: 'if I had to go out their [sic] again I'd run away. He said their [sic] was nothing he could do for me so I ran away again and I'll run away again if I have to go out their [sic].' Slovik's regimental commander advised him that this written confession would be very damaging for Slovik and gave him the option of taking it back and destroying it. Slovik refused to do so. He was then placed in the divisional stockade.[96]

Slovik was determined to avoid combat. He appeared to be under the somewhat valid impression, given the US Army's disciplinary policy up to that time, that the worst punishment he would face as a deserter would be imprisonment. He reasoned that he would be released from prison after the end of the war and be reunited with his wife. The threat of prison held no fears for Slovik and so he rejected an unofficial offer, made by the division's judge advocate, whereby his desertion would be overlooked if he voluntarily returned to his unit. Slovik's court martial took place on 11 November 1944. He pleaded not guilty and did not testify. No witnesses were produced in his defence. The board of the court martial found Slovik guilty of two counts of violating the 58th Article of War – desertion to avoid hazardous duty – and sentenced him to be 'shot to death with musketry'.[97] The presiding officer of the court martial later commented that Slovik was a:

healthy-looking soldier in open defiance of the authority of the United States. There was his confession: he had run away from his duties as a rifleman ... and *he would run away again*. Given the circumstances of a division locked in bloody battle and taking heavy casualties, I didn't think I had a right to let him get away with it.[98]

The presiding officer also commented that although he felt 'for the good of the division, he [Slovik] ought to be shot ... I don't think a single member of that court actually believed that Slovik would ever be shot.'[99] He assumed that the sentence would be reduced to imprisonment and Slovik released two to three years after the war ended – a belief also held by Slovik. The divisional commander, Major General Norman Cota, confirmed Slovik's death sentence.[100] Cota's attitude to Slovik's case was virtually identical to that of the presiding officer. Cota later commented: 'Given the situation as I knew it in November 1944, I thought it was my duty to this country to approve that sentence. If I hadn't approved it – if I had let Slovik accomplish his purpose – then I don't know how I could have gone up to the line and looked a good soldier in the face.'[101]

On 23 December 1944, Eisenhower confirmed Slovik's death sentence. But before the execution could be carried out, the case had to undergo an extensive legal review to ensure it was legally sufficient to support the sentence. The staff judge advocate who reviewed Slovik's case for Eisenhower wrote:

He was obstinately determined not to engage in combat ... There can be no doubt that he deliberately sought the safety and comparative comfort of the guardhouse. To him and to those soldiers who may follow his example, if he achieves his end, confinement is neither deterrent nor punishment. He has directly challenged the authority of the government, and future discipline depends upon a resolute reply to this challenge. If the death penalty is ever to be imposed for desertion it should be imposed in this case, not as a punitive measure nor as retribution, but to maintain that discipline upon which alone an army can succeed against the enemy.[102]

A month later, on 23 January, Eisenhower ordered that the execution be carried out in the 109th's regimental area. On the morning of 31 January

1945 Slovik was bound to a post in a courtyard enclosed by a high wall (to deter curious French onlookers), all insignia of the US Army having been ripped from his uniform. He was shot by a firing party of twelve soldiers in front of 42 witnesses representing the various units of the 28th Division. He was 24 years old. His wife was informed by telegram that her husband had died in the European theatre of operations but not that he had been executed. Only when she was denied the payout on his GI life insurance policy did she discover that he had died 'under dishonourable circumstances'. For the rest of her life she sought to have her husband pardoned.[103]

The underlying issue of Slovik's execution is the conflict between the rights of an individual and their responsibilities to their nation. Slovik was an unwilling soldier who was executed for a crime of omission – refusing to serve on the front line – rather than commission. In this instance, social expectations of service clearly outweighed individual rights, a factor encapsulated in a message to the troops of Slovik's regiment from the regimental commander following the execution: 'The person that is not willing to fight and die, if need be, for his country has no right to life.'[104]

Approximately 40,000 American soldiers 'deserted before the enemy' during the Second World War. Many of these men were tried by lesser courts martial and received dishonourable discharges or were incarcerated in theatre disciplinary centres. Almost 3,000 deserters were tried by General Courts Martial and awarded sentences ranging from twenty years' imprisonment to death. Although 130 US soldiers were sentenced to death for desertion during the Second World War, the usual practice was for the death sentence to be commuted to a lengthy prison term.[105] A fact that was no doubt evident to Slovik.

Slovik's biographer, William Bradford Huie, concludes that a combination of three factors resulted in Slovik being the only US soldier to be executed for a military-specific offence during the Second World War. The first was Slovik's written confession, which represented a blatant challenge to the authority of the US military. The second was that Slovik had a prison record, Huie commenting: 'if you are going to shoot one man, it's easier to shoot one with a "record".'[106] The divisional judge advocate recalled that:

I never expected Slovik to be shot. Given the common practice up to that time, there was no reason for any of us to think that the Theatre Commander [Eisenhower] would ever actually execute a deserter. But I thought that if ever they wanted a horrible example, this was one. From Slovik's record, the world wasn't going to lose much.[107]

The third and perhaps most crucial factor was timing. The American commanders were alarmed by the high number of deserters and felt that an example needed to be made, particularly as the Allied forces were then reeling under the German Ardennes offensive (Battle of the Bulge). But the deterrent value of Slovik's execution was limited, as details were not released outside the 28th Division, though General Cota noted: 'the execution did have in my division the deterrent effect visualised by the theatre commander.'[108]

The post-war decline of the death penalty

The French Army retained the death penalty for military offences after the Second World War but it was rarely if ever imposed. During the battle for Dien Bien Phu in 1954, Colonel Lalande, the commander of strongpoint Isabelle, decided to execute some Algerian soldiers for cowardice to set an example for the others, who he felt had not pursued a counter-attack against the Viet Minh with sufficient vigour. He ordered the platoon commanders of the company that took part in the failed attack to designate two men from each platoon to be executed that evening. Remonstrations by the Algerians' officers, who argued that the members of the French Foreign Legion had also failed to break through, resulted in Lalande proposing a compromise. The trial would go ahead but the designated men would be acquitted: this subsequently occurred.[109]

Although it had not been applied since the end of the Second World War, the British military retained the death penalty for the offences of mutiny and treachery until 1998. The death penalty, however, remains a disciplinary option for the US military. The 2000 edition of the United States' *Manual for Courts-Martial* lists fourteen capital offences. Many of these offences are enacted only in time of war, such as desertion (Article 85); assaulting, wilfully disobeying a superior commissioned officer (Article 90); misbehaviour of sentinel or lookout (Article 113); and misbehaviour before the enemy (Article 99), which incorporates provisions pertaining to cowardice. Others, such as murder (Article 118) and rape (Article 120) are constantly in force.

The US Army has executed 135 soldiers since 1916. The last to be executed was Private John A. Bennett, who was hanged on 13 April 1961 following his conviction for the rape and attempted murder of an eleven-year-old girl in 1955. As at January 2009, nine servicemen were incarcerated on death row at the US Disciplinary Barracks, Fort Leavenworth, Kansas. All nine men had been convicted of premeditated murder or felony murder.

The campaign to abolish military executions was largely championed by civilians, particularly politicians, and paralleled the movement away from all forms of capital punishment in the Western world (with the exception of the United States). But the attitude of soldiers towards military executions is often markedly different to that of civilians. Some soldiers empathise with those executed. A soldier of the American Civil War commented: 'How impossible it was for us to fathom the agony of their souls, as they marched to their own funerals, saw their own coffins, and their very graves before their eyes. *Oh, it was terrible.*'[110] Others may justify the execution on the basis that the condemned man deserted his comrades in battle and so deserves this punishment. Such soldiers have little sympathy for men who refuse to face the dangers of the battlefield alongside them. One of the soldiers in Slovik's firing squad stated the night before the execution: 'I got no sympathy for the sonofabitch! He deserted us, didn't he? He didn't give a damn how many of us got the hell shot out of us, why should we care about him? I'll shoot his goddam heart out. If only one shot hits him, you'll know it's mine.'[111]

As officers are responsible for maintaining discipline and keeping soldiers in the front line, they generally adopt a pragmatic view of the need for the death penalty. The German field marshal Erich Von Manstein wrote in his memoirs:

The man who commands an army is also its supreme arbiter, and the hardest task that can ever confront him is the confirmation of a death sentence. On one hand it is his inexorable duty to maintain discipline and, in the troops' own interest, to inflict severe penalties for delinquency in action. On the other, it is a grim thought to know that one can snuff out a human life by a mere signature.[112]

Some officers believed that only the fear of a disgraceful death by execution could overcome the soldier's natural instinct of self-preservation. In *War as I Knew It*, General George S. Patton Jr. drives home this point:

One of the great defects in our military establishment is the giving of weak sentences for military offences ... I am convinced that, in justice to other men, soldiers who go to sleep on post, who go absent for an unreasonable time during combat, who shirk in battle, should be executed.[113]

* * *

Although the peak number of executions by the British Army was reached during the First World War, the death penalty was frequently imposed by the German and Soviet armies during the Second World War. In the latter cases, this was a reflection of their totalitarian regimes, and the imposition of the death penalty for desertion was not a disciplinary option for the British generals of the Second World War, despite the fact that a number of them felt it was only the fear of this shameful death if they fled that would keep men at the front. But as the general population became more informed about the nature of war, and the moral ambiguities of executing soldiers was debated, civil support for the military's use of the death penalty waned. They reasoned that war was terrifying enough without soldiers also having to worry about being tied to a stake and shot at dawn.

Chapter 8

KILLING YOUR OWN
– FRIENDLY FIRE

O n 28 March 2003, a United States Air Force (USAF) A-10 pilot (call sign POPOV36) fired on a patrol of British armoured reconnaissance vehicles, believing them to be Iraqi. The transcript of the cockpit video clearly reveals the pilot's distress when told he had killed and wounded friendly troops:

> Dammit. Fucking dammit. (POPOV36)
> God dammit. Fuck me dead (weeping). (POPOV36)
> You with me. (POPOV35 – the attacking pilot's wingman)
> Yeah. (POPOV36)
> They did say there were no friendlies. (POPOV35)[1]

Adrien Bourgogne of Napoleon's Imperial Guard tells of an incident that occurred on the retreat from Moscow in 1812. A mounted grenadier mistook one of Napoleon's orderly officers for a Cossack and ran him through with his sword. According to Bourgogne, 'The unhappy Grenadier, on seeing his mistake, endeavoured to get killed. He flung himself amongst the enemy, striking to right and left, but everyone fled before him.'[2] The Soviet soldier Mansur Abdulin recalled that following a clash between Soviet armies encircling the Germans at Stalingrad, 'We also cried: for there were dead and wounded on both sides, and each person blamed himself. Later, when clearing the battlefield of bodies, we avoided each other's eyes.'[3] Robert Crisp, who served with the British Royal Tank Regiment during the North African campaign of the Second World War, mistook a British tank for a German Panzer and fired four rounds into it. Another British tank suddenly appeared waving a red flag (the signal to cease fire). 'I looked again, my heart coming into my mouth in sheer horror ... I knew immediately what I had done,' Crisp wrote in his memoirs of the campaign. He ran over to the shattered tank, hoping that no one had been killed, where he was confronted by a young officer: 'You bloody fool. You've killed my gunner.' The surviving crew members placed a rope around the figure slumped in the turret.

> I had neither the strength nor the will to help. Perhaps he wasn't dead. Perhaps he was just unconscious. The officer might have made a mistake. There was still a little hope.
>
> Out of the turret-top they hauled a lad's body – red hair, fair skin, freckled face. As they pulled him out, the head rolled sideways and two, wide-open, empty eyes looked straight into mine. In that moment I touched the rock-bottom of experience.[4]

The emotional anguish of killing a comrade is evident in these accounts, as well as in Ron Kovic's memoir of his service in Vietnam and its aftermath, *Born on the Fourth of July*. In October 1967 Kovic shot and killed a nineteen-year-old fellow Marine during a firefight near the Demilitarised Zone (DMZ) with North Vietnam. It was dark and the Marines were withdrawing when a figure suddenly appeared before Kovic and he fired three shots, hitting the man in the throat. Kovic was driven back to base with the body of the dead Marine sprawled at his feet. After reporting the incident to his superiors, Kovic retired to his tent and sat down on his bed. He went over the scene repeatedly in his head. Consumed by guilt, he pointed the barrel of his rifle at his head. 'Oh Jesus God almighty, he thought. *Why*? Why? Why? He began to cry slowly at first. *Why*? I'm going to kill myself, he thought. I'm going to pull this trigger.'[5] But he did not pull the trigger and fell into a troubled sleep. The next morning he tried to convince himself it was an accident or that maybe someone else had fired the fatal shot, but in the end he knew that was not the case. Even confessing his actions to the battalion executive officer did not provide Kovic with the absolution he sought. He began to feel alienated from his fellow Marines, who he believed blamed him for the man's death and who he felt talked about the incident behind his back. Kovic lamented that:

> The next few weeks passed in a slow way, much slower than any time in his whole life. Each day dragged by until the night, the soft soothing night, when he could close himself off from the pain, when he could forget the terrible thing for a few hours. Each night before he slept he prayed to his god, begging for some understanding of why the thing had happened, why he had been made into a murderer with one shot … What kind of god would give him these terrible feelings and nightmares for what seemed to be the rest of his life?[6]

'A sinister and malignant stroke'

The anguish caused by friendly fire incidents is not confined to those who fired the fatal shot. Guy Chapman served for most of the First World War as an officer in a battalion of the Royal Fusiliers. In his memoir of the war, *A Passionate Prodigality*, Chapman wrote of an incident where a mortar shell fell short and exploded a few yards away from a British raiding party, killing a young officer and wounding ten soldiers.

> The catastrophe wrenched many of us as no previous death had been able to do. Those we had seen before had possessed an inevitable quality, had been taken as an unavoidable manifestation of war, as in nature we take the ills of the body. But this death, at the hands of our own people, though a vagary of the wind, appeared some sinister and malignant stroke, an outrage involving not only the torn body of the dead boy but the whole battalion.[7]

A German officer, Walter Bloem, similarly recalled the devastating effect on the morale of his men when mistakenly shelled by their own artillery as they advanced across France in the opening campaign of the First World War. Bloem stated that: 'Death coming from the enemy's shells is expected, part of the bargain of war, but coming from the mistaken fire of one's own artillery it is beyond the pale – utterly devilish.'[8] Eugene Sledge, who served with the US Marine Corps in the Second World War, felt very much the same when a US tank mistook his mortar party for Japanese soldiers and opened fire.

> A surge of panic rose within me. In a brief moment our tank had reduced me from a well-trained, determined assistant mortar gunner to a quivering mass of terror. It was not just that I was being fired at by a machine gun that unnerved me so terribly, but that it was one of ours. To be killed by the enemy was bad enough; that was a real possibility I had prepared myself for. But to be killed by mistakes by my own comrades was something I found hard to accept. It was just too much.[9]

Evgeni Bessonov, who served with the Soviet 4th Tank Army, wrote in his memoirs of the Second World War of an incident where the misdirected fire of a regiment of Katyusha rocket launchers fell on some units of his brigade, killing 30 to 35 soldiers. 'In one month of fighting we did not have such high casualties, as from one Katyusha salvo! It was painful to see the dead soldiers — young, healthy and needed for further battles.'[10] Joseph Owen was more explicit when recounting an incident where a friendly artillery barrage fell among the Marines trying to break out of the encirclement by Chinese troops near the frozen Chosin Reservoir in North Korea. 'Friendly fire. The worst thing that can happen in combat,' wrote Owen. 'Our dismal day turned to horror. Cries and moans and agonised screams pierced the black smoke that drifted over the broken ground.'[11] When the smoke dissipated, four Marines were dead and three wounded.

Hal Moore tells us of a friendly fire incident in his account of the Vietnam War where a 105mm artillery round dropped short and exploded among one of his platoons. The round killed nineteen-year-old Private

First Class Richard Clark. Clark was part of a platoon that had been cut off for 26 hours and had survived several determined attacks by the NVA during the battle at Landing Zone X-Ray in the Ia Drang Valley of South Vietnam in November 1965. One of Clark's comrades plaintively remarked: 'I just couldn't believe this. He was right next to me in Ia Drang when we were trapped. We made it through that, we got out of there, and then he gets killed by our own artillery. Why did Richard Clark have to die that way?'[12]

As his company prepared to assault their final objective on Wireless Ridge in the closing stages of the Falklands War, Major Philip Neame told his Forward Observation Officer to put down some artillery fire on the Argentine position. Tragically the wrong target number was called up and the artillery fire fell on Neame's soldiers, wounding one and killing another. 'This lad had already been injured and casevac'd; he'd recovered, returned to us and now, of all things, he'd been killed by our own artillery. It seemed a complete waste,' Neame bitterly recounted. 'These things do happen in war, far worse has happened in the past and far worse will probably happen in the future, but it made me really mad.'[13]

Anthony Swofford wrote of an incident during the 1991 Gulf War where American tanks attacked his unit's supply convoy. Four vehicles were destroyed during the attack, resulting in two soldiers killed and six wounded. Swofford felt that the rounds fired by the US tanks were

> more mysterious and thrilling and terrifying than taking the fire from the enemy, because the enemy fire made sense but the friendly fire makes no sense – no matter the numbers and statistics that the professors at the military college will put up on transparency, friendly fire is fucked fire and it makes no sense and cannot be told in numbers.[14]

The senselessness of friendly fire casualties weakens morale among soldiers, not because these deaths are unexpected – fratricide has always been a feature of war – but because they seem so tragically avoidable. The soldiers lament that if only their comrades had been better trained or the weapons better designed then these casualties would not have been incurred. But the unavoidable fact is that if soldiers are sent off to war some of them will most likely be killed by their comrades.

Covering up the truth

Deaths due to friendly fire are particularly difficult for the families of the deceased soldiers to accept. An illustrative example of the angst caused by fratricide is the death of Corporal Michael Mullen, a member of Charlie Company, US 6th Infantry Regiment. On the night of 17 February 1970, Charlie Company established a night defensive position on a jungle hilltop near Tu Chanh, South Vietnam. They requested that their supporting artillery battery – four 105mm howitzers – register some targets on the trails near their position to be fired upon if the Viet Cong attempted a night attack. These targets were sited about 400 metres out from the company perimeter. The first registration round was on target but the second, a high-explosive round, hit a tree and exploded directly above the 1st Platoon area, spraying the men below with shrapnel. Mullen and another soldier were killed and six men wounded. A subsequent investigation revealed that an artillery officer at the fire control centre, who was responsible for calculating the trajectory of the rounds, had forgotten to take into account the height of the trees on the hilltop. A round that was intended to sail over Charlie Company's position had instead hit the tree and exploded.[15]

Mullen's parents found the circumstances of their son's death very difficult to accept, particularly because it was his own army that had killed him. His mother became an outspoken member of the anti-war movement and used her son's $2,000 death gratuity to place a half-page advertisement in her state's paper (the *Des Moines Register*). The advertisement depicted 714 tiny crosses (representing the number of Iowans that had been killed in the Vietnam War) and asked how many more lives needed to be sacrificed before the Iowan people spoke out against the war.[16] The radicalisation of the Mullens later became the subject of C.D.B. Bryan's bestselling book *Friendly Fire*.

The Mullens' distrust of the military arose because army officials were initially evasive about the circumstances of their son's death. Indeed, deaths by friendly fire may be covered up, with relatives told only that their loved one was killed in action and no precise details provided. Eugene Sledge recalled an incident that occurred during the battle for Peleliu (Second World War), where a Marine was mistaken for a Japanese infiltrator and shot by his comrades. Sledge was told by his company commander never to discuss the incident or mention the slain man's name again. Furthermore, the death was recorded as 'killed in action against the enemy (wound, gunshot, head)'.[17] The journalist Rick Atkinson noted that despite the US Army and Marine Corps knowing within a month that a number of the casualties during the 1991 Gulf War were caused by friendly fire, in most instances they waited four more months before informing the men's families of the actual circumstances of their deaths.[18] In many cases, this covering-up of fratricide incidents occurred because the military authorities were concerned about the effect this news would have on the families of the deceased, but also less altruistically because friendly fire incidents may be detrimental to a commander's career – or they may even end it.

Cases of mistaken identity

On 2 May 1863, the Confederate officer Lieutenant General Thomas 'Stonewall' Jackson was mistakenly shot by his own soldiers during the Chancellorsville campaign of the American Civil War. Jackson's army had broken the right wing of General Joseph Hooker's Army of the Potomac just before dusk with a daring flanking assault. When night fell, Jackson rode out into the woods with a few of his staff to reconnoitre a side track from which he hoped to launch a night attack. Jackson was therefore between the Confederate and Union lines. Earlier that evening he had warned his men to exercise caution, as the Union troops were still to their front. Jackson finished his reconnaissance around 9pm and, accompanied by his personal staff, rode back towards the Confederate line. The soldiers of the 18th North Carolina Regiment, hearing the galloping horses coming from the direction of the Union retreat, assumed the Union soldiers were mounting a counter-attack and opened fire. The volley killed several of Jackson's staff, with Jackson himself struck three times. The extent of his wounds led to his left arm being amputated the following morning. Jackson appeared to be recovering from his wounds but was diagnosed with pneumonia on 7 May and died three days later. He was 39 years old.

A similar incident occurred in the closing stages of the siege of Sevastopol during the Crimean War. Timothy Gowing had led a patrol forward of the British lines to provide early warning of a Russian attack. The British were attacked by an overwhelming Russian force and retired. 'But during our absence from the trench it had been filled with men of various regiments; and, not knowing that there was anyone in front but the enemy, they opened a regular file fire, and we were in a pretty mess between two fires,' Gowing recounted. Twenty out of 30 men were killed. New sentries were posted and the men manning the front trench were told there were friendly troops to their front. Gowing bitterly observed: 'Had that been done before, the greater portion of my men would not have died, as they were nearly all shot by our own people; these are some of the "blunders" of war.'[19]

Some 60 years later, on the night of 13 July 1918, a patrol was returning to the British lines after carrying out a raid on the German trenches. The patrol was led by Siegfried Sassoon. Unfortunately Sassoon was carrying the distinctive British helmet in his hand instead of wearing it, and his silhouette was mistaken for that of a German, resulting in Sassoon being shot in the head by one of his own soldiers. Fortunately it was only a scalp wound and he recovered.[20] Edmund Blunden recalled an incident when he was returning to the British trenches from a patrol in no man's land. 'We had now been out in the open long enough for our sentries to forget us; and as we came back to our wire one did forget us, firing excitedly.' Illumination flares were sent up and Blunden and his two companions began to attract German machine gun fire. Blunden yelled out his name, the sentry's fire eased up and they scrambled back into the British trenches.[21]

As Jackson, Gowing, Sassoon and Blunden discovered, being fired upon by your own troops is most likely to occur when a patrol is returning to a defended locality. For this reason patrols will usually exit a defended position through an adjacent unit and return to the position through their own troops, so that the incoming patrol has the greatest chance of being rec-ognised and not shot at by their own comrades. However, 'There's always the one guy who don't get the word and that's the guy who shoots you up coming home. Always,' observed a sergeant in the US Army following a friendly fire incident in Vietnam in 1965.[22] In this instance, a private was asleep in his foxhole when the alert was sent out that the patrol was coming in and nobody woke him to tell him. Consequently, when he woke up and saw the silhouetted figures to his front, he assumed they were NVA soldiers and opened fire, emptying a whole magazine into them. One American soldier was shot in the hip and two others in the legs. It was only when the private stopped firing to reload that they were able to alert him to the fact that he was firing on friendly troops.[23]

The major contributing factors to fratricide incidents among ground troops include battlefields with limited visibility, poor coordination measures, and inexperienced, nervous or ill-disciplined troops. A salient example of what can occur when these factors are combined is the Allied assault on the Aleutian island of Kiska during the Second World War. Over the period 15–16 August 1943, 35,000 inexperienced American and Canadian troops landed on Kiska, which had been seized by the Japanese in June 1942. The Allied forces, expecting fanatical Japanese resistance, staged a full-scale assault on the fog-shrouded island. By nightfall on the 16th, 28 men were dead and 50 wounded. Although four men were killed and several wounded by booby-traps and mines left behind by the Japanese, the vast majority of these casualties were caused by friendly fire. The Japanese had abandoned the island on 28 July – two weeks before the Allied assault.[24]

Many friendly fire incidents can be attributed to inexperienced sol-diers but even highly trained 'elite' troops experience fratricide. During the Falklands War, a British Special Air Service (SAS) patrol ambushed a British Special Boat Service (SBS) patrol, killing the patrol commander. The SBS troopers had strayed into the SAS patrol's operational area after having been dropped off at night by a naval helicopter some miles from their intended location. The SBS patrol leader, a sergeant, was attempt-ing to confirm his position when he was killed. The SAS detected the SBS patrol through their night-vision equipment and set an ambush, having been briefed that there were no other friendly units in the area. The commander of the SAS element was lying alongside the patrol's machine gunner and urged him to wait until the approaching men were confirmed as the enemy before opening fire. When the SBS patrol had approached to within ten metres or so of the men lying in ambush they were challenged by the SAS officer in English. The SBS commander halted, held his rifle well away from his body and stood still, and the two men immediately behind him also froze. But the last man in the patrol, who possibly had not heard the challenge, tried to creep away. The machine gunner perceived this action to be a threat to the SAS troopers and opened fire, killing the SBS patrol commander.[25]

On 21 May 1982, Charlie Company of the British Army's 3rd Battalion of the Parachute Regiment fired on a patrol from the battalion's Alpha Company returning to the battalion position in the hills bordering the San Carlos settlement in the Falkland Islands. The two patrols had spotted each other and each assumed the other patrol was Argentine. Both patrol commanders then requested a fire mission from battalion HQ. The confusion was further compounded when the Alpha Company patrol commander gave a grid reference that placed his patrol about a kilometre from their actual position. Battalion HQ therefore assumed its companies had two separate contacts with the enemy, rather than with each other. The machine gunners of Charlie Company opened fire and wounded nine men from Alpha Company. Elements of the Alpha Company patrol then attempted to reach the comparative safety of the top of the slope but were hit by the artillery fire requested by Charlie Company. Some 40 shells landed around the patrol and two more men were wounded, both with serious head injuries (they survived). As the reports of the wounded came in, battalion HQ realised that Alpha and Charlie companies were fighting each other and ordered the troops to stop firing.[26]

The effect of fear

The prevailing confusion of the battlefield that resulted in the death of the SBS sergeant and the wounding of the British paratroopers is a major contributing factor to many fratricide incidents. This confusion is intertwined with and often inseparable from fear. In most cases, if the plane, tank, ship, or soldier you are engaging is in range of your weapon systems, then you are probably also in range of theirs. So targets tend to be engaged at ranges that make positive identification problematic. Soldiers rationalise that the target might be the enemy and probably is. Hugh Dundas recalled that when the Typhoon was being introduced into service with the Royal Air Force, a number of these aircraft were shot down by Spitfires. He theorised that 'the unfamiliar Typhoon was all too similar to ... the familiar Fokker Wulf [sic] 190. And as the Spitfire pilots saw far more 190s than Typhoons they tended to attack first and ask questions later. Nevertheless the situation caused great bitterness.'[27] From a similar case of mistaken identity, Paul Fussell notes that eventually the Allied pilots of the Second World War learnt that on no account were they to fly over London on their way to bomb the European mainland because of the inherent risk to their planes posed by the trigger-happy London anti-aircraft gunners. He quotes a Canadian pilot who stated: 'the jittery army gunners always cut loose at you, despite the fact that we were flying north to south and there were 800 of us. We could hardly be Germans to the most unimaginative mind and yet they always pounded up the flak.'[28]

Royal Air Force pilots also had cause to be somewhat wary of the Royal Navy. A pilot from the Second World War recalled: 'If the Royal Navy were in the Channel and your aircraft went anywhere near them, they shot at you ... I take a very dim view of the fact they couldn't recognise the difference between a Hurricane and a Messerschmitt.'[29] Guy Gibson wrote that as his squadron prepared for the famous Dambuster raid, 'Many times out at sea the boys were getting fired on by his Majesty's ships, who have notoriously light fingers ...'[30] Arthur Hadley tells us of serving with an American unit in the Second World War that had shot down so many of its own planes that its anti-aircraft guns all had a little tag on the trigger that read: 'This gun will only be fired under command of an officer.'[31] Moreover, Jim Bailey wrote of an incident befalling a fellow pilot in the Royal Air Force where 'Our own aerodrome defences succeeded in putting three bullets into his aircraft, when he had flaps and wheels down, navigation lights on, and was coming in to land.'[32] But it was not only Allied aircraft that were brought down by friendly fire.

On 28 June 1940, an anti-aircraft gun crew on an Italian warship shot down an Italian bomber over Tobruk harbour. On board was the governor of Libya and commander of the Italian African Army, Air Marshal Italo Balbo, who was returning to his HQ at Tobruk. British fighters had subjected the cruiser to repeated air attacks since dawn. When the crew of the cruiser's anti-aircraft gun detected two planes flying the route previously used by the British planes, they opened fire with 20mm tracer. Balbo's plane was hit and crashed. There were no survivors.[33]

A similar fratricide incident occurred on the evening of 7 December 1941, when a flight of six F4F Wildcats (fighter aircraft) from the US aircraft carrier *Enterprise* was fired upon by American anti-aircraft guns while attempting to land at Hickam Field on the Hawaiian island of Oahu. Specific instructions were issued to all ships and anti-aircraft batteries that six American aircraft were coming in to land, but it was not enough to prevent the gunners, still jittery from the Japanese attack on Pearl Harbor that morning, from opening fire. At 9pm, as the planes approached Oahu, a torrent of fire was directed skyward from ship- and shore-based guns. Only one aircraft managed to land on an airfield (Ford Island Naval Air Station), while another set down on a golf course. The remaining aircraft were shot down. One pilot was killed instantly. Another two pilots died from their wounds the following day (one of whom had been machine-gunned from the ground after parachuting from his stricken plane).[34]

Sicily, 1943 – ground-to-air fratricide

The Anglo-American invasion of Sicily in July 1943 was also marred by a series of tragic fratricide incidents involving ground forces firing on friendly aircraft. The invasion plan included four separate airborne operations using the paratroopers of the American 82nd Airborne Division and the British 1st Airborne Division. All airborne operations were to be conducted at night. The commander of the 82nd Airborne, Major General Matthew Ridgway, expressed a concern during the planning phase that the airborne troops might be engaged by friendly fire when they passed over the massive Allied invasion fleet. But Admiral Cunningham (Royal Navy), the commander-in-chief of naval operations for the invasion, refused to guarantee that his ships would not fire on the transport aircraft. He stated that he could not risk the safety of his ships by allowing unidentified aircraft to pass close overhead at night. Despite the reservations of Ridgway, the airborne operations went ahead.[35]

On the evening of 9 July 1943, the British 1st Air Landing Brigade took off from Tunisia in a fleet of 147 gliders towed behind American C-47 (Dakota) aircraft. As they approached the coast of Sicily, the C-47s were engaged by the Italian gunners ashore as well as by the Allied gunners on board the ships of the invasion fleet. The heavy flak caused many of the inexperienced C-47 pilots to panic, and this factor, coupled with a stronger than expected offshore wind, resulted in many of the gliders being released too far out to sea, the C-47s then returning to the safety of Tunisia. Sixty-nine of the 147 gliders crashed into the sea, drowning 326 British soldiers. Of the remaining gliders, two were shot down, several were towed back to Tunisia and 59 landed in Sicily, though they were spread out over an area of some 25 square miles.[36]

The invasion fleet reached Sicily on 10 July, and throughout that day and the next the ships were under almost constant attack from German and Italian planes, as well as from the enemy artillery ashore. The ships' anti-aircraft gunners, having been subjected to air raids throughout the day, were understandably jittery, as were the hundreds of anti-aircraft gun crews ashore protecting the disembarked invasion force. Around 9pm on the evening of the 11th, some 2,000 men of the American 504th Regimental Combat Team were to be flown over Sicily by the C-47s at a height of only 1,000 feet above the massed ships of the invasion fleet and the landed troops of the US 7th Army. Ridgway's concerns about friendly fire would be tragically fulfilled.[37]

Ridgway continued to press for a guarantee that the ships and the troops ashore would not fire on the relatively slow-moving C-47s. He was informed by the invasion HQ that the gunners would be briefed on the flight path of the American planes and ordered not to fire. Ridgway anxiously waited as the American planes reached the Sicilian coast and turned northwest. They then flew at low altitude for 35 miles over the friendly troops and ships. The lead flight passed overhead without incident. But as the second flight neared the final checkpoint a lone anti-aircraft gun fired. A few more desultory rounds arced skyward, then suddenly hundreds of anti-aircraft guns (both ashore and afloat) began firing into the night sky. When the gunners were later questioned about why they had fired, they stated that they thought the transports were German bombers, which earlier that night had attacked the invasion force, or conversely that they were German transports dropping paratroopers onto the invasion beaches.[38]

Several planes were shot down before they could drop their paratroopers, while others attempted to escape the flak by turning out to sea. The paratroopers, crammed inside the low-flying transports, were being massacred. Many planes crashed into the sea or onto the beaches. Naval gunners, believing they were firing at German paratroopers, shot at the American soldiers as they slowly descended after exiting the burning C-47s. A US Navy destroyer continued to fire at a C-47 that had ditched into the sea until someone recognised it as an American plane. Ashore, tank crews fired on the transports with their machine guns and ground troops fired on the paratroopers when they landed. Twenty-three of the 144 C-47s that took part in this operation were shot down. Another 57 aircraft were badly damaged. The paratroopers suffered 81 dead, 132 wounded and sixteen missing, while the 52nd Troop Carrier Wing, which flew the C-47s, reported seven dead, 30 wounded and 53 missing – a total of 319 casualties due entirely to friendly fire.[39] Perhaps the best summation of what brought about this tragedy was given by Eisenhower's deputy, Air Chief Marshal Arthur Tedder (Royal Air Force), who wrote:

Even if it is physically possible for all the troops and ships to be duly warned, which is doubtful, any fire opened up either by mistake or against enemy aircraft would almost certainly be supported by all troops within range. AA firing at night is infectious and control almost impossible.[40]

The last phase of the ill-fated airborne assaults, the drop of the British 1st Parachute Brigade, was also largely non-effective. Four gliders were shot down by friendly fire and the British paratroopers were so scattered that fewer than 300 men out of an emplaned force of over 1,800 actually landed on the drop zone.[41] Of the four airborne operations, only that of the American 505th Regimental Combat Team managed to avoid friendly fire casualties.

The German forces also experienced incidents of ground-to-air fratricide during the Second World War, the most costly of which occurred during the Ardennes offensive (also known as the Battle of the Bulge). On 1 January 1945, the Luftwaffe launched a massive attack involving 900 fighters and fighter-bombers – virtually its entire operational fighter force. The attack was directed against Allied airfields in Belgium, France and the Netherlands. The dawn attack surprised the Allied forces and more than 200 of their aircraft were destroyed, with a further 150 seriously damaged. However, as the Luftwaffe's planes were returning to their airbases they were engaged by German anti-aircraft guns. The attack had been planned in such secrecy that the German gunners had not been informed of the operation. Consequently, they shot down 184 German planes, in the process killing some of the Luftwaffe's most experienced pilots. The Luftwaffe would never again mount an attack of this magnitude.[42]

Two-way exchanges

The fratricide incidents described thus far have mostly involved a one-way exchange of friendly fire, whereby troops have realised that they have been engaged by friendly troops and have not responded with deadly force. When neither force realises that they have been fired upon by friendly troops and they fight each other, the probability of friendly fire casualties increases. Such incidents may arise because of a lack of coordination measures or due to poor tactical schemes of manoeuvre. On 9 July 1944, a tank company from the US 3rd Armoured Division was ordered to capture Hill 91 at Hauts-Vents, Normandy. Because of a navigation error, whereby upon reaching a crossroads the eight Sherman tanks turned right (north) rather than left (south), the tankers ended up attacking Charlie Company of the 823rd Tank Destroyer Battalion. The two American units fought each other for 25 minutes, the confusion added to by the limited visibility of the battlefield. Two tanks were destroyed and the tank company suffered six casualties. Charlie Company had one man killed and three wounded. Eventually the tankers realised they had been engaging friendly forces, reversed direction and set off for Hill 91, which they reached at dusk – just in time to be strafed by American fighters in an attack that had been requested earlier but was delayed because of bad weather (fortunately no further friendly fire casualties were incurred).[43] A similar incident occurred on 3 November 1956 at Um Katef during Israel's 1956 campaign in the Sinai against Egyptian forces. Proceeding on different axes of advance, the tanks of the Israeli 7th Armoured Brigade engaged the tanks of their comrades in the 37th Armoured Brigade, destroying eight tanks in five minutes.[44]

Another incident of this nature occurred in late 1967 during the Vietnam War. A US artillery unit conducting a harassment and interdiction fire mission applied the wrong charge and the rounds landed in an American base camp, killing one man and wounding 37. The victims' unit responded with accurate counter-battery fire that killed twelve men and wounded 40 at the initiating firebase. The incident lasted 23 minutes and resulted in 90 casualties.[45] The responding artillery unit was most likely unaware that their counter-battery fire was directed towards their own comrades, though this is not always the case. Fratricide incidents have occurred where forces have deliberately fired upon friendly forces, usually because this is their only option to save their lives.

One such incident took place just before dawn on 19 April 1953 during the Korean War. Two companies (King and Love) of the US 31st Infantry Regiment were ordered to attack 'Pork Chop Hill', but from different sides of the feature. Consequently, two soldiers from Love Company engaged the

King Company soldiers with a machine gun as soon as they reached the crest of the hill, having mistaken the American soldiers for Chinese troops. The soldiers from King Company soon realised they were being engaged by American troops, but all attempts to get the two Love Company soldiers to cease fire proved futile, the fire stopping only when the two men were shot and killed by their comrades in King Company. Moreover, a Canadian soldier, who fought in Normandy during the Second World War, confessed to deliberately killing an American soldier: 'He was shooting at me as if I was a German. It's him or me. So I shoot the boy even though I know he's a Yank.'[46] But occasionally troops will fire on friendly troops not to save their own lives, but because they are angry and seek revenge.

During the Sicilian campaign of August 1943, the US Army Air Force earned the epithet of the 'American Luftwaffe' because of their tendency to bomb friendly troops, despite ground elements taking precautions such as stretching luminescent panels across the hoods of their vehicles to indicate Allied troops. General Omar Bradley wrote in his memoirs of an occasion when he was strafed by US planes three times in one day. In another instance, a flight of American dive-bombers continued to strafe an American tank column even after the ground forces had let off yellow smoke bombs to indicate that they were friendly troops. The tank crews lost patience with the pilots and shot down one of the aircraft.[47] A similar incident occurred in the Ardennes in December 1944. Two US P-47 Thunderbolt fighter aircraft made a strafing run on an American infantry company, despite the appropriate air recognition panels being displayed. Fortunately no one was hurt. But when one of the P-47s made a second run and dropped a 100-pound bomb on a mess tent, causing many friendly casualties, this proved too much for one of the American officers.

I put 75 rounds from a truck-mounted 50 cal. MG [machine gun] through his motor. He went up in flames. He rolled his plane and parachuted. We jumped in a jeep and were fully intent on beating him to death. He came down, fortunately for him, next to the division MPs. We were told he was cashiered.[48]

The Germans also had cause to shoot down their own aircraft. The German general Heinz Guderian described in his memoirs an incident where the Luftwaffe attacked his corps headquarters. 'It was perhaps an unfriendly action on our part, but our flak opened fire and bought down one of the careless machines. The crew of two floated down by parachute and were unpleasantly surprised to find me waiting for them on the ground.'[49]

The changing nature of the threat

Although death by friendly fire has always been a danger faced by soldiers, the nature of this danger has changed as weapons have evolved. The close-order formations of the Napoleonic period were particularly susceptible to friendly fire. Marshal Saint-Cyr estimated that one quarter of all French infantry casualties during this period were due to the men in the forward ranks being accidentally shot by those arrayed in the ranks behind them.[50] John Keegan suggests that the obsession with drill among the soldiers of this period came from their desire to avoid friendly fire: 'For among close-packed groups of men equipped with firearms, one's neighbour's weapon offers one a much more immediate threat to life than any wielded by an enemy.'[51]

The mass armies of the Napoleonic period, like those of the American Civil War, fought in constrained areas with limited visibility, due to the plumes of smoke generated by the firing of thousands of cannons and muskets. In such an environment, even the best-drilled and most dis-ciplined troops would inevitably engage their comrades in musket fire when the battle broke down into desperate localised struggles. Narrative accounts of this period abound with references to friendly fire incidents of this nature. Edward Costello recalled an incident that occurred during the attack on San Sebastian (August 1813) in the Peninsular War, when his company mistakenly engaged their comrades: 'fire they did, and did some mischief too, for the Major bringing up their left shoulders more than he should have done, they fired a volley into the 6th Scotch, who were some distance on our right front, and badly wounded fifteen or sixteen of their men.'[52]

In addition to being shot by the muskets of their comrades, the soldiers of this period also had to contend with the misdirected shells of their own artillery. Deaths from friendly artillery would result either from the shells falling short or because of misidentification of the target by the friendly gunners. A typical account of such an incident is provided in the diary of Edmund Wheatley: 'On ascending [higher up the hill] a 9 lb. shot from [our own] artillery [across the river] fell short and nearly cut off half of us.'[53] William Fletcher had a similar experience during the battle of Gettysburg in the American Civil War.

> I had slept but a few minutes when our batteries behind opened fire on the enemy's supposed line, to confuse, and then we would be ordered to charge. The guns were not elevated enough and were doing fine work on our position. The bursting and flying pieces of shell and rock put us in a panic condition – we could not drop to the front and protect ourselves, for we would be exposed to the enemy.[54]

The limited concentration of artillery fire on these battlefields, however, restricted the number of deaths due to friendly artillery fire – not so during the massive artillery barrages of the First World War.

Indirect fire

The greatest agent of fratricide during the First World War was indirect fire weapons. Rudimentary communication systems meant that headquarters often lost contact with their subordinate elements advancing on the battlefield. This loss of communications meant that the headquarters no longer knew the exact location of their forces when giving fire control orders to the artillery. When this factor was coupled with the sheer volume of artillery fire that characterised warfare on the Western Front, casualties from friendly artillery fire became so commonplace as to be unremarkable. Charles Shrader records in his study of fratricide that when estimating the probable casualties from an offensive during the war, 'the thorough staff planner usually included an allowance for casualties due to a friendly barrage'.[55] After the war, the French general Alexandre Percin wrote in *Le Massacre de notre Infanterie, 1914–1918* that friendly artillery fire resulted in an estimated 75,000 French casualties during the First World War.[56]

Among the many friendly fire episodes detailed in his book, Percin recounted an incident on 22 August 1914 when the 1st Regiment of Colonial Infantry encountered three regiments of German soldiers in a dense forest near Rossignol, Belgium. An accompanying French battery set up their guns several hundred yards from the forest and, although they were not given any orders to do so, fired blindly into the trees – supposedly in support of the French infantry. Several hours later when the battle had ended, of the 3,200 French soldiers who entered the forest, approximately 2,000 were dead with a further 1,000 or so wounded or taken prisoner. Percin estimated that the French artillery fire had caused a third of the French casualties, including nearly 700 deaths.

Another tragic friendly fire incident was to occur in the early hours of 24 February 1916 during the battle of Verdun. A battalion of the French 72nd Division was ordered to hold the village of Samogneux. The battalion repeatedly fought off determined German attacks but a number of panic-stricken soldiers fled to the rear, claiming that the village had fallen. A breakdown in communications strengthened the belief of the French commanders that the Germans had indeed captured the village and the inevitable counter-attack was ordered. A preparatory artillery barrage was arranged and for over two hours French shells rained down on Samogneux, in the process killing a number of Germans but also decimating the remnants of the French battalion still holding the village. In vain the defenders fired off the green 'cease fire' rockets but these were either not seen or ignored. Once the barrage finished, the Germans moved in and occupied the position and the remnants of the French battalion were taken prisoner.[57]

The British troops of the First World War also came to fear their own artillery. A typical fratricide incident befell the British Army's 66th Division during the Passchendaele campaign of mid-1917. At dawn on 9 October 1917, after a gruelling eleven-and-a-half-hour trek in darkness and rain through an almost impassable landscape pocked with shell holes filled with water and mud, the division's lead elements reached the start line for the assault. The exhausted soldiers plunged straight into the attack, only to fall victim to German machine guns and British artillery fire. Because of the slow progress of the exhausted men, the lead elements of the 66th Division arrived at the start line twenty minutes late and had fallen far behind their supporting creeping artillery barrage. Their divisional commander, aware of the slow progress of the infantry, ordered the British gunners to bring the barrage back. Tragically, the gunners did not establish the location of the forward elements of their troops before opening fire. Consequently, the artillery fire caused heavy casualties among the British soldiers as they struggled across no man's land.[58]

Indirect fire weapons were also a major cause of friendly fire casualties during the Second World War. The mobility of the armies that fought that war produced irregular frontages as units pushed forward or withdrew. This mobility made it difficult to keep track of the exact location of friendly units and complicated the delivery of accurate supporting fires. As a result, positions were often shelled by their own comrades.

Even if the precise location of friendly troops was known, human error could still result in troops falling victim to friendly indirect fire. The incorrect charge bag could be used or there might be an error in the shell trajectory calculations. Friendly fire casualties also happened because of faulty ammunition, which could result in rounds falling short. On 7 August 1944, elements of the US 124th Infantry Regiment had one man killed and three wounded by the Japanese in a battle near the Driniumor River in New Guinea. But in that same battle, eight American soldiers were killed and fourteen wounded by faulty mortar ammunition that dropped short.[59] Yet by the latter stages of the Second World War, the greater destructive power of air-delivered ordnance and the concentration in which it was dropped on the battlefield resulted in aerial bombing overtaking artillery as the greatest cause of friendly fire casualties.

Air-to-ground fratricide

Mansur Abdulin recounted in his memoir of his service with the Red Army a friendly fire attack by Soviet fighters as his regiment advanced towards the Ukraine following the battle of Kursk. The Soviet planes overflew the marching column and then wheeled about and dived towards the troops.

> Are they crazy? Now I can see only the wings and the noses of the air-craft, pointing straight at me. Unwilling to believe that our planes are about to attack us, I – just in case – plunge through the basement floor window of the nearest house. Immediately, I hear the sound of explosions in the street above my head. Earth, broken glass, plaster and bitter lime dust fall on my head. I thought that the thunder [and] smoke would never cease. And the planes, having fired all their missiles, started shooting us with their guns. They tore our regiment to pieces and then safely returned to their airfield … The street was like a trench turned into a communal grave, into which someone had dropped a mass of dead men and horses …[60]

The soldiers of the Western democracies also had cause to fear their air forces. In early 1944, fierce German resistance at Cassino was holding up the Allied advance on Rome. Two previous assaults on the main German position atop Monte Cassino had been repulsed by the defenders. On the morning of 15 March 1944, in preparation for a renewed assault by the troops of the 2nd New Zealand and 4th Indian divisions, 435 Allied bombers dropped over 1,000 tons of high-explosive bombs on the Cassino area. Regrettably, some of the bombs fell on friendly troops, resulting in 28 Allied soldiers killed and 114 wounded. A subsequent investigation determined that a malfunction in a bomb rack on one of the lead aircraft resulted in 40 bombs being dropped on friendly positions. Obscuration of the target by smoke and dust, and the resulting lack of specific aiming points, also contributed to some of the bombs going astray.[61] The fratricide at Cassino was to prove a prelude for the much higher number of friendly fire casualties arising from the close air support for the breakouts from the Normandy beachheads.

Operation Cobra was the codename given to the breakout of General Omar Bradley's US 1st Army from the Normandy beachhead near St Lô in July 1944. The breakout was to be preceded by an immense aerial carpet-bombing of the surrounding area by the planes of the Eighth and Ninth US Air Forces. The bombing would be conducted in three waves. Initially fighter-bombers would bomb the forward German positions. These would be followed by the less accurate heavy bombers, who would bomb the German defences for 60 minutes. Medium bombers would follow, and would attack the German rear areas for 30 minutes. The air effort represented the largest close air support operation yet attempted.[62]

General Bradley and the air planners reached an impasse during the planning for the breakout. The air planners wanted the American ground forces withdrawn 3,000 yards from the edge of the target area to minimise friendly fire casualties. Bradley wanted his ground troops to remain as close as possible to the German positions so they could advance upon the Germans before they had time to recover from the bombing. Bradley stated that 800 yards was the maximum he would ask his men to fall back. Eventually a compromise was reached: the ground troops would pull back 1,200 yards. But another problem had arisen.[63]

Bradley was concerned about the potential for friendly casualties due to the planes dropping their bombs short of the target and wanted the bombers to fly parallel to the St Lô to Périers road, so that they did not overfly his ground troops. The air planners disagreed. To achieve the required concentration of heavy bombers within the 60 minutes allocated, they insisted that the bombers would need to approach at right angles to the target area. This approach would minimise the effect of German counter-air fire and speed up the passage of the aircraft over the target, but it meant the bombing run-in would be over friendly positions. The risk to friendly ground troops would be increased because the bombing method used by the air force was for formations of twelve to fourteen aircraft to follow the flight path of the lead aircraft and conform to the bomb drop of the lead bombardier. If the lead bombardier of a formation dropped his bombs short, invariably the remainder of the formation would bomb short as well. In the end, the air force got its way – the bombers would approach perpendicular to the target area.[64]

After several postponements due to unsuitable weather, the attack was scheduled for the afternoon of 24 July. However, heavy cloud over the target area resulted in the bombers being recalled mid-flight. Many of the aircrews did not receive the abort message and most of the fighter-bombers and several hundred of the heavy and medium bombers continued on to the target. Because of poor visibility over the target area, coupled with aircrew errors, a large number of bombs were dropped short, killing 25 American soldiers and wounding 131, mainly from the US 30th Division. The attack was rescheduled for the following day.[65]

Clearer weather enabled the full aerial attack to go in on the 25th, yet inaccurate bombing again resulted in US casualties. Visibility over the target area was reduced by a huge column of dust and smoke from the damage caused by the preceding fighter-bomber attack, blown into the flight path of the heavy bombers by a wind coming from the south. Reference points were obscured, so the bombardiers assumed the dust clouds from the previous attack marked the target.[66] Tragically for the friendly troops on the ground, the wind was blowing the dust clouds towards the American positions. The war correspondent Ernie Pyle, located in the forward area with the American assault troops, provided a graphic account of the gripping fear felt by the ground troops, helpless to do anything about the approaching cataclysm.

As we watched, there crept into our consciousness a realisation that the windrows of exploding bombs were easing back towards us, flight by flight, instead of gradually easing forward, as the plan called for. Then we were horrified by the suspicion that those machines, high in the sky and completely detached from us, were aiming their bombs at the smoke line on the ground – and a gentle breeze was driving the smoke line back over us! An indescribable kind of panic comes over us. We stood tensed in muscle and frozen in intellect, watching each flight approach and pass over us, feeling trapped and completely helpless.

And then all of an instant the universe became filled with a gigantic rat-tling as of huge ripe seeds in a mammoth dry gourd. I doubt that any of us had ever heard that sound before, but instinct told us what it was. It was bombs by the hundred, hurtling down through the air above us.[67]

In addition to the misidentification of the target area, other aircrew errors also endangered the friendly troops on the ground. In one instance the lead bombardier of a flight failed to synchronise his bombsight and so dropped his bombs short, with the other eleven bombers in his flight following suit. Because of this act of carelessness, twelve B-24s dropped almost 500 high-explosive bombs on friendly troops.[68]

The short bombing resulted in the American 9th and 30th Divisions suffering 111 killed and 490 wounded.[69] Numbered among the friendly casualties was Lieutenant General Lesley McNair, who was an official observer for the Normandy campaign. McNair, who had previously been responsible for the expansion and training of the stateside US Army, had joined a battalion taking part in the attack to view the results of this training.[70] He was the most senior US officer to be killed in action in the European theatre during the Second World War.

A few weeks after Operation Cobra, on 8 August during Operation Totalise, it was the turn of the Canadian and Polish troops to be bombed by Allied aircraft. Sixty-five soldiers of the 1st Polish Armoured Division were killed and 250 wounded by the Allied air attack, while the Canadian North Shore (New Brunswick) Regiment suffered around 100 casualties.[71] A week or so later, on 14 August during Operation Tractable, the follow-up operation to Totalise, the Canadians again fell victim to inaccurate Allied bombing. On this occasion the Royal Air Force's Bomber Command attempted to provide close air support to a ground attack by Canadian and Polish troops on German positions near Falaise, resulting in 150 deaths due to friendly fire.[72]

The ground troops that fought the Korean War also had cause to fear the approach of friendly aircraft. On the morning of 23 September 1950, two companies from the 1st Battalion of the Argyll and Sutherland Highlanders captured Hill 282 from the North Koreans. The North Koreans, however, held the adjacent Hill 388, which overlooked the Argylls' position. The North Koreans began to shell the British positions and mounted a fierce counter-attack. The British called for UN artillery support but none was available. However, the Argylls were able to identify the forming-up points from which the enemy were launching their attacks and requested a UN air strike. The British troops ensured that their own positions were clearly marked from the air by laying out crimson and gold air recognition panels. Shortly after noon, three US Air Force (USAF) P-51 Mustangs responded to the request for air support. The pilots ignored the air recognition panels and, after circling Hill 282 three times, began their attack run, strafing with 20mm cannons and then dropping napalm onto the helpless British soldiers. The air strike killed seventeen British soldiers and wounded a further 76, most of whom received extensive burns.[73]

Fifteen years later, in November 1965, Lieutenant Colonel Hal Moore was facing a similar situation to the Argylls, with soldiers of the North Vietnamese Army (NVA) threatening to overrun his battalion during the battle of Ia Drang in the Vietnam War. Moore and his troops also experienced the inherent danger of close air support when a USAF F-100 Super Sabre jet fighter released two napalm canisters over the battalion command post. 'When the flames died down we all ran out into the burning grass,' recounted Joe Galloway, a reporter covering the battle. 'Somebody yelled at me to grab the feet of one of the charred soldiers. When I got them, the boots crumbled and the flesh came off and I could feel the bare bones of his ankles in the palms of my hands. We carried him to the aid station. I can still hear their screams.'[74] The burnt soldier died two days later.

Helicopter gunships accounted for a number of friendly fire casualties during the Vietnam War. David Hackworth wrote in his memoir, *About Face*: 'Unfortunately, the feeling of power most chopper drivers felt could work both for the infantry and against us: overenthusiastic pilots sometimes got a little carried away, firing first and only then ascertaining whether the target was friend or foe.'[75] Hackworth backed up this statement by describing a fratricide incident during Operation Hawthorne in June 1966, when helicopter gunships from the 1st Cavalry Division, responding to a call for fire support, strafed American troops rather than the NVA, resulting in 21 American soldiers wounded and one killed.[76]

The US Coast Guard also fell victim to American pilots during the Vietnam War. On the night of 11 August 1966, the Coast Guard cutter *Point Welcome*, which was patrolling near the DMZ between North and South Vietnam, was bombed and strafed by four USAF F-4 Phantoms and a B-57 bomber. Two men were killed and eleven seriously wounded in the attack (total losses for the US Coast Guard during the war were seven men killed and 59 wounded).[77]

The deadliest air fratricide incident of the Vietnam War was to occur a year later. On 19 November 1967, the 2nd Battalion of the American 503rd Regiment, 173rd Airborne Brigade assaulted Hill 875 near Dak To, South Vietnam. The NVA had turned Hill 875 into a formidable fortress, complete with deep bunkers connected by tunnels. The attacking American paratroopers were subjected to withering rifle and machine gun fire from the NVA positions. Furthermore, the NVA counter-attacked towards the battalion's reserve company, the remnants of which were driven up the hill towards the two assaulting companies. The besieged Americans formed a defensive perimeter and attempted to hold off the attacking NVA with the assistance of indirect fire and close air support. Tragically, around 8pm, a USAF jet dropped a 500-pound bomb inside the US perimeter, killing 42 American soldiers and wounding 45. This single friendly fire incident caused almost half of the total fatalities suffered by the 2nd Battalion during the battle of Dak To.[78]

Although these incidents were horrific, deaths attributed to friendly fire represented only a small number of the 58,000 Americans killed in the Vietnam War. Fratricide would briefly flare up as an issue before being lost in the general clutter of tragic stories flowing out of the war. The proportionally high number of fratricide casualties during the 1991 Gulf War, however, would thrust the issue of friendly fire into the public spotlight.

The Gulf War, 1991: 'We can shoot farther than we can see'

The Pentagon officially acknowledged 28 incidents of American-on-American fratricide during the 1991 Gulf War, most of which were attributed to human error. At least 35 of the 148 American deaths in combat during the war were due to friendly fire. Additionally, the injuries of 72 out of 467 Americans wounded in combat were ascribed to friendly fire.[79] The American forces suffered their first fratricide casualties two weeks before the start of the ground offensive.

The evening of 29 January 1991 saw the first engagement between the US and Iraqi forces. Tragically, the opening salvo of the Americans would kill their own comrades rather than the enemy, when a Marine Light Armoured Vehicle (LAV) of Task Force Shepherd's Delta Company was struck by a TOW (anti-tank) missile fired by one of its fellow vehicles. The missile sliced through the rear troop hatch of the vehicle and detonated the fourteen TOW missiles stacked in its storage rack. The LAV exploded in a massive fireball, killing instantly the four Marines inside. Regrettably, these were not to be the only deaths from friendly fire that Delta Company suffered that night.[80]

That evening, as an Iraqi force was advancing into Saudi Arabia heading for the town of Khafji, Delta Company was fighting a desperate rearguard action, with close air support provided by the A-10s (anti-tank aircraft) of the USAF 355th Tactical Fighter Squadron. One of the A-10s acquired an Iraqi tank with its targeting system and released a Maverick anti-tank missile. But instead of locking automatically onto the tank, the missile malfunctioned and dived straight down, striking one of Delta Company's LAVs. The warhead punched through the thin armour of the LAV's roof and detonated inside the troop bay. Miraculously, the force of the explosion flung the driver clear of the vehicle and although badly burnt, he survived. Seven of his comrades were not so fortunate. Delta Company had been in combat for four hours and eleven Marines had been killed – all by friendly fire.[81]

Despite the high percentage of American casualties due to friendly fire, the odds of being killed by the Americans were even greater if you were a member of the British contingent sent to the Gulf. Over a third of the British fatalities during the ground combat phase resulted from friendly fire (nine of 24). All nine fatalities came from a single fratricide incident on 26 February 1991, when two USAF A-10 Thunderbolts shot up a column of 37 Warrior Infantry Fighting Vehicles from Charlie Company, 3rd Battalion of the Royal Regiment of Fusiliers Battle Group. In addition to the nine deaths, eleven British soldiers were seriously wounded in the American attack.[82]

The A-10 pilots had detected what they thought were about 50 Iraqi T54/55 tanks and support vehicles moving north. The lead aircraft made two passes, at 15,000 and 10,000 feet, and observed the vehicles through binoculars. As the pilot did not identify any friendly vehicle markings, the A-10s began their attack. Both aircraft fired a Maverick missile, each of which destroyed a vehicle.[83] The misidentification occurred despite the fact that the Iraqi tanks have a long gun barrel (100mm calibre) that protrudes well past the main body of the vehicle, while the Warrior's main armament is a 30mm cannon that doesn't even reach the front edge of the vehicle. Additionally, the Warrior is two-thirds the size of a T54/55 tank. At the time, there was no Iraqi air threat and armoured vehicles posed little direct danger to the A-10s.

The British 4th Brigade, to which the 3rd Battalion of the Royal Regiment of Fusiliers belonged, was in contact with the enemy for 54 of the 100 hours of ground combat in the 1991 Gulf War. The brigade advanced 350 kilometres and destroyed 60 Iraqi tanks, 90 Iraqi armoured personnel carriers and 37 Iraqi artillery pieces. The brigade suffered no losses to enemy action. All of its fatalities arose from the USAF attack on the Warriors.[84]

A subsequent British Board of Inquiry determined that no blame or responsibility for the incident could be attributed to the Fusiliers or the Assistant Divisional Air Liaison Officer, who requested the attack. The A-10 pilots stated that they had attacked on the basis of the information passed to them by a previous flight and because of their positive identification of the target as enemy vehicles. The board concluded that the Warriors were displaying the correct inverted 'V' recognition symbol and the required fluorescent panels but that open hatches or equipment carried outside the vehicles could have obscured some of these panels. Furthermore, the board noted that although a reconnaissance flight had observed the panels at 6,000 feet, this was below the operating height of the A-10s. The board did not firmly establish whether the A-10 pilots were at fault. However, its report concluded:

> On the basis of the evidence before it, the Board was unable to establish why the attacked Warrior vehicles were misidentified by the A-10 pilots as enemy T54/55 tanks, particularly in view of the previous identification runs at 8,000 and 15,000 feet … The Board remarked that it was clear all UK and USAF personnel involved were striving to achieve their individual tasks to the best of their abilities in a fast-moving battle. The Board thought it inevitable that, at some stage, difficulties may arise when individuals are under such pressure.[85]

The Coalition ground forces were provided with an unprecedented level of close air support (CAS) during the 1991 Gulf War. An analysis of all CAS missions conducted during the war credits CAS aircraft (generally A-10s and F-16s) with the destruction of 3,500 Iraqi tanks, 2,400 armoured vehicles (other than tanks) and 2,600 artillery pieces. During the brief ground combat phase, approximately 4,500 CAS sorties were flown, only a few of which resulted in fratricide.[86]

The military power of the Coalition, particularly air power, overwhelmed the Iraqi forces and brought about their rapid collapse. The lack of sustained enemy resistance, and consequently the short duration of the ground conflict phase, helped minimise friendly casualties. Studies published by the Pentagon contend that the brief period of ground combat and the low number of casualties to enemy fire meant that the proportion of fratricide casualties was much higher in the 1991 Gulf War than in previous wars; the Pentagon's supposition being that if the ground combat phase had lasted longer, the experience level of the troops would have increased and therefore the fear and inexperience that leads to fratricide would have correspondingly decreased.[87]

Troop inexperience was undoubtedly a contributing factor for a number of the fratricide incidents that occurred during the 1991 Gulf War. But the principal reason for the high number of friendly fire deaths in this conflict was the nature of the combat. The speed and manoeuvrability of modern armoured forces results in a much more fragmented battlefield than the linear battle lines of the past. Consequently, Coalition units would unexpectedly come upon other Coalition units, or they would become intermingled with Iraqi forces. When this factor is combined with decreased visibility because of dust, darkness or smoke, and a tendency to shoot first and verify identity later, it is little wonder that there were fratricide incidents.

Another factor in the high number of such incidents during the 1991 Gulf War was that Coalition armoured vehicles and attack aircraft were able to engage at ranges that often precluded precise identification of friend or foe. As one US Army report into the fratricide incidents put it, the basic problem was that: 'We can shoot farther than we can see.'[88] For example, Abrams tanks using thermal sights are able to engage targets up to three kilometres away. At such a range, while state-of-the-art sights can identify general targets, they often lack the resolution to determine vehicle type. Indeed, positive identification beyond 700 metres is difficult using thermal sights.[89] Additionally, standoff weaponry, when coupled with the fear that pervades the battlefield, can bring about a tendency to shoot before positive identification has been made – but, more importantly, before the enemy can fire at you. The Pentagon's final report to Congress on the 1991 Gulf War noted that tank crews were routinely held to a standard of achieving a round on target in less than ten seconds after the target was detected, with a well-trained crew having the initial round downrange in about six seconds. The report commented that the history of tank-on-tank combat demonstrates that 'when opponents have equally sophisticated fire control and equally lethal munitions, success usually belongs to the crew that fire first.'[90] In such an environment, with extreme pressure to fire quickly, ambiguous targets are likely to be assumed to be hostile until proved friendly. A report to Congress on the fratricide incidents noted that target misidentification was responsible for more of these incidents than any other factor in the war.[91]

Tragic errors of target misidentification are not limited to wartime. On 14 April 1994, two USAF F-15s shot down two US Army Black Hawk transport helicopters, killing the 26 passengers and crew. The soldiers killed comprised the command group of Operation Provide Comfort, tasked with providing humanitarian aid to over 1 million Kurds who had fled their homes in northern Iraq. The UN had designated all areas north of the 36th parallel in Iraq as a 'No Fly Zone' for Iraqi aircraft to shelter the Kurds from further air attacks. This area was patrolled by mainly British and American planes as part of Operation Northern Watch. The USAF principally employed Airborne Warning and Control Systems (AWACS) aircraft and F-15 fighter aircraft to perform this mission.

The US Secretary of Defense, William Perry, convened an Aircraft Accident Investigation Board to determine what caused this tragedy. The board found that four key errors, omissions and failures caused the incident: the F-15 pilots misidentified the Black Hawks; the AWACS crew failed to intervene; the Black Hawks' flight plan had not been sufficiently coordinated with the Operation Northern Watch taskforce; and the Identification Friend or Foe system failed.[92] Again, as noted for the attack on the British Warriors, the destroyed vehicles (in this case the helicopters) presented little, if any, threat to the F-15s.

Friendly fire in the Global War on Terror

Notwithstanding the lessons supposedly learned during the 1991 Gulf War, and the subsequent funding directed towards developing technological solutions to minimise casualties from friendly fire, fratricide incidents beset the forces engaged in the Global War on Terror.

Just before 2am on 18 April 2002, four Canadian soldiers were killed and eight wounded when a USAF F-16 dropped a 500-pound bomb on a night-time live-fire military exercise. The incident occurred at a firing range about nine miles south of the Kandahar airfield in south-eastern

Afghanistan. The soldiers belonged to the paratroop company of the 3rd Battalion of the Princess Patricia's Canadian Light Infantry Battle Group.

The Canadian soldiers were killed in a designated training area which, for reasons of safety, had restrictions in place prohibiting military aircraft from dropping below 10,000 feet when passing over it. Additionally, the Canadian HQ had informed the Americans that a company of Canadian soldiers would be conducting an overnight live-firing exercise in this area. The two F-16s were returning from an unrelated mission when they observed what they assumed to be enemy fire near Kandahar. What the pilots had actually seen was the Canadian soldiers firing small arms and anti-tank weapons at inert targets. The lead pilot reported seeing what looked like fireworks and assumed it was ground-to-air fire (in actuality probably ricocheted tracer rounds) directed at his aircraft (flying at 23,000 feet, the danger of the planes being hit by small-arms fire was very low). He requested and received permission from the sector's AWACS aircraft to determine the exact coordinates of the fire and both planes wheeled around. The second F-16 pilot (wingman) requested permission to fire his 20mm cannon at the target but was instructed to 'hold fire' while the target was verified. The lead pilot then said: 'Let's just make sure that it's, uh, that it's not friendlies is all.' Under the USAF rules of engagement, pilots who believe they are in danger are authorised to fire back. On a second pass over the area, the wingman saw what looked to him like ground fire being directed at the lead jet. He reported: 'I've got some men on a road and it looks like a piece of artillery firing at us. I am rolling in self-defense.'

The wingman could have taken evasive action by leaving the area when he saw the firing below, but instead he invoked his right of self-defence. He dropped his plane down to 10,000 feet and, without receiving permission to do so, dropped a 500-pound bomb. Less than ten seconds later, the Combined Air Operations Center advised the controlling AWACS aircraft that they were friendly troops being targeted. The F-16 pilots stated that they thought a Multiple Launch Rocket System (MLRS) was firing at them – the Canadians did not use any artillery pieces during the live-fire exercise. The Canadians had no knowledge of the impending danger until the moment the bomb struck.[93]

The deaths triggered a national outpouring of grief and anger in Canada, particularly because these were the first Canadian deaths in combat since the Korean War some 50 years earlier. Separate Canadian and American boards of inquiry were convened to investigate the incident. In a statement to the press, Canada's Chief of Defence Staff, General Ray Henault, commented: 'My understanding is that there was no hostile activity in the area that would have created this incident. How this sort of thing could happen is a mystery to us.'[94] The inquiries determined that the two American F-16 pilots were solely at fault for the death of the Canadians and stated that the pilots had violated the rules of engagement and engaged in an inappropriate use of lethal force.[95]

The death of the four Canadians on 18 April 2002 was not the first fratricide incident to involve the USAF during the Afghan campaign, nor was it the deadliest. The first such incident occurred on 26 November 2001, when five US Special Forces soldiers were wounded and six allied Afghan fighters killed by an F/A-18 air strike near Mazar-e-Sharif. The aircraft responded to a request for close air support from a Forward Air Controller on the ground and dropped a precision-guided bomb that exploded near the friendly troops. The Pentagon concluded that the incident was the result of 'procedural errors in the transmission and application of friendly and enemy coordinates'.[96]

The deadliest fratricide incident occurred on 5 December 2001, when a USAF B-52 bomber dropped a 2,000-pound precision-guided bomb about 100 metres from a position manned by members of the 3rd Battalion, US 5th Special Forces Group. At the time, the US soldiers were directing air strikes against Taliban forces to the north of Kandahar. Three Special Forces soldiers were killed and nineteen wounded by the blast. The US soldiers were fighting alongside members of the Afghan opposition forces, five of whom were killed and eighteen wounded, including the Afghan Interim Authority leader, Hamid Karzai (elected President of Afghanistan in October 2004), who suffered minor facial lacerations. The bomb had a satellite-based guidance system; therefore US defence officials theorised that either the weapon system malfunctioned or the wrong coordinates were sent by the ground force or entered by the B-52 crew.[97] The US

Secretary of Defense, Donald Rumsfeld, in a comment that probably did not provide much solace to the relatives of the dead and wounded, rationalised that: 'A very smart weapon, a good weapon might work 85 to 90 per cent of the time. The rest of the time it doesn't work right. Now that's a very good percentage. But it means that there is one out of 10 that is not going to do what it was intended to do.'[98]

There were a number of similarities between the 1991 Gulf War and 2003 Iraq War. The enemy had not changed, nor had the major partners in the coalition. Tragically, another similarity was that during the combat phase several British solders were killed by the Americans – their allies. Mid-afternoon on 28 March 2003, a USAF A-10 aircraft (part of a two-aircraft patrol) attacked two Scimitar armoured reconnaissance vehicles belonging to Delta Squadron of the Blues and Royals, Household Cavalry Regiment. The Scimitars were part of a five-vehicle patrol conducting a recce along the Shatt al-Arab river, about 25 miles to the north of the Iraqi city of Basra. The patrol had paused when, without warning, the two lead vehicles were attacked by an A-10. Both vehicles were hit by 30mm rounds and caught fire. One British soldier was killed and four wounded in the attack. A sixth soldier escaped the burning vehicles without injury.[99]

The attack generated an irate public response. A number of reports in the British press stated that some of the wounded soldiers were calling for the A-10 pilot to be prosecuted for manslaughter. The commander of one of the Scimitars, Lance Corporal Steven Gerrard, speaking from his hospital bed, angrily remarked:

I can command my vehicle. I can keep it from being attacked. What I have not been trained to do is look over my shoulder to see whether an American is shooting at me … He [the pilot] had absolutely no regard for human life. I believe he was a cowboy. There were four or five that I noticed earlier and this one had broken off and was on his own when he attacked us. He'd just gone out on a jolly.[100]

The A-10 was flying at low altitude and should have been able to differentiate between Iraqi and Coalition armoured vehicles, Gerrard noting that the aircraft was only about 50 metres off the ground when the pilot began a second attack run. One of the pilots misidentified the British vehicles as Zil 157s – Russian-made trucks used by the Iraqi Army, which, incidentally, are wheeled vehicles while the Scimitars are tracked. The Scimitars were all marked with an inverted 'V' (the Coalition symbol for a friendly vehicle) and had the standard NATO fluorescent orange signal panels tied across their turrets to assist identification from the air, and flying from the back of one of the vehicles was a Union Jack flag measuring eighteen inches by twelve inches. Additionally, one of the wounded soldiers exited his vehicle and waved frantically at the A-10 pilot in an attempt to halt the second attack run. Significantly, the disabled vehicles did not pose any threat to the A-10.[101] Trooper Christopher Finney, who ironically the day before had received a letter from his father that cautioned: 'Be careful, come home soon and watch out for those damn Yanks', was awarded the George Cross for rescuing under fire some of his comrades who had been wounded in the A-10 attack. Finney commented: 'All the wagons have markings to say they are Coalition. I don't know why he shot a second time, he was that close. To be honest, I think they are just ignorant. I don't know if they haven't been trained or are just trigger happy.'[102]

No disciplinary action was taken against the two pilots involved in the attack, an official US military investigation concluding that their actions were not reckless and that they reasonably believed they were engaging the enemy.[103] A transcript of the cockpit video for the attack revealed that the pilots were repeatedly informed by a US Marine Corps Forward Air Controller attached to a British unit on the ground that there were no friendly vehicles in the target area. Although the pilot of the attacking A-10 initially identified the orange panels, when informed that there were no friendly vehicles in the area he convinced himself the panels were in fact rocket launchers.[104]

The attack on the Scimitars was unfortunately not the only fratricide incident the British forces suffered during the 2003 Iraq War. Five days earlier, on 23 March, a US Patriot missile battery near the Kuwaiti border shot down a British Tornado aircraft from 9 Squadron, Royal Air Force, killing the two crew. The Tornado had completed its mission and was returning to base when the Patriot battery targeted it. A US Army spokesperson later stated that a software error in the Patriot system had caused it to misidentify the Tornado as an incoming missile. Additionally, a British soldier from the 2nd Royal Tank Regiment was accidentally shot dead by a comrade on 24 March while trying to quell a riot in Al Zubayr (near Basra) and two British soldiers from the Queen's Royal Lancers were killed on 25 March when their Challenger II tank was engaged by another British Challenger to the west of Basra. Thirty-three British soldiers died between the start of the invasion on 20 March and the end of major combat operations on 1 May 2003 – six (18 per cent) were victims of friendly fire.

The US forces in Iraq also suffered casualties from friendly fire. The deadliest fratricide incident involving US troops occurred at Nasiriyah on 23 March 2003, when USAF A-10 aircraft fired on Charlie Company of the 1st Battalion, 2nd Marine Regiment. The 1st Battalion had been tasked with securing two bridges on Highway 8, which linked the southern Iraqi cities with Baghdad. One bridge rose over the Euphrates river on the south-eastern edge of Nasiriyah, the other over the Saddam Canal, located 4.5 kilometres to the north of the first bridge. As the Marines were involved in heavy fighting against Iraqi irregulars at the time of the A-10 attack, US military investigators found it difficult to determine who was killed by enemy fire and who by friendly fire. Of the eighteen Marines killed in this action, potentially up to ten were killed by the A-10, nine of whom died when a Maverick missile slammed into the rear of a Marine amphibious assault vehicle. Of the seventeen Marines wounded, thirteen were verified as being wounded solely by enemy fire, the others by a combination of enemy and friendly fire. A subsequent military investigation cleared the pilots of any blame, attributing the incident to a loss of situational awareness because of 'deviations from the planned scheme of manoeuvre, the urban environment, and problematic communication links'.[105] Some of the surviving Marines later expressed their frustration that the USAF A-10 pilots did not recognise the distinctive Marine armoured assault vehicles, noting that the Marine Corps has its own air arm that provides close air support for Marine ground forces. One Marine commented: 'This was my second time being strafed by an A-10. First Gulf War I was strafed. If I can't work with Marine air, I don't want to work with anything.'[106]

The attack on the Marines at Nasiriyah was, tragically, not the deadliest fratricide incident to occur during the 2003 Iraq War. Around midday on 6 April 2003, a US Navy F-14 aircraft attacked a vehicle convoy of Kurdish Peshmerga guerrillas that was accompanied by US Special Forces soldiers. The incident occurred about 30 miles to the south-east of the northern Iraqi city of Mosul. The Kurdish convoy was heading towards a town recently captured by Kurdish forces and had joined up with a group of US Special Forces, who were also mounted in vehicles. The Iraqi Army had recently abandoned the area, leaving behind tanks, armoured personnel carriers and artillery pieces. One of the US Special Forces soldiers spotted an Iraqi tank about a mile away and called for close air support. Two F-14s responded, one of which dropped a bomb directly onto the convoy. Most of the Kurdish and US soldiers were outside their vehicles when the bomb hit. Seventeen Kurdish fighters and a Kurdish translator working for the BBC were killed in the explosion. Forty-five others were wounded, including three US Special Forces soldiers and Wageeh Barzani, commander of the Kurdish Peshmerga guerrillas and brother of the leader of the Kurdistan Democratic Party.[107] Travelling with the Kurds was a team from the BBC, one of whom reported:

> It was an American plane that dropped the bomb right beside us. I saw it land about 10, 12 feet away, I think. This is just a scene from hell here. All the vehicles are on fire. There are bodies burning around me, there are bodies lying around, there are bits of bodies on the ground. This is a really bad own goal by the Americans.[108]

24 per cent casualties?

For all modern armies, well aware of the horrific incidents described in this chapter, reducing the number of fratricide casualties is an abiding goal. But a technical panacea remains elusive. Developments in the areas of enhanced optics and more sophisticated Identification Friend or Foe devices, while promising, will probably always lag behind advances in weapon ranges and standoff capability. Another complicating factor is the growing preference for coalition operations. The various national elements of a coalition may have different equipment, and some elements may even have similar equipment to that used by the enemy, thereby complicating the process of battlefield identification. Realistic training should help reduce the number of friendly fire casualties, but the inescapable fact is that the 'fog of war' will always remain and consequently fratricide incidents will continue to occur. The key question is how many casualties should be expected?

The oft-quoted figure of friendly fire accounting for less than 2 per cent of all battlefield casualties is based on a study completed in 1982 by Lieutenant Colonel Charles R. Shrader of the US Army. However, he admitted that the figure of 2 per cent was his best guess, acknowledging the paucity of accurate data on historical friendly fire casualties. Furthermore, he concluded:

> [C]ommanders at various levels may be reluctant to report instances of casualties due to friendly fire either because they are afraid of damaging unit or personal reputations, because they have a misplaced concern for the morale of surviving troops or the benefits and honours due the dead and wounded, or simply because of a desire to avoid unprofitable conflicts with the personnel of supporting or adjacent units.[109]

The difficulty in apportioning deaths to friendly or enemy fire is illustrated by the 23 March 2003 incident at Nasiriyah, Iraq – and this incident was investigated using modern forensic techniques! It therefore stands to reason that the number of casualties arising from friendly fire may indeed be greater than 2 per cent. Colonel David Hackworth, who conducted an internal Pentagon study into fratricide during the Vietnam War, concluded that 15 to 20 per cent of all US casualties in Vietnam were caused by friendly fire.[110] Furthermore, a study conducted by the US Congressional Office of Technology Assessment in 1993, which drew upon casualty surveys from the Second World War and the Vietnam War, suggested that these conflicts had fratricide rates of 15 to 20 per cent.[111] So the 24 per cent figure for friendly fire victims during the 1991 Gulf War may be less of an anomaly and closer to the norm than most modern militaries would care to acknowledge.

What is surprising in modern warfare is not that so many soldiers are killed by friendly fire, but that so few are. For example, during the 2003 invasion of Iraq, elements of the 1st Reconnaissance Battalion of the US Marine Corps had an artillery round fall short and pepper their vehicles with shrapnel, they were engaged with small-arms fire from another task-force of Marines, and they had their convoy shot up at night by a Navy reserve medical unit. Furthermore, one of their foot-mounted patrols was engaged with a .50-cal machine gun by one of their fellow companies from within the battalion, and a USAF F-15 aircraft dropped a 500-pound bomb a kilometre short of its intended target and within 200 metres of the Marines. Remarkably, no casualties were caused by these friendly fire incidents, which in itself is extremely fortunate given the massive killing power of the modern American military.[112]

✻ ✻ ✻

There is probably no greater tragedy for a military force than to suffer casualties from friendly fire, because in most cases these are avoidable casualties. Some of the fratricide incidents discussed in this chapter could rightly be attributed to the so-called 'fog of war', but certainly not all of them. In certain cases the firer was in direct danger, where a split-second reaction could mean the difference between life or death. But even in these instances, invariably some human error had been made, such as inadequate coordination measures, or someone had simply got lost. A different standard of responsibility, however, should be applied to those fratricide incidents where the firer was in little or no personal danger, where a suspected desire to achieve a 'kill' overrode the need to first determine if they were engaging a legitimate target. This attitude of shooting first and verifying the target later will probably remain a feature of war, and the tragedy of this is that the enemy will not be the only threat a soldier must face on the battlefield. He must also fear his comrades.

Chapter 9

THE MILITARY VERSUS THE MEDIA

At the heart of the often turbulent relationship between the military and the media is the dichotomy between the military's need for operational security and the media's right to inform the public of the nature and purpose of the military operations conducted by the nation's armed forces. But enabling media access to the combat zone increases the burden of the soldier in two areas. When the media reveals operational details, it increases the personal risk faced by the soldier in combat. Furthermore, a ubiquitous media presence makes the soldier's job more difficult, as he needs to consider whether his actions will be recorded, taken out of context, and served up by the media to be judged in the court of public opinion. But the public, recognising the burdens that a media presence imposes on soldiers, is, not surprisingly, generally more supportive of military censorship of the press than journalists themselves. Phillip Knightley notes in his history of war journalism, *The First Casualty*, that the public knew that the news from the 1991 Gulf War was being censored, and almost 80 per cent of poll respondents thought this was a good idea – indeed, nearly 60 per cent of respondents thought the authorities should exert more control over the war coverage.[1]

The military and the media have a symbiotic (though some may say parasitic) relationship. During the First World War, one of Lord Northcliffe's editors was asked what sells newspapers. He replied: 'The first answer is "War". War not only creates a supply of news but a demand for it. So deep rooted is the fascination in a war and all things appertaining to it that ... a paper only has to be able to put up on its placard "A Great Battle" for sales to go up.'[2] Further evidence of the profitability of war to the media industry is provided by Trevor Royle, who notes that in May 1982, during the height of the Falklands War, sales of the *Guardian* went up by 50,000 copies a day and the ITN *News at Ten* attracted an audience 20 per cent higher than average.[3] Yet the military also needs the media.

The media has a key role to play in keeping the public informed and ensuring that vital public support for staying the course during a conflict is maintained. Soldiers also need the media to tell their stories, for as Max Hastings noted when he accompanied the British taskforce to the Falklands in 1982: 'when men are risking their lives for a cause, they passionately want to know that what they are doing is being noticed – and that means written about and broadcast to the world.'[4] Dan Mills recounted in *Sniper One*, his memoir of his deployment to Iraq in 2004, that the initial absence of press coverage of the efforts of the British soldiers at Al Amarah was particularly galling. 'I've been through every single fucking one of these papers, and there's not a fucking word about the PWRR [Princess of Wales's Royal Regiment] in any of them,' complained one of his soldiers. 'Lads like H needed to know they were getting respect back home for being in that shit hole. They believed in Queen and Country and they wanted to go home heroes, so everyone in their local pub would want to buy them a drink,' concluded Mills.[5] This sentiment was also encountered by Peter Arnett, who quoted in his memoirs a letter sent to him by some American soldiers following the publication of an article he had written profiling their unit, which was stationed near the border with Laos during the Vietnam War and was besieged by the North Vietnamese Army. The letter stated: 'You wouldn't believe how high our morale went, to know that the people back home knew the hell we were going through, to know that we were no longer the forgotten soldiers.'[6]

One of the first official war correspondents was William Howard Russell, a 'special correspondent' for *The Times* (London) during the Crimean War. Previously the London newspapers had largely relied on letters sent to them by junior officers for their coverage of campaigns. Russell's initial dispatches praised the efforts of the British, French and Turkish troops in their victory over the Russians at the Battle of the Alma. But as the number of soldiers dying from disease and wounds grew, and the expected breakthrough failed to materialise, Russell's reports became much more critical of the conduct of the war, in particular the tactical ineptitude of the senior British officers and their lack of concern for the well-being of the men. Such reporting, although it helped bring about much-needed change to the British Army, earned Russell the condemnation of the authorities (both civil and military) and set a precedent for the ongoing acrimony which, although varying in intensity, characterises the relationship between the military and the media to this day.

The major London papers sent a total of three correspondents to the Crimean War. Less than a decade later, at least 500 correspondents and war artists covered the American Civil War for the Union (the Confederacy mainly relied on serving officers to provide its newspapers with updates on battles and campaigns). At the peak of the Boer War there were almost 300 correspondents providing accounts for British, European and American newspapers. Ninety correspondents were accredited to accompany the American Expeditionary Force of the First World War and 558 journalists covered the Normandy landings during the Second World War. The largest number of accredited reporters during the Korean War was 270, with a peak in accredited journalists during the Vietnam War of 637.[7] The exponential growth in the number of news organisations, including the emergence of 24-hour news networks such as CNN, resulted in an estimated 1,600 journalists and support personnel being based in Saudi Arabia during the 1991 Gulf War, with hundreds more covering the conflict from neighbouring countries – some even reporting directly from Baghdad. Of this multitude, 192 were permitted by the Americans to accompany combat units. When NATO troops entered Kosovo in June 1999 they were accompanied by an estimated 2,750 journalists and support staff. The war correspondent Martin Bell, who was part of this throng, noted that 'Satellite vans clogged the highway, and patrols of the Parachute Regiment in Pristina would go out with more photographers than soldiers.'[8] During Operation Iraqi

Freedom in early 2003, over 600 reporters and photographers, from 220 media organisations, were embedded by the US Department of Defense with its Army, Navy, Air Force and Marine Corps units. Eighty-three of those journalists and photographers were embedded into a single formation: the 3rd Infantry Division.

The growth of the media presence on the battlefield challenged the military's efforts to control the flow of information. The military responded with varying degrees of censorship (the infamous blue pencil). During the First and Second World Wars, the sense of a common purpose resulted in a largely compliant media acting as a propaganda arm of the government. Details of military disasters were heavily censored because the government felt that broadcasting the full extent of such events would adversely affect the all-important 'home front' morale. For example, neither the number of casualties incurred nor the number of ships sunk in the Japanese attack on Pearl Harbor was released to the press. Official communiqués initially stated that only one 'old battleship' had been sunk, some other ships were damaged and heavy casualties had been inflicted on the Japanese. In actuality, five battleships had been sunk and three damaged, three cruisers and three destroyers had sustained extensive damage, 200 planes had been destroyed and 2,344 men were killed. The Japanese had lost only 29 planes.[9] Furthermore, it was not until January 1943, over a year after the event, that newsreel footage of the attack was finally shown to the public.

Vietnam: the free press war

The nadir of the relationship between the military and the media was reached during the Vietnam War. Vietnam was the first American conflict in which there was virtually no censorship of the press, though reporters agreed to abide by a set of fifteen ground rules which were principally concerned with maintaining operational security. The American military commander in Vietnam from 1964 to 1968, General William C. Westmoreland, had initially considered some form of censorship of the media but rejected the idea as it would have been unenforceable because of the large number of foreign correspondents in-country, along with the difficulties of coordinating such a measure with the Vietnamese. 'The situation was ripe for dispute and misunderstanding and resentment,' wrote Peter Arnett, 'creating a climate of discord that would permanently tarnish our relations with the military and designate the press a handy whipping boy for the disasters that were to come.'[10]

Once reporters had received their accreditation through the American Military Assistance Command, Vietnam and their corresponding Bai Chi credential from the Republic of South Vietnam which admitted them to the Vietnamese Press Corps, they were given unrestricted access to operational areas. Joe Galloway recalled: 'Vietnam was the most free press exercise in the history of this country. You had that press card, you agreed to a simple list of rules. The press card would take you anywhere. You could go anywhere and stay as long as you wanted to as long as you had the [gumption] to go.'[11] Reporters were given virtual free rein to jump on and off military transport. If there was space on a US military helicopter they could even be flown straight into battle.

But from the military's perspective, this journalistic freedom came at a price. The reporters out in the field tended to file stories that reflected the viewpoint of the combat soldiers they walked alongside through the rice paddies and the jungle. This reporting often conveyed the frustration felt by drafted soldiers who were trying to defeat an elusive and ruthless enemy. The dispatches emanating from the field tended to clash with the overly optimistic daily official briefings held at the American military headquarters in Saigon, the infamous '5 o'clock follies'. A credibility gap arose between the official assessment of the progress of the war and what the reporters were exposed to in the field. The reporters became increasingly distrustful of the military and this was reflected in a burgeoning array of stories that portrayed the American soldiers as 'drugged-out baby killers' (to reflect one extreme) or as misguided: doomed to fight a war they could not win.

The divergence of viewpoints between the American authorities (both military and civil) and those of the Fourth Estate was most pronounced during the Tet offensive of February 1968. The military's claim that they had inflicted a tactical defeat on the NVA and their Viet Cong allies by successfully countering the uprising was broadcast on television across America, superimposed over images of the desperate fighting inside the American Embassy in Saigon. The military's claim of a tactical victory was valid: the anticipated nationwide uprising of the South Vietnamese did not occur and the local Viet Cong infrastructure, which had been built up over many years, was destroyed. But at the same time the United States had suffered a strategic defeat. The American public, who had been led to believe that victory was close at hand, now began to sense that the war was unwinnable, and their support for the war began to wane. Increasingly negative press coverage after the Tet offensive continued to turn public opinion against the war and the soldiers who had to fight it.

The Vietnam conflict generated a lasting and widespread animosity between the American military and the media. A significant portion of the military blamed the media for eroding public support for the war. The media countered by claiming that public support was lost because of mounting casualties and a lack of political leadership. Henry Gole, an army officer who had served two tours of duty in Vietnam and who was a faculty member of the US Army's War College during the 1980s, observed that the officers he encountered at the War College

> almost unanimously despised journalists and made no effort to conceal their attitude. Those students held Congress and the civilian leadership in low regard and resented the apathy and ignorance of the American public about national security and matters martial. But they reserved a special venomous attitude for 'the media', a term more sneered than spoken ... Some 20 years after their experience in Vietnam, student attitudes towards the media were overwhelmingly negative and seemingly permanent, at least in that generation of embittered officers.[12]

The military's enduring distrust of the media

These American army officers of the 1980s were perpetuating the soldier's traditional distrust of the motives of the media. Robert Graves, an officer in the British Army during the First World War, listed in descending order the status of various segments of the population as viewed by the soldiers in the trenches:

> [T]he trench-soldiers themselves and those who had gone home wounded … the staff, Army Service Corps, lines of communication troops, base units, home-service units, and then civilians down to the detested grades of journalists, profiteers, 'starred' men exempted from enlistment, conscientious objectors, members of the Government.[13]

Somewhat ironically, while the soldiers of the Vietnam era blamed the media for eroding public support for the war, the contempt which the soldiers of the First World War felt towards war correspondents was due to their enthusiastic promotion of the war. The war correspondents fed the public a steady diet of heroism and glory, largely ignoring the squalor and futility of the trenches. Alistair Horne in *The Price of Glory* remarks of the French soldier of the First World War that:

> [M]ore than the muddlers, the *embusqués* [shirkers], the profiteers and the defeatists, of all the excrescences growing at the rear, the fighting men probably resented most of all the so-called *bourreurs de crâne* [brainwashers]. These were the writers and newspapermen, paid hacks of the propaganda machine … who from their comfortable offices in Paris wrote of the nobility of war.[14]

Such was the loathing of the media among the troops that one of Graves's fellow British officers, Siegfried Sassoon, wrote in his poem, 'Fight to a Finish', of the 'Yellow-Pressmen' lining the streets during a victory parade to cheer 'the soldiers who'd refrained from dying'. Giving form to what was perhaps a relatively common fantasy among the troops, the soldiers taking part in the parade fixed bayonets and ran the cold steel through the pressmen, the poem's narrator commenting that the pressmen gave a satisfactory 'grunt and squeal' as the bayonet found its mark.[15] Martin Bell put the matter quite succinctly when he wrote: 'It is generally accepted now that, of the men who wrote contemporary accounts of the war, it was the reporters who told the lies and the poets who told the truth.'[16]

Although contemporary military officers readily acknowledge the vital role played by the media in maintaining public support for a conflict, and realise that the military can use them to get their viewpoint across to the public, the traditional distrust of the motives of the press remain. Admiral Sir Jeremy Black, who was the captain of HMS *Invincible* during the Falklands War, wrote of the tensions that arose between his men and the journalists on board his ship.

> At the root of our difficulties lay one fundamental fact. The Services are taught from their earliest days to be team players and that the results of the team working together are the only results that matter, regardless of individuals who subordinate themselves to the team. Press men are the very antithesis of this. They operate as individuals, they don't understand the word team even when their lives depend on it and the only other individual in their firmament is their Editor. They will pursue their own ends at any price, having never heard of the word loyalty, and will trample on anyone or anything that stands in their way.[17]

Some twenty years later, Nathaniel Fick, who served as an officer in the US Marine Corps during the 2003 invasion of Iraq, recalled that his initial impression of the journalist who would be embedded in his platoon was 'a clueless opportunist chasing a Pulitzer Prize on the backs of men he wouldn't speak to on the street at home'.[18]

On occasion, the military's general dislike of the press becomes focused on a specific individual. On 1 May 2004, Piers Morgan, the editor of the *Daily Mirror* (UK), published photos purporting to show the abuse of an Iraqi detainee by members of the 1st Battalion of the Queen's Lancashire Regiment. Following in the wake of the Abu Ghraib scandal, where it was revealed that members of the US military had abused and humiliated Iraqi prisoners, the effect of the pictures' publication was explosive. Chris Hunter, serving with the British Army in the southern Iraqi city of Basra at the time the photos were made public, wrote of the immediate fallout, as the militia army of Muqtada al-Sadr rampaged across the city. Hunter's convoy of three vehicles was ambushed as they returned to a British base. 'I still can't believe how close we came to checking out last night. None of us can,' wrote Hunter in *Eight Lives Down*, his account of his service in Iraq. 'Piers Morgan has now accelerated past Muqtada al-Sadr, right to the top of my shit list. There's a rumour that his brother is a serving officer, yet even that didn't dissuade him from putting the lives of our guys at risk. What sort of man would do that?'[19] The pictures were later proved to have been faked, as the type of vehicle shown in the photographs had not been deployed to Iraq. The *Daily Mirror* subsequently issued an apology and sacked Morgan.

The threat to operational security

Major media companies tend to package news as entertainment, with a resulting emphasis on conflict, dramatic footage and melodrama. This factor, coupled with an increase in the number of 24-hour cable news stations and the growth of internet news services, has fundamentally changed the dynamics of the media business. The widespread use of the telegraph in the mid-19th century meant that editors could publish information from their correspondents in the field the day after an event; and this led to the media preoccupation with the 'scoop'. At the time, papers were usually printed daily and had set deadlines. The current cut-throat nature of media competition, however, has given rise to an incessant demand for breaking news, encouraged sensationalism and resulted in less reflection and verification. Ever-looming deadlines encourage reporters to 'go with what you have got' rather than investigate further and risk being pre-empted by a competitor. A veteran reporter commented in 2001: 'Not only has this proliferation increased the likelihood that leaked information could compromise operational security, but the increased media competition detracts from careful fact checking.'[20]

The need to maintain operational security has always been foremost in the military's dealings with (and tolerance of) the media. During the Peninsular War, the then Sir Arthur Wellesley, commander of the British

Army in Spain and Portugal, wrote to the Secretary of State for War and the Colonies, Lord Liverpool, to protest against the release of operational information by the press.

> I beg to draw your Lordship's attention to the frequent paragraphs in the English Newspapers, describing the position, the numbers, the objects, the means of attaining them possessed by the armies in Spain and Portugal. In some instances the English Newspapers have accurately stated, not only the regiments occupying a position, but the number of men fit for duty of which each regiment was composed; and this intelligence must have reached the enemy at the same time as it did me, at a moment at which it was most important that he should not receive it.[21]

The opposing French forces also bemoaned the release of operational information by the press. 'The liberty of the press is certainly an excellent thing; but more patriotism ought to prevail in the use that is daily made of it,' admonished Elzéar Blaze, an officer in Napoleon's army, in his account of his campaigns. He continued: 'we ought not to tell our neighbours what is so important to them to know. The enemy formerly kept spies among us; he now obtains his intelligence at a much cheaper rate – he subscribes to our journals.'[22]

A similar complaint was made by Field Marshal Lord Raglan, the commander-in-chief of the British Army forces in the Crimea, who felt that William Howard Russell's dispatches had breached operational security and thereby afforded assistance to the enemy. Indeed, in an article published in *The Times* on 23 October 1854, that paper's 'special correspondent' (Russell) revealed the number and type of artillery pieces the British had dispatched to the Crimea, the location of the main powder magazine, the shortage of roundshot, the site of Lord Raglan's HQ, and the disposition of the forces on the ground.[23] Raglan wrote to the Secretary of War, the Duke of Newcastle, pointing out the details contained in *The Times*' article of 23 October, noting that the main powder magazine was shortly afterwards the object of a heavy cannonade, and requesting that something 'be done to check so pernicious a system at once'.[24] In a subsequent letter to Newcastle, Raglan bemoaned that: 'The enemy at least need spend nothing under the head of Secret Service.'[25] Christopher Hibbert notes in his biography of Raglan that the enemy shared Raglan's view of the damage done by the release of operational information by the newspapers, the Czar reportedly stating: 'We have no need of spies, we have *The Times*.'[26]

On 25 February 1856, in the closing stages of the Crimean War, the newly appointed commander-in-chief, Lieutenant General Sir William John Codrington, issued a general order pertaining to war correspondents. The order forbade the publication of details that could prove of value to the enemy, authorised the expulsion from the theatre of war of any correspondent who, it was alleged, had published such details (no doubt Codrington had Russell in mind when including this clause), and threatened future transgressors with the same punishment.[27] This order represented the opening salvo in the military's ongoing campaign to control the press during war. The next battlefield of this campaign was the American Civil War.

General William Tecumseh Sherman displayed considerable animosity towards the press throughout the American Civil War, on one occasion having a reporter for the *New York Herald* court-martialled on charges of giving intelligence to the enemy. The reporter in question, Thomas W. Knox, had drawn Sherman's ire by reporting on the failed attempt by the Union Army to seize the Chickasaw Bluffs, to the north of the Confederate stronghold of Vicksburg. Although it was a fact that the Union forces had been rebuffed, suffering some 1,700 casualties in the process, Knox's implication that the Union defeat was a direct consequence of Sherman's insanity and inefficiency was unlikely to place him in the general's good graces. Incidentally, the court martial, while finding Knox not guilty of the more serious charges, found him guilty of wilfully disobeying Sherman's prohibition on accompanying the army on this campaign and of causing a dispatch to be printed without the sanction of the general in command. The court ordered Knox 'to be sent without the lines of the army, and not to return under penalty of imprisonment'.[28]

Sherman's intolerance of the press was not merely in response to the personal attacks they frequently made upon him. Rather he believed that the press directly endangered the lives of his men by revealing the direction of Union attacks and the components of the attacking force. For example, on 17 July 1861, the *New York Times* reported: 'The army in Virginia today took up the line of march for Richmond, via Fairfax and Manassas. The force starting today was fully fifty thousand strong ... about three thousand Regular Infantry, Cavalry and Artillery, and fifty thousand volunteers.'[29] Four days later the Union and Confederate forces fought the First Battle of Bull Run (First Manassas), resulting in a resounding defeat of the Union Army. In a letter to Senator Thomas Ewing, Sherman's foster father (as well as his father-in-law), Sherman wrote:

> I regard newspaper correspondents as spies ... They publish without stint positive information of movements past and prospective, organisations, names of commanders, and accurate information which reaches the enemy with as much regularity as it does our People ... Being in our very midst, catching expressions dropped by officers, clerks, and orderlies, and being keen expert men they detect movements and give notice of them. So that no matter how rapidly we move, our enemy has notice in advance. To them more than to any other cause do I trace the many failures that attend our army. While they cry about blood and slaughter they are the direct cause of more bloodshed than fifty times their number of armed Rebels. Never had an enemy a better corps of spies than our army carries along, paid, transported, and fed by the United States.[30]

A different army, a different continent and a different century would provide the best-known manifestation of Sherman's concerns about the press endangering the lives of soldiers by revealing operationally sensitive information.

The Falklands, 1982: a battle over censorship

The relationship between the press and the British Ministry of Defence during the Falklands War of 1982 was combative, with the press accusing the Ministry of excessive censorship and the Ministry responding by pointing out the dangers to the troops that arose from the media's apparent inability to exercise self-censorship. The British media openly speculated about potential dates for the landing of British forces on the Falkland Islands, as well as possible tasks of SAS and SBS patrols. The media also reported that Argentine bombs had failed to detonate because the munition technicians were setting the safety fuses too long (these fuses arm after a predetermined interval to stop the bombs detonating too close to attacking aircraft). The members of the taskforce were urged not to mention this information in their letters home because once alerted it would be a simple matter for the enemy to reset the safety fuses. 'The night after the commander of the *Intrepid* had asked us not to mention the fuses in letters home, the World Service of the BBC announced it for us,' bitterly recalled Hugh McManners in his memoir of the war. 'The freedom to print the truth is obviously a cornerstone of democracy, but it would seem that editorial common sense practised by professionals is concerned with selling copy rather than any other implications the broadcast story might have. Suffice it to say that we were not impressed.'[31] Another serious breach of operational security was to occur in relation to the attack by the 2nd Battalion of the Parachute Regiment (2 Para) on the Argentine positions near the settlements of Darwin and Goose Green.

At 10am local time on 27 May 1982, the soldiers of 2 Para were sheltering in the abandoned Camilla Creek House and outbuildings preparing for an attack on the Argentine positions at Darwin that evening. Some of the men were listening to the BBC World Service and were shocked to hear the BBC's defence correspondent state: 'there is something quite big going on. They're saying for example that the 2nd Parachute Regiment has moved south towards Darwin area.' The British official history of the Falklands campaign records that the Argentine commander had decided to reinforce the Darwin/Goose Green position on the 26th, the day prior to the BBC announcement. A hundred men were taken from Port Stanley to Goose Green by helicopter, arriving during the initial stages of the battle on the 28th. Another 140 reinforcements and some artillery pieces from the Argentine positions on Mount Kent arrived on the afternoon of the 28th, too late to influence the outcome of the battle, which was then coming to an end. Despite this conclusion, whether the Argentine reinforcements were prompted by the BBC announcement is still a matter of debate. What is not questioned is that the announcement represented a serious breach of operational security. The commander of the land forces in the Falklands, Brigadier Julian Thompson, immediately expressed his dismay about the leak to Fleet Headquarters in the UK. Admiral John Fieldhouse replied: 'Equal horror expressed here. Have made representations at highest level.'[32]

At Camilla Creek House the BBC announcement was greeted with shock and then anger, as 2 Para felt the vital element of surprise was now lost. The battalion commander, Lieutenant Colonel H. Jones, ordered his men to leave the shelter of the buildings, disperse and dig-in, in case the Argentines acted upon this information by launching an air or artillery attack. This dispersal interrupted the preparation for battle and complicated the required coordination measures. Any remaining doubt that the Argentines may have held about Goose Green being the target of a British attack was soon dispelled when, shortly after the BBC announcement, two British forward patrols were spotted and disengaged under fire.

The media's announcement of the impending attack on Goose Green was as much due to political impatience as it was to the media's inability to exercise self-censorship. The British politicians were anxious for some positive news from the Falklands to offset the negative press arising from the loss of ships among the taskforce (by 25 May five ships had been sunk). At the War Cabinet meeting on the 26th, the breakout of the land elements from the beachhead at San Carlos was discussed. The Secretary of State (John Nott) told the House of Commons that afternoon: 'Our forces on the ground are now poised to begin their thrust on Port Stanley.' There were only two possible directions for this advance: north towards Teal Inlet and the main Argentine positions, or south towards the less well-defended positions on the Goose Green isthmus. Journalists openly speculated that Goose Green seemed the most likely target. On the 27th, the Prime Minister, Margaret Thatcher, refused to go into details of the impending attack during parliamentary questions, merely stating: 'Our ground forces are now moving from the beachhead.' The news of the breakout was duly broadcast on the 1pm news and was picked up and rebroadcast on the BBC World Service an hour later. Although the direction of the breakout had not been officially announced, the media were speculating that it would be towards Goose Green and that is what the press in London broadcast.[33]

Although no Falklands-based journalist played any part in the BBC's announcement, 'at San Carlos rage towards the media was unbounded', recalled Max Hastings. 'The reporters, it was felt, had shown that they simply could not be trusted with operational information. For the rest of the war, some units would have nothing to do with any of us.'[34] Later in the campaign, the already tenuous relationship between the military and the media was frayed even further when two journalists conversed with some colleagues at another settlement on the civilian landline about the forthcoming British attacks towards Port Stanley and were overheard by a Royal Marines officer, who reported the matter to Brigadier Thompson. Although the initial fear that this conversation could have been monitored by the Argentines proved unfounded, as the line was cut well short of Stanley, the damage was done. Hastings was barred from attending Thompson's briefing on the forthcoming battle: 'journalists as a group were at that moment regarded with deep suspicion, even animosity, in some military circles,' Hastings noted.[35]

The risks to security from modern media

While the BBC's premature announcement of the attack on Goose Green by 2 Para is one of the best-known and most cited examples of an instance where the press have compromised operational security, it is by no means the only example. In April 1999, the Pentagon's chief spokesman, Kenneth Bacon, commented:

[W]e live in an incredibly competitive media age. We now have three, twenty-four-hour-a-day cable networks all competing for scoops, all competing to get on the air as soon as possible with new details. And I think ... that the press is much less restrained in the use of operational information today than they used to be.[36]

By way of example, Bacon stated that in 1995 a television network reported that the US military was about to launch a cruise missile against a Serbian surface-to-air missile site at Banja Luka, Bosnia. Because of this disclosure, the missile site was moved and could have potentially been used to shoot at American planes. Bacon also mentioned that in late March 1999, during the Kosovo conflict, the *Washington Post* announced on its front page that the Allied forces were going to target the Serbian Interior Ministry and the Ministry of Defence in Belgrade. The targets were struck the following week. Bacon claimed that the forewarning by the media allowed the Serbians to take 'various actions to reduce the impact of those strikes'.[37]

There was also a notable breach of operational security by a member of the press during the 2003 invasion of Iraq which showed the inherent risk of the military's practice of retrospective censorship in the age of instant communication from anywhere in the world. On 1 April 2003, Geraldo Rivera, a journalist representing the Fox News Channel, who was embedded with the 101st Airborne (Air Assault) Division, voluntarily left Iraq – on a US Army helicopter – after Coalition officials threatened to expel him. Rivera was allowed to leave voluntarily only because a compromise had been reached between Fox News and the Pentagon. During a live television broadcast Rivera had sketched a map of Iraq in the sand and indicated the current and planned future locations of the 101st as they prepared to assault Baghdad. A military official commented that Rivera 'gave away the big picture stuff. He went down in the sand and drew where the forces are going.'[38]

Although the vast majority of correspondents observe the military guidelines that seek to maintain operational security and would never intentionally put their own soldiers in danger, nevertheless, there is an attendant risk that breaches of operational security will occur if the media is permitted into the combat zone. This risk is increased by technology that enables instant connectivity from virtually anywhere and the proliferation of media outlets, particularly those of opposing or neutral nations. Such outlets will usually be less inclined than domestic media organisations to observe military guidelines on press coverage. As an example of official impatience with this situation, during a White House meeting with the British Prime Minister, Tony Blair, on 16 April 2004, the American President, George W. Bush, allegedly outlined an intention to bomb the headquarters of the Arab satellite TV station al-Jazeera, located in the Qatari (a US ally) capital of Doha. Bush felt that al-Jazeera's reporting was fuelling the ongoing Iraqi insurgency and was opposed to them airing footage showing strikes on American forces by insurgents and images of dead American soldiers and contractors, as well as recorded messages from Osama bin Laden. Significantly, on 13 November 2001, al-Jazeera's office in Kabul was destroyed by a US missile (there were no casualties) and on 8 April 2003, al-Jazeera's Baghdad station was hit by a US missile, killing a cameraman.[39] Fearing such an action, in February 2003 al-Jazeera had advised the US military of the exact coordinates of their Baghdad office in the hope of avoiding an 'accidental' missile strike.[40]

Fighting with an embedded media

The Bush administration's frustration with al-Jazeera is indicative of the difficulty of trying to control the flow of information generated by a ubiquitous media presence equipped with satellite communications. In such an environment, the only viable means of control is a self-censorship code. The basis of self-censorship is that journalists are provided access to an operational area and ongoing support from the military in return for agreeing to follow a set of defined ground rules on what can and what can not be reported. While the media has the technological means of releasing a story violating the ground rules agreement, the likely outcome would be the expulsion of the offending journalist from the operational area. In this way, the military makes it clear that the media's presence in operational areas continues only because the military chooses to tolerate it – but that this tolerance is conditional.

A defining element of the relationship between the military and the media is the interplay between two conflicting objectives. The media espouses the public's 'right to know' about the actions of state-sponsored military forces. But this factor should be somewhat counterbalanced by the soldier's right to undertake a difficult and hazardous duty without having to unduly concern himself with whether or not his actions will be subjected to 'trial by media'. The soldier should not have to worry that his conduct on the battlefield may be sensationalised, presented out of context and served up to be judged by those whose understanding of the conditions of war is limited and whose interest in the welfare of the soldier is minimal. Nathaniel Fick wrote that:

> As a citizen, I supported the Pentagon's much touted embedded media campaign as a way to give Americans an uncensored look at the war and the warriors. As an officer, I dreaded dealing with information leaks, distraction to my Marines, and constant moral oversight of people who knew little about our culture and the demands of combat decision making.[41]

Soldiers function in an environment characterised by great stress and confusion, where life-or-death decisions have to be made instantaneously. When the commanding general of the 1st Marine Expeditionary Force, Lieutenant General James Conway, addressed the officers of the 1st Marine Division about the rules of engagement prior to them moving into Iraq in 2003, he reassured them that 'a commander would be held responsible not for the facts as they emerged from an investigation, but for the facts as they appeared to him in good faith at the time – at night, in a sandstorm, with bullets in the air.'[42]

Yet despite the best intentions of the military to ensure that apparent breaches of the laws of armed conflict by soldiers are examined in context, the ubiquitous media presence may sensationalise such incidents and hold them up to be judged in the court of public opinion. At the same time, this media presence may uncover or even prevent wartime atrocities, as Dave Grossman notes in *On Combat*: 'the modern soldier is likely to have his every act videotaped and reported on national TV, and there is no tolerance for any deviation from the rules of War. Today our soldiers are held to the highest standards, and that is a good thing.'[43] But the media have a tendency to judge the actions of soldiers against a benchmark of normal codes of behaviour, rather than the kill-or-be-killed nature of the battlefield. So the ever-increasing media presence on the battlefield imposes an additional burden on the soldier: that his every action may be judged by the public, safe at home and spared the terror and confusion of the battlefield.

Breaking the rules; and trial by media

A notable (because filmed) case of an alleged breach of the rules of engagement occurred on 13 November 2004, during the US-led offensive targeting insurgent strongholds in the Iraqi city of Fallujah. The incident began when an unidentified corporal, belonging to the 3rd Battalion, 1st Marine Regiment, entered a mosque with his section. The day before, the mosque and an adjacent building had been the scene of a fierce firefight between insurgents and Marines. The Marines secured the mosque, killing ten insurgents and wounding five who, after being treated by Navy corpsmen, were left in the mosque to be recovered by another element when time and circumstances allowed (fighting was ongoing and there were American casualties that needed to be moved first). Intelligence reports indicated that insurgents had reoccupied the mosque during the night, resulting in the dispatch of a force of Marines to reclear the building. The insurgents inside were engaged by machine guns mounted on the Marines' tanks. After the tanks ceased fire, gunshots were heard coming from inside the mosque. A squad of Marines then approached the mosque through a breach in the outer wall and encountered other Marines already in the complex. One of the Marines already present in the mosque informed the soldier commanding the newly arrived squad that there were five insurgents in the building and he had shot them. Both parties of Marines then moved inside.[44]

When the Marines entered the main building of the mosque they found the body bags containing the dead from the previous day's attack, as well as the five wounded and apparently unarmed insurgents who had been left behind in the mosque. It appeared one of these men was now dead and that three others were bleeding from new gunshot wounds. The fifth insurgent was partially covered by a blanket and was lying where he had been left the previous day. He had not been shot again. The Marine corporal noticed the apparently dead Iraqi and realised he was breathing. The corporal shouted out: 'He's fucking faking he's dead – he's faking he's fucking dead.' He then shot the apparently unarmed and critically wounded man in the head at close range. Another Marine commented: 'Well he's dead now.' The Marines then continued with their clearance of the building.[45]

The Hague Convention of October 1907 codified the Laws and Customs of War on Land. Article 23 states that it is 'especially' forbidden 'To kill or wound an enemy who, having laid down his arms, or having no longer means of defence, has surrendered at discretion; and to declare that no quarter will be given.'[46] Yet despite this prohibition of the killing of prisoners of war, there is ample evidence that it occurs in combat. Guy Chapman wrote of a conversation he had with a fellow British officer during the First World War. The officer had come to Chapman for advice about how he should discipline one of his soldiers who had calmly shot a surrendering (and unarmed) German officer in the head. Chapman commented: 'I don't see that S___'s really to blame. He must have been half mad with excitement by the time he got into that trench. I don't suppose he ever thought what he was doing. If you start a man killing, you can't turn him off again like an engine.'[47] The officer revealed to Chapman that another of his soldiers had also executed a prisoner of war. They agreed it was too late to do anything about these incidents and that the best course of action was to overlook them.[48]

The killing of the unarmed German officer was by no means an isolated incident during the First World War. 'Prisoners we are not troubled with now for we kill every bosche at sight,' an Australian officer on the Western Front blithely recounted in a letter home.[49] Robert Graves noted that nearly all of the officers he served with at Le Havre during the First World War knew of specific cases where prisoners of war had been murdered:

> The commonest motives were, it seems, revenge for the death of friends or relations, jealousy of the prisoner's pleasant trip to a comfortable prison camp in England, military enthusiasm, fear of suddenly being overpowered by the prisoners or, more simply, not wanting to be bothered with the escorting job. In any of these cases the conductors would report on arrival at headquarters that a German shell had killed the prisoners; no questions would be asked.[50]

The practice of executing prisoners of war was not restricted to the First World War. In *Citizen Soldiers*, Stephen Ambrose comments that of the 1,000-plus combat veterans of the Second World War he had interviewed, only one admitted to having personally shot a prisoner of war; the man adding that while he had felt some remorse for this act, he would do it again. But as many as a third of the veterans Ambrose interviewed had witnessed other GIs shooting surrendered German soldiers.[51] In *The Men of Company K*, Harold Leinbaugh and John Campbell wrote of an incident that occurred near the German village of Kogenbroich. Their men had captured a dozen Germans and were escorting them to the rear when 'One man whose best friend had just been killed took revenge on four prisoners. He said they had jumped him and tried to escape – maybe they had, but he got no more escort duty.'[52] Moreover, General George S. Patton Jr. suggested in his memoirs that: 'Prisoner of war guard companies ... should be as far forward as possible in action to take over prisoners of war, because troops heated with battle are not safe custodians.'[53]

A related issue to the killing of surrendering soldiers is the considerable personal risk entailed in providing medical assistance to a wounded enemy soldier because he may be seeking to get his foe to come closer before striking. Timothy Gowing recalled that in the Crimean War some of the wounded Russian soldiers '[shot] down our men just after they had done all they could for them. Our comrades at once paid them for it either by shooting or bayoneting them on the spot.'[54] A British report on the atrocities committed by the Germans when they occupied Belgium in 1914 commented:

> In dealing with the treatment of the wounded and of prisoners and the cases in which the former appear to have been killed when helpless, and the latter at, or after, the moment of capture, we are met by some peculiar difficulties, because such acts may not in all cases be deliberate and cold blooded violations of the usages of war. Soldiers who are advancing over a spot where the wounded have fallen may conceivably think that some of those lying prostrate are shamming dead, or, at any rate, are so slightly wounded as to be able to attack, or to fire from behind when the advancing force has passed, and thus they may be led into killing those whom they would otherwise have spared.[55]

An example of the inherent danger of accepting the surrender of an enemy soldier on the battlefield is provided by Denis Winter, who in his book *Death's Men* quotes a soldier from the First World War.

> Lying on his stomach, he turned his head and asked for mercy but his eyes said murder. I plunged my bayonet into the back of his heart and he slumped with a grunt. I turned him over. There was a revolver in his right hand under his left armpit. He had been trying to get a shot at me under his body. As I withdrew the bayonet, I pressed the trigger and shot him to make sure.[56]

Günter Koschorrek described how one of the NCOs in his unit was killed when a Soviet officer, who was lying wounded on the ground and whom the NCO had just finished bandaging, pulled out a pistol and shot him in the back. The Soviet officer was immediately killed by a burst from a German sub-machine gun.[57] David Hackworth recalled an incident during the Korean War when his unit was conducting a sweep of the battlefield following the repulse of a Chinese attack. 'Many [Chinese soldiers] played possum, lying motionless on the ground, pretending they were dead; when this was discovered, every single "corpse" got a slug in the head – if blood pumped out you knew you'd gotten a live one.'[58]

Yet no soldier was ever court-martialled for any of the incidents described above, nor was the morality of these actions debated in editorials or analysed on talk shows across the globe. But tellingly, the Marines who entered the mosque in Fallujah were accompanied by an embedded journalist, who captured the shooting incident on film and, as part of a media pool, was obliged to share this with other networks. Footage of the incident was soon being shown across the world, particularly on al-Jazeera, which repeatedly aired the unedited footage (the actual killing was edited out when it was shown on the American news networks). Once the footage had been aired, the Marine corporal who had killed the insurgent in the mosque was removed from his unit and questioned by the US Navy's Criminal Investigative Service. Significantly, CNN reported that on the same day, and a mere block away from the shooting in the mosque, a Marine was killed and five others wounded by a booby-trapped body they had found in a house following a shootout with insurgents.[59]

The report of the Criminal Investigative Service was reviewed by the Commanding General of the 1st Marine Division, Major General Richard F. Natonski. Prior to the assault on Fallujah the Marines were informed that the rules of engagement allowed the use of deadly force against men of military age deemed to be displaying hostile intent, even if these men had not fired upon the Marines.[60] Natonski determined that 'the actions of the Marine in question were consistent with the established rules of engagement, the law of armed conflict and the Marine's inherent right of self defense.'[61] A Marine Corps press release stated that careful consideration was given to the impact of the enemy's known tactic of feigning death or surrender and then attacking. The press release revealed that the Marine corporal had in fact shot and killed three insurgents, not one, while clearing the mosque. Autopsy reports revealed that the three insurgents had died from multiple gunshot wounds and that all three men had been hit by bullets fired from the corporal's M-16 (although it is not clear in the press release, presumably some of the insurgents' wounds were inflicted during the battle of the previous day). In a sworn statement provided to investigators, the corporal admitted he had shot the three men in self-defence as he believed they posed a threat to himself and his colleagues. It was noted in the investigators' report that the videotape showed the wounded insurgent was concealing his left arm behind his head.[62] The press release concluded with the statement that:

> [I]t was reasonable to believe that the corporal fired on the [insurgent] after reasonably believing that the individual was committing a hostile act by exhibiting a known enemy [tactic] (feigning death and subsequently moving his concealed arm). Based on all the evidence in the case, and the rules of engagement that were in effect at the time, it is clear that the corporal could have reasonably believed that the [insurgent] shown in the videotape posed a hostile threat justifying his use of deadly force.[63]

Tellingly, the press release did not provide any details about the two other insurgents killed by the Marine corporal in the mosque. Nor was any information provided on the circumstances of the death of a fourth wounded Iraqi insurgent, who was killed in the mosque by another Marine during the same patrol. Apparently justification was required only for the killing captured on video, which is indicative of the role played by the media in this incident. The key issue here is not one of morality or justification for the actions of the Marine in question. That is a separate debate. As noted, similar incidents involving the killing of unarmed prisoners of war have occurred with an alarming frequency in previous conflicts. But what made this particular case so different to the others was the presence of the media, which transformed a localised incident into one played out across the world's television screens. A comment recorded by Stephen Ambrose in *Band of Brothers* is particularly relevant to this narrative.

Tom Gibson of the 101st Airborne Division recounted to Ambrose an incident involving the murder of ten German prisoners of war by an American army officer during the fighting in Normandy in the Second World War. The officer shot the Germans while they were digging a roadside ditch and were under guard by other American soldiers. Gibson affirmed that:

> I firmly believe that only a combat soldier has the right to judge another
> combat soldier. Only a rifle company combat soldier knows how hard it
> is to retain his sanity, to do his duty and to survive with some semblance
> of honour. You have to learn to forgive others, and yourself, for some of
> the things that are done.[64]

A generation later, James McDonough, who served with the 173rd Airborne
Brigade during the Vietnam War, when reflecting upon the decisions he
had to make while leading his men in combat, concluded:

> War is not a series of case studies that can be scrutinized with objectiv-
> ity. It is a series of stark confrontations that must be faced under the
> most emotion-wrenching conditions. War is the suffering and death of
> people you know, set against a background of the suffering and death
> of people you do not. That reality tends to prejudice the already tough
> choices between morality and pragmatism.[65]

Another contemporary example of the media's tendency to sensationalise
incidents, and the resulting impact on soldiers, arose out of the intense
media coverage of Israeli soldiers tasked with containing the Second
Palestinian Intifada (September 2000 to November 2006). These soldiers
were subjected to severe criticism from their superiors as they responded
with progressively more violent tactics to Palestinian provocation. Martin
Van Creveld in his history of the Israeli Defence Force, *The Sword and the
Olive*, comments:

> [T]he *chariggim* (excesses) were the result of stress as the troops
> panicked and used 'excessive force' in situations they considered life-
> threatening – but that their superiors, prompted by the growing pres-
> ence of reporters and TV cameras, thought should have been handled
> with greater care ... The upshot is that those who use 'excessive' force
> and 'unnecessarily' kill or wound Palestinian demonstrators find

themselves at risk of being treated as criminals; those who did not use enough force and lost lives, as fools.[66]

The propensity of the media to sensationalise incidents will result in the actions of soldiers in combat being subjected to greater scrutiny and will potentially make a challenging and dangerous job more difficult. Shortly after the shooting of the wounded insurgent at the mosque in Fallujah, the Marines were informed by the journalist present that these were the same men who had been wounded, disarmed and given medical treatment from the previous day's attack. 'At that point the Marine who fired the shot became aware that I was in the room,' the journalist recalled. 'He came up to me and said, "I didn't know sir – I didn't know". The anger that seemed present just moments before turned to fear and dread.'[67]

What produced this 'fear and dread' in the Marine? Was it the realisation that he had just shot and killed an unarmed and helpless man, or was it the knowledge that his actions would now be broadcast across the world's television screens, potentially taken out of context and subjected to intense scrutiny?

In early 2002, Andrew Exum's platoon of US soldiers were accompanied by two reporters while they conducted a patrol in the Shak-e-Kot valley of Afghanistan. During the patrol Exum and his men shot and killed an al-Qaeda fighter. Reflecting soldiers' concerns about the media's tendency to distort and present out-of-context actions on the battlefield, Exum noted that:

> After we returned from that mission, the higher-ranking officers and sergeants major fretted over what the reporters might write in their stories. They worried that the reporters might not understand what we had done, why we had fired so many bullets, and what my soldiers had done after the man was dead, when Corporal Littrell cut the clothes off the body with a knife to search him.[68]

Later, Exum and his men were required to make sworn statements to an officer investigating the incident. They were told that this action was just a precaution. 'One of the reporters worked for a national magazine, and the officers weren't sure how he would interpret our actions. In case the reporter's story branded us as anything less than heroic, the officers just wanted to have our statements on record for use in a rebuttal.'[69] Exum bitterly recalled:

> It was complete bullshit. If the reporter's story was negative, the officers could use our statements as proof that the army had already taken action and was on its way toward prosecuting the offenders. With our sworn statements, they could put together an investigation to find us guilty of 'excessive violence' or something equally ridiculous.[70]

The need to counter media sensationalism, particularly when combating an insurgency where allegations of the abuse of prisoners/civilian casualties/disrespecting holy sites and so on could inflame the local populace, has been incorporated into the contemporary military's preparation for battle. During the November 2004 assault on Fallujah, Major General Jim Molan, an Australian Army officer serving as the Chief of Operations for the US-led Multi-National Force–Iraq, was directed to ensure that within one hour of an allegation appearing in the media, a response was 'fired back'. Molan noted that: 'We were not to deny anything immediately, nor investigate everything in such detail that we could only reply five days later, far too late for the media cycle.'[71] Despite the efforts of Molan and his staff, he surmised: 'Yet for the millions in the world who gained their knowledge of the war from the media, the courageous and moral actions of thousands of these troops [members of the Multi-National Force–Iraq] were overlooked. A video of one US Marine shooting a wounded enemy gave the battle its enduring image.'[72]

* * *

The relationship between the military and the media is multifaceted and reflects the often diverging objectives of each organisation. The military's emphasis is on completing the mission, but the armed forces of a democratic nation will find it difficult to successfully prosecute a campaign without the support of the people. The media has a vital role to play in gaining or even losing this public support. But the media is a business, and like any business it will not survive if it does not fulfil a need. Intense competition among media outlets, in particular the insatiable demands for new material driven by the 24-hour cable news channels, tends to compel journalists to 'go with what they have got', which often represents a triumph of style over substance and sensationalism over analysis. The military's stance is that they need to ensure operational security, both to maximise the potential for completing the task and to protect the soldiers who must carry out the mission. So the military seeks to restrict the flow of information to protect their soldiers from the enemy. But the media also has a role to protect soldiers from the state. This role was perhaps best expressed by Associate Justice Hugo L. Black of the US Supreme Court in his opinion in the case of *New York Times Co. v. United States* (1971) over whether the paper had the right to publish sections of the so-called 'Pentagon Papers'. Black wrote:

> In the First Amendment the Founding Fathers gave the free press the protection it must have to fulfill its essential role in our democracy. The press was to serve the governed, not the governors … And paramount among the responsibilities of a free press is the duty to prevent any part of the government from deceiving the people and sending them off to distant lands to die of foreign fevers and foreign shot and shell.[73]

But that is a different topic altogether and will be examined in the final chapter of this book.

Chapter 10

IS THERE A NEED FOR WAR?

hile we may aspire to the idealism of living life in peace, 3,500 years of military conflict (dated from the battle of Megiddo in 1480 BCE) indicates that the American philosopher George Santayana was probably pretty close to the mark when he wrote: 'Only the dead have seen the end of war.'[1] History has shown us repeatedly that the threat of force is generally countered only by an equal or greater force. Well may we note the warning given to us in the 4th century CE by the Roman writer Flavius Vegetius Renatus: 'Let him who desires peace, prepare for war.'[2] Two examples, separated by some 12,500 miles and over 2,000 years, will serve to illustrate the application of Vegetius' maxim.

Might vs. morality

At the time of the Peloponnesian War between Athens and Sparta, Melos was a sparsely populated island in the Sea of Crete (part of the Cyclades). It was originally a Spartan colony but had remained neutral during the Peloponnesian War. Its misfortune was to be surrounded by other small islands that were part of the Athenian empire. In 416 BCE a large Athenian force landed on Melos. The Athenian generals sent representatives to speak before the Council of the Melians, where they demanded that Melos become a tribute state of Athens. The Melians declared that they wished to remain neutral. The Athenians responded by stating: 'The standard of justice depends on the equality of power to compel and that in fact the strong do what they have the power to do and the weak accept what they have to accept.'[3] The Athenian position was clear: become a tribute state or be destroyed.

The Melians replied with a moral viewpoint: 'We are standing for what is right against what is wrong'; but the Athenians countered with a Machiavellian rationalisation: 'This is not a fair fight, with honour on one side and shame on the other. It is rather a question of saving your lives and not resisting those who are far too strong for you.' The Melians stated that they were 'not prepared to give up in a short moment the liberty which our city has enjoyed from its foundation for 700 years.' The Athenian representatives then left the council. The Athenian army built a wall completely around the city of Melos and left a garrison behind to put the city to siege, while the majority of the force returned to Athens. The siege lasted throughout the summer and into the following winter, the Melians occasionally staging limited raids into the Athenian siege lines. The intransigence of the Melians resulted in another military force being dispatched from Athens. The siege was renewed with vigour and eventually the city capitulated. The Athenians put to death all men of military age and enslaved the women and children.[4]

The Chatham Islands are located 500 miles to the east of New Zealand. They are an archipelago of ten small islands of which only two (Chatham and Pitt) are currently inhabited. The islands were discovered by the same wave of Polynesian colonists who settled the nearest landmass, New Zealand. By the late 18th century, the inhabitants of the Chatham Islands, the Moriori, had not had any contact with their ancestral forefathers, the Maori, for over 800 years. Initially, internecine warfare had led to a dangerous decline in the Moriori population. A Moriori chief, Nunuku Whenua, brought an end to this warfare and decreed that henceforth all disputes were to be resolved through a ritualised contest – using as weapons sticks no thicker than a thumb and no longer than an arm – that was to cease as soon as blood was drawn. Adherence to Nunuku's law led to the Moriori becoming a pacifistic society.[5]

On 29 November 1791, the Royal Navy brig *Chatham* was blown off course on a voyage from New Zealand to Tahiti and chanced upon the islands. Following in the wake of the *Chatham* were sealers and later whalers, who began to visit the islands in the first decade of the 19th century. The Moriori warmly greeted the foreigners and performed the ceremony of peace. Working on board the visiting ships were some Maori who returned to New Zealand with tales of fertile islands whose peaceful occupants would offer little resistance to invaders.[6]

On 19 November 1835, a British trading vessel arrived at the Chatham Islands carrying 500 Maori armed with guns, clubs and axes. On 5 December the ship returned, carrying another 400 Maori. The party of 900 men, women and children comprised two tribes who had been displaced from their traditional lands on New Zealand's North Island. Although the Moriori initially welcomed the Maori, the Maori soon made clear their intention to seize the islands. A few Moriori resisted, but as they did not possess firearms the Maori invaders quickly put down these localised insurrections. The Moriori gathered to discuss what could be done. Some advocated attacking the Maori but others favoured obeying Nunuku's law. The latter group prevailed. The Moriori decided to return to their villages and were prepared to live in peace with the Maori and share the resources of their islands, though this accommodation would most likely have been interpreted as cowardice by the warlike Maori. At the time, the Moriori outnumbered the Maori invaders by almost two to one. Before this offer of peaceful coexistence could be relayed, the Maori attacked in force, having mistaken the Moriori gathering for a council of war. The Maori killed 226 men and women – who were then ritually eaten – out of a Moriori population of approximately 1,600.[7]

In a subsequent Native Land Court hearing into the dispossession of the Moriori, a Maori participant in the massacre stated:

[W]e took possession ... in accordance with our customs and we caught all the people. Not one escaped. Some ran away from us, these we killed, and others we killed – but what of that? It was in accordance with our custom ... I am not aware of any of our people being killed by them.[8]

A Moriori survivor recounted that the Maori 'commenced to kill us like sheep … [We] were terrified, fled to the bush, concealed ourselves in holes underground, and in any place to escape our enemies. It was to no avail; we were discovered and killed – men, women and children indiscriminately.'[9]

The Moriori survivors were enslaved, becoming little more than chattels of the Maori. To reinforce their lowly status, the Moriori women were not taken as wives by the Maori, as was the usual Maori tradition. Furthermore, Moriori men and women were prohibited from marrying or cohabiting. Deaths from influenza and other infectious diseases hastened the decline of the Moriori population. In 1862, the surviving Moriori petitioned the British governor of New Zealand, Sir George Grey, for the restoration of their lands. They stated that since their enslavement, 1,336 Moriori had died of 'despair', the survivors then numbering only 101. Although most of the Maoris had left the Chathams by 1870, the Moriori population never recovered from the invasion. The last full-blooded Moriori died in 1933. A later investigation into the fate of the Moriori stated:

[The] Moriori were enslaved in appalling conditions. They were housed in inadequate whare [dwellings], poorly fed, compelled to undertake extreme labour, brutalised, forbidden to marry or to have children, and made to respond to everyone's bidding, including even that of Maori children. For a time before 1842, and possibly for a short while thereafter, they were gratuitously killed at whim.[10]

These are but two of the many historical examples where morality has succumbed to might. In the 5th century BCE, Thucydides, in 'The Melian Dialogue', captured the essence of the moral foundation of international politics.

In matters that concern themselves or their own constitution the Spartans are quite remarkably good; as for their relations with others, that is a long story, but it can be expressed shortly and clearly by saying that of all the people we know the Spartans are most conspicuous for believing that what they like doing is honourable and what suits their interests is just.[11]

The credibility of power

However, it is also worth reflecting upon the four words etched onto the wall of the Korean War Monument in Washington, DC: 'Freedom is not free'. A capable, well-equipped and proven defence force is perhaps the best guarantor of a state's continued existence. For a state's military to constitute a credible deterrent to foreign aggression, certain quantitative and qualitative elements must be present. Quantitatively, the key elements are the number of personnel and the quantity and type of its military platforms, both of which are direct reflections of the willingness of the state to apportion its revenues to military expenditure (the so-called 'guns versus butter' budgetary debate). The qualitative elements must be developed rather than purchased/recruited, and include such factors as professionalism, morale and willingness to fight. While the relative quantitative strength of armies is a matter of force ratios, the relative qualitative strengths are largely a matter of conjecture, though this conjecture is heavily influenced by a state's military reputation from previous conflicts. Put simply, a credible military must not only have the capability to fight, it must also have done so in the recent past. Jonathan Schell expressed this idea when he wrote of the Vietnam War: 'Or was it our goal not so much physically to stop an enemy as to preserve our reputation all around the world as a mighty nation ready and able to use its power to advance its interests and beliefs – to preserve what four presidents called the "credibility of our power".'[12]

In the latter half of the 20th century many nations moved away from mass conscript armies to smaller professional armies staffed by volunteers. This fundamental change in the composition of the military paralleled a decreasing possibility of armies being used for self-defence, as the security of many nation states was now largely guaranteed by the possession of nuclear weapons (or via an alliance with a nuclear power). Consequently, militaries evolved to become more expeditionary in nature and thus more likely to be used to achieve foreign policy objectives, rather than in the direct defence of the nation against an aggressor.

Throughout the last quarter of the 20th century, democratic countries that retained compulsory military service (such as Sweden, Switzerland, Norway, Finland, Germany, Italy and Singapore) were less likely to engage in expeditionary warfare than countries with a professional, volunteer military. From a politically pragmatic viewpoint, leaders are more likely to commit their nations to war if the burden of the actual fighting is borne by marginalised elements of society. It is also worth considering the effect of a country's political leaders having experienced military service, which is more likely to have occurred in countries that have retained conscription.

Military service is not a prerequisite to being an effective political leader. History is rife with triumphant generals who were failures as politicians, but there are also plenty of examples of successful wartime leaders who had no military experience. Previous military service does, however, provide the political leadership with a better understanding of the burden that will be borne by their soldiers if the nation is committed to war. Although he survived the slaughter of the 30-day campaign to capture the small Pacific island of Peleliu, which cost the lives of almost 1,800 American Marines and soldiers – along with 11,000 Japanese soldiers – Eugene Sledge felt that something had died within him. 'Possibly I lost faith that politicians in high places who do not have to endure war's savagery will ever stop blundering and sending others to endure it.'[13] An American soldier fighting his way up the Italian peninsula in May 1944 wrote in a letter to his brother that he was certain there would be another war because 'peace will be settled by men who have never known combat and ... hold no dread of another war.'[14] A veteran of the Falklands War bitterly suggested that: 'If our leaders knew what actually happened in wars, then wars wouldn't happen.'[15]

The majority of the recent political leaders of the major Western democracies have no direct experience of war. A salient example of the dearth of military service among the leaders of democratic nations was the group of neo-conservatives holding key positions in the administration of President George W. Bush who were the principal proponents of an invasion of Iraq in 2003. Republican Senator Chuck Hagel, who had served in Vietnam in 1968 as an infantry squad leader with the US Army's 9th Infantry Division, commented:

> It is interesting to me that many of those who want to rush this country into war and think it would be so quick and easy don't know anything about war. They come at it from an intellectual perspective versus having sat in jungles or foxholes and watched their friends get their heads blown off. I try to speak for those ghosts of the past a little bit.[16]

The Iraq conflict was America's deadliest war since Vietnam. Yet over 60 years ago, a number of the world's elder statesmen set out to establish an organisation that was intended to make the death and destruction experienced in Iraq an anachronism belonging to humankind's violent past. The concept of the United Nations arose out of the devastation of the Second World War. One of its principal aims, as stated in the preamble to the United Nations Charter, is 'to save succeeding generations from the scourge of war'. In the words of its second Secretary-General, Dag Hammarskjöld, the United Nations was created 'not to bring mankind to heaven, but to save it from hell'. But the absence of a military force of its own requires the UN to be dependent on member states to enforce its judgments, and the interests of member states may, and often do, conflict with the aims of the United Nations. The failure of supranational bodies such as the UN to bring about lasting peace has condemned us to live in a world where military might, rather than morality, is often the arbitrator of international disputes.

'The continuation of policy by other means'

This failure to achieve a lasting peace is a political failure, which is compounded by another key fact, succinctly expressed by General William Westmoreland who, when reflecting on the three wars he had fought, concluded: 'The military don't start wars. Politicians start wars.' This sentiment was echoed by the German general Heinz Guderian in his memoirs:

[P]olicy is not laid down by soldiers but by politicians. This has always been the case and is so today. When war starts the soldiers can only act according to the political and military situation as it then exists. Unfortunately it is not the habit of politicians to appear in conspicuous places when the bullets begin to fly. They prefer to remain in some safe retreat and to let the soldiers carry out 'the continuation of policy by other means.'[17]

Guderian is of course referring to a dictum of Carl Von Clausewitz, who had stated some 120 years earlier that 'war is nothing but the continuation of policy with other means'.[18] Clausewitz postulated that subordinating the political point of view to military objectives would be absurd, for it is policy that has created war, therefore military considerations must be subordinate to political objectives. He elaborated further: 'No one starts a war – or rather, no one in his senses ought to do so – without first being clear in his mind what he intends to achieve by that war and how he intends to conduct it. The former is its political purpose, the latter its operational objective.'[19] Yet a related issue is the extent to which the understanding of what is militarily possible affects the political objective.

Clausewitz's view was that 'the political object is the goal; war is the means of reaching it.'[20] However, this axiom presumes a degree of rationality that may be absent in the emotional tumult when a country teeters on the precipice separating peace and war. Logically, no leader would commit his nation to war knowing that it would inevitably lead to the defeat of its armed forces and national humiliation. Yet, as John Keegan emphasises in *The Face of Battle*:

Any objective study quickly reveals, however, that most wars are begun for reasons which have nothing to do with justice, have results quite different from those proclaimed as their objects, if indeed they have any clear-cut result at all, and visit during their course a great deal of casual suffering on the innocent.[21]

The disparity between the intended and actual consequences of fighting a war arises from several factors. Wars are usually fought with but not by machines. The human factor dispels any pretence of mathematical certainty, that is, the numerically superior side (assuming equality in the amount of armaments and capabilities of weapon platforms) will not always be victorious. On the field of battle, fighting spirit, morale or just plain luck may prove to be the decisive factor on the day, rather than the quantity of soldiers.

Moreover, war generates extremes of human behaviour, from selfless bravery to cowardice, from compassion to cruelty and from love to hate. These behavioural extremes defy rationality and increase the unpredictability of war. Furthermore, technological superiority in weapons may be minimised or even negated by an adaptable enemy who, where possible, will choose to fight on ground, or in a manner of, his own choosing. For the enemy also has a seat at the table and will be doing his utmost to disrupt and thwart the war plans of his opposition. Hence the comment by Lieutenant General William S. Wallace, commander of the American V Corps during Operation Iraqi Freedom in March 2003 who, having become frustrated by the ruthless fanaticism of the Fedayeen opposition, let slip that: 'The enemy we're fighting is different than the one we'd war gamed against …'[22]

This human factor affects politicians as well as soldiers. Politicians are subject to self-doubt, intractability or baseless optimism. Clausewitz noted the short-term perspective of leaders who, confident in the ability of their country's military to vanquish their opponent, embrace war as an instrument of state policy. He commented that some politicians when advocating war 'evade all rigorous conclusions proceeding from the nature of war, bother little about ultimate possibilities, and concern [themselves] only with immediate possibilities.'[23] Clausewitz cautions politicians 'not to take the first step without considering the last' when considering committing their country to war.[24]

This brings us back to the paradigm of war serving a political purpose. The primary goal of any political party is to gain and then retain power. All subsequent actions of any ruling clique, whether it attained its political ascendancy via the ballot or via the bullet, need to be examined through this prism. That is not to say that politicians never act in the public good, but rather it is self-evident that you need to be in power to implement your agenda. Politicians' actions can also accord with the principal-agent problem, whereby having been elected to serve the interests of their constituents they may choose instead to further their own interests, even if these are contrary to the best interests of the nation. Consequences of the principal-agent problem are excessive anxiety about their reputation and an inability to admit mistakes. This factor can cause politicians to persist with a course of military action, eschewing the notion of sunk costs and unable to admit that the potential benefits of continuing are greatly outweighed by the cost of doing so. Though in fairness, the actual cost (and consequences) of a war can often be determined only with hindsight.

Clausewitz argued that the political basis for war modified war's nature and prevented it from degenerating to its absolute form:

> Since war is not an act of senseless passion but is controlled by its political object, the value of this object must determine the sacrifices to be made for it in *magnitude* and also in *duration*. Once the expenditure of effort exceeds the value of the political object, the object must be renounced and peace must follow.[25]

But the key issue here remains one of political objectives. In the absence of any great unifying cause, such as the threat of invasion, domestic issues generally dominate politics. As domestic issues affect the day-to-day lives of constituents, the impact of these issues can be readily quantified. For example, voters easily understand the implication of increasing unemployment and rising prices and they often have a personal stake in the outcome of political decisions concerning these issues. Foreign policy, however, is outside the direct experience of all but a few specialists, such as diplomats, journalists, academics and soldiers. Psychologists have observed that people are apt to magnify the likelihood of sensationalised though remote possibilities, while minimising or discounting much more probable scenarios. Such is often the case with foreign policy, where ignorance and stereotyping, coupled with media sensationalism, results in a distorted picture of the threat posed by foreign adversaries or the value to the nation of foreign policy objectives. Therefore, politicians may emphasise an external threat to divert public attention from domestic to foreign affairs.

The political objective of a war may change over the course of the conflict. This factor was recognised by Clausewitz, who observed that 'the original political objects can greatly alter during the course of the war and may finally change entirely *since they are influenced by events and their probable consequences*.'[26] American forces invaded Iraq in March 2003 to seek out and destroy Saddam's weapons of mass destruction and because of alleged Iraqi links to al-Qaeda, thereby linking the invasion to the proclaimed Global War on Terror. It was only after no such weapons were found, and the links to al-Qaeda had proved extremely tenuous, that the stated political objective became that of bringing democracy to the Iraqi people. It was the prerogative of the Bush administration to modify the political object it sought to achieve from the military's invasion and occupation of Iraq. The key question, however, was whether the value of the revised political objective still justified the cost (both sunk and ongoing) of embarking on this enterprise. Returning to Clausewitz, if 'in the end his [the leader's] political object will not seem worth the effort it costs [he] must then renounce his policy.'[27] Two questions arise: how is the cost of war measured, and who must bear the cost?

The true cost of war

The cost of war may be measured in simplistic terms, such as the number of soldiers killed or wounded and the amount of matériel destroyed. Conversely, the cost may be measured in increasing levels of complexity: the growth of the military budget, ongoing medical costs for wounded veterans, extent of regional destabilisation brought about by the conflict and so on. The issue of who must bear this cost is also multifaceted, but one factor is clear: even in the most militarised of countries, the responsibility for the fighting and the dying is not shared equally. A small minority of men and women bear the burdens of war so that the rest of their society will not have to. This book has sought to reveal the nature of these burdens. Political leaders commonly exhort a nation's citizenry to 'get behind the war effort and support the troops', but what they are really seeking is for the citizens to support the political decisions that committed the soldiers to war. Rather than succumbing to jingoism, citizens could better support their soldiers by attempting to better understand the tasks that society asks of them and ensuring that they are not committed to war because of ill-conceived political decisions. It is possible to oppose the war but still honour the sacrifices of the soldier.

Political power in a democracy is ultimately dependent on the consent of the governed. The greatest demand any society can make upon its members is to send them off to war. For the majority, who will not have to endure the terror, the frustration and the loneliness of the battlefield, the very least we owe those proxies whom we have dispatched in our place is to hold our governments to the highest standard of responsibility. War should always be the last resort to resolve a dispute.

Political leaders should resist succumbing to the seductiveness of a 'surgical military strike' to solve a seemingly intractable foreign policy dilemma, for while it is a relatively straightforward task to blow something up, what then? War often has its own inescapable momentum. Once a nation has embarked on a course of military action, no matter how ill-advised, it will find it very difficult to disengage its troops. The propaganda that served to sanctify the 'cause' and generate public support for the initial deployment of troops now drags behind the nation's political leaders like an anchor, restricting their freedom of action. International treaty obligations also impinge on the options available to wartime leaders; for example, the Hague Convention of 1907, which sought to mitigate the excesses of war on land, states:

> The authority of the legitimate power having in fact passed into the hands of the occupant, the latter shall take all the measures in his power to restore, and ensure, as far as possible, public order and safety, while respecting, unless absolutely prevented, the laws in force in the country.[28]

These obligations, coupled with the reluctance of politicians to admit they made a mistake and underestimated the requirements to accomplish a mission, will result in an armed intervention continuing to its bloody conclusion, often wrapped up in sanctimonious rhetoric about ensuring that the soldiers killed thus far have not died in vain.

In *Knights in White Armour*, the journalist Christopher Bellamy comments that the Russian term *negodnos*, which he translates as the 'unusability or inappropriateness of military force for solving certain problems', best encapsulates the fact that not all foreign policy challenges have a military solution. Bellamy emphasises that: 'However military technology advances, however precise the means of delivery or "surgical" the aims, war remains bestial. There are no nice ways of killing people.'[29]

The decision to go to war is often a case of 'bounded rationality', whereby judgements are based on imperfect information and awkward facts and possibilities are rationalised away during the planning process. Hence the inadequate preparation by the American forces to contain the Iraqi insurgency once the major combat phase of the campaign was over. Yet while some factors will always be unknowable, the past can provide a guide.

Any leader who sets their country upon the road to war should first consider the historical record. Would Adolf Hitler have invaded Poland in September 1939 if he had foreseen that this decision would ultimately lead to his suicide in the ruins of his Berlin bunker, the almost total devastation of Germany and its partition for almost half a century? Likewise, would General Hideki Tojo and the Japanese High Command have authorised the attack on Pearl Harbor knowing that this attack would ultimately lead to the Japanese losing all of their overseas possessions, two of their major cities being devastated by atomic bombs and the rest reduced to ruin by conventional bombing, and the occupation of Japan by foreign forces? A military adage is that 'no plan survives the first shot' – well might it be said that war aims are equally vulnerable.

We should not be bound by the lessons of history but neither should we ignore them. Shelford Bidwell writes in *Modern Warfare*:

The Vietnam War is an example of the appalling dangers and difficulties which arise from rashly embarking on war without a careful appreciation of costs and benefits and of the possibility of it running out of control as the dams holding up the reservoirs of human aggression are breached.[30]

Bidwell's analogy of a dam holding back the reservoir of human aggression is particularly apt. The inherent violence of a military attack invariably generates in its wake animosity, aggression and a desire for revenge. 'A feeling of profound gratitude that I was in a position to get revenge for 9/11 surged through me,' stated Nathaniel Fick, an officer in the US Marine Corps. 'Its intensity was startling. It wasn't just a professional interest in finally doing what I'd trained so long to do. It was personal. I wanted to find the people who had planned the attack on America and put their heads on stakes.'[31]

The animosity towards the enemy aroused by the initial attack generally ignores one salient fact: those who planned and then ordered the attack are usually insulated from direct retribution by formations of men who had little or no direct responsibility for this action. It is these men who must face the initial retaliatory strike. As the conflict develops, national objectives are obscured by a self-sustaining cycle of violence, as each localised action leaves in its wake a desire for 'pay-back'. As Fick advanced into Iraq he saw medevac helicopters flying dead and wounded Marines to the rear. He recalled: 'emotion began to creep in. I was angry. I wanted revenge. For the first time, my blood was up.'[32]

Yet it appears that this particular lesson of human behaviour in war must be continually relearnt. Three decades after the tragedy of the Vietnam War, men who had largely avoided serving in this war committed their nation to another drawn-out conflict by invading Iraq – their glib proclamations to the contrary notwithstanding. Once again, the political leaders professed surprise, if not indeed dismay, as the casualties in Iraq – both civilian and military – continued to mount and the country slipped further towards anarchy.

Perhaps the greatest paradox of war is how quickly the spontaneous public outbursts of enthusiasm and support for soldiers marching off to war are subsumed by a deepening awareness of the horror that has been unleashed. It is not without irony that we choose to commemorate the ending of wars rather than their beginnings, as if to reaffirm our willingness to embrace peace. Yet we are cursed with a short memory, and it is dismaying to note how quickly we fall into step when the war drums sound. In *War Is a Force That Gives Us Meaning*, Chris Hedges, reflecting upon the numerous conflicts that he had covered as a foreign correspondent, surmised that:

> [W]ar is a drug ... It is peddled by mythmakers – historians, war correspondents, filmmakers, novelists, and the state – all of whom endow it with qualities it often does possess: excitement, exoticism, power, chances to rise above our small stations in life, and a bizarre and fantastic universe that has a grotesque and dark beauty. It dominates culture, distorts memory, corrupts language, and infects everything around it ...[33]

* * *

This book has not sought to contribute to the philosophical and moral debate about whether a war can be considered 'just'. Clearly, in some circumstances aggression can be countered only by aggression and the consequences of not taking up arms outweigh the cost of doing so. The second half of the 20th century would have been very different indeed if the totalitarian regimes of Germany, Japan and Italy had not been opposed by the Western democracies and the Soviet Union during the Second World War. As one who has stood in the remnants of the death camp at Auschwitz II-Birkenau, I am not so idealistic, nor so historically short-sighted, as to state that war never produces positive outcomes. But I feel that I agree with Bill Mauldin that even these positive outcomes arise out of embracing a lesser evil to see off a greater evil. 'I didn't feel we had accomplished anything positive,' Mauldin reflected some 40 years after the end of the Second World War. 'We had destroyed something negative: Hitler.'[34]

Rather than debate the justification put forward, often retrospectively, for various wars, this book has tried to reveal the burden of the soldier, for it is he (and she) who must bear a disproportionate amount of the suffering unleashed by war. We would all do well, particularly those leaders who commit their nations to wars of choice rather than wars where national survival is at stake, to dwell upon the musings of General Dwight D. Eisenhower, who commented in 1946, looking back on a lifetime spent in the military: 'I hate war as only a soldier who has lived it can, only as one who has seen its brutality, its stupidity.'[35]

ACKNOWLEDGEMENTS

Extracts from Mansur Abdulin, *Red Road from Stalingrad* reproduced by
permission of Pen and Sword Military.

Extracts from Lawrence M. Baskir and William A. Strauss, *Change and
Circumstance: The Draft, The War, and the Vietnam Generation* reproduced by
permission of Random House, Inc.

Extracts from Martin Bell, *Through Gates of Fire* reproduced by permission of
Weidenfeld and Nicolson, an imprint of The Orion Publishing Group, London.

Extracts from Andrew Exum, *This Man's Army* (© 2004 by Andrew Exum)
reproduced by permission of Gotham Books, an imprint of Penguin Group
(USA) Inc.

Extracts from Nathaniel Fick, *One Bullet Away: The Making of a Marine Officer*
reproduced by permission of Weidenfeld and Nicolson, an imprint of The
Orion Publishing Group, London.

Extracts from Paul Fussell, *The Great War and Modern Memory* reproduced by
permission of Oxford University Press, Inc.

Extracts from Guy Gibson, *Enemy Coast Ahead – Uncensored* reproduced by
permission of David Higham Associates.

Extracts from Robert Graves, *Goodbye to All That* reproduced by permission of
Carcanet Press Limited.

Extracts from J. Glenn Gray, *The Warriors: Reflections on Men in Battle* reproduced
by permission of Lisa Gray Fisher and Sherry Martin.

Extracts from General Heinz Guderian, *Panzer Leader* reproduced by permission
of Penguin Books Limited.

Extracts from Colonel David Hackworth and Julie Sherman, *About Face*
reproduced by permission of Eilhys England Hackworth and Pan Macmillan
Australia Pty Ltd.

Extracts from Max Hastings, *Going to the Wars* reproduced by permission of Pan
Macmillan.

Extracts from Joshua Key, *The Deserter's Tale: Why I Walked Away From the War in
Iraq* reproduced by permission of Anansi Press.

Extracts from Ron Kovic, *Born on the Fourth of July* reproduced by permission of
Akashic Books.

Extracts from John Lawrence and Robert Lawrence MC, *When the Fighting is Over*
reproduced by permission of Bloomsbury.

Extracts from Harold P. Leinbaugh and John D. Campbell, *The Men of Company K*
reproduced by permission of HarperCollins Publishers.

Extracts from Gary McKay, *In Good Company: One Man's War in Vietnam* reproduced by permission of Allen and Unwin Pty Ltd.

Extracts from James M. McPherson, *The Battle Cry of Freedom* reproduced by permission of Oxford University Press, Inc.

Extracts from William Manchester, *Goodbye Darkness: A Memoir of the Pacific War* reproduced by permission of Don Congdon Associates Inc.

Extracts from Robert Mason, *Chickenhawk* reproduced by kind permission of Robert Mason and Penguin Group (USA) Ltd.

Extracts from Dan Mills, *Sniper One* reproduced by permission of Penguin Books Ltd.

Extracts from Lord Moran, *The Anatomy of Courage* (© 1945 & 2007; this edition Constable & Robinson Ltd, 2007) reproduced by kind permission of the publisher.

Extracts from John Peters and John Nichol, *Tornado Down* reproduced by permission of Penguin Books Ltd.

Extracts from Lewis B. Puller, Jr., *Fortunate Son: The Healing of a Vietnam Vet* copyright © 1991 by Lewis B. Puller, Jr. Used by permission of Grove/Atlantic, Inc.

Extracts from Colin P. Sisson, *Wounded Warriors* reproduced by permission of Colin P. Sisson.

Extracts from Eugene B. Sledge, *With the Old Breed at Peleliu and Okinawa* (© 1981 by E.B. Sledge) reproduced by permission of Presidio Press, an imprint of The Ballantine Publishing Group, a division of Random House Inc.

Extracts from Kayla Williams, *Love My Rifle More Than You: Young and Female in the US Army* reproduced by permission of Weidenfeld and Nicolson, an imprint of The Orion Publishing Group, London.

Extracts from Cecil Woodham-Smith, *The Reason Why* reproduced by permission of Constable & Robinson Ltd.

Every effort has been made to contact copyright holders for works quoted in this book. If notified, the publisher will be pleased to acknowledge the use of copyright material in future editions.

BIBLIOGRAPHY

Mansur Abdulin, *Red Road from Stalingrad: Recollections of a Soviet Infantryman*, Pen and Sword Military, Barnsley, South Yorkshire, 2004

Patsy Adam-Smith, *The Anzacs*, Nelson, Melbourne, 1985 reissue

Mark Adkin, *Goose Green: A Battle is Fought to be Won*, Leo Cooper, London, 1992

Alexander Aitken, *Gallipoli to the Somme: Recollections of a New Zealand Infantryman*, Oxford University Press, London, 1963

Don E. Alberts (editor), *Rebels on the Rio Grande: The Civil War Journals of A.B. Peticolas*, University of New Mexico Press, Albuquerque, 1984

Hervey Allen, *Toward the Flame: A War Diary*, University of Pittsburgh Press, 1968 edition

Stephen E. Ambrose, *Citizen Soldiers: The US Army from the Normandy Beaches to the Bulge to the Surrender of Germany June 7, 1944–May 7, 1945*, Touchstone edition, Simon & Schuster, New York, 1998

—— *Band of Brothers: E Company, 506th Regiment, 101st Airborne from Normandy to Hitler's Eagle's Nest*, Simon & Schuster, New York, 2001 edition

Ellery Anderson, *Banner Over Pusan*, Evan Brother Ltd, London, 1960

James Anton, *Royal Highlander: A soldier of HM 42nd (Royal) Highlanders during the Peninsular, South of France & Waterloo Campaigns of the Napoleonic Wars*, Leonaur, 2007 edition

Peter Arnett, *Live from the Battlefield: From Vietnam to Baghdad, 35 Years in the World's War Zones*, Touchstone edition, Simon & Schuster, New York, 1995

Max Arthur, *Above All, Courage – The Falklands Front Line: First-Hand Accounts*, Guild Publishing, London, 1985

—— *Forgotten Voices of the Great War*, Ebury Press, London, 2003 edition

—— *Forgotten Voices of the Second World War*, Ebury Press, London, 2004

Major Phil Ashby, *Unscathed: Escape from Sierra Leone*, Pan Books, London, 2003

Tony Ashworth, *Trench Warfare 1914–1918: The Live and Let Live System*, Pan edition, London, 2000

Rick Atkinson, *Crusade: The Untold Story of the Gulf War*, HarperCollins, London, 1993

—— *In the Company of Soldiers: A Chronicle of Combat in Iraq*, Little, Brown, London, 2004

Jim Bailey, *The Sky Suspended: A Fighter Pilot's Story*, Bloomsbury, London, 2005 edition

Horace L. Baker, *Argonne Days in World War I*, University of Missouri Press, Columbia, Missouri, 2007

Jean-Baptiste Barres, *Chasseur Barres*, Leonaur, 2006 edition

Lawrence M. Baskir and William A. Strauss, *Change and Circumstance: The Draft, The War, and the Vietnam Generation*, Alfred A. Knopf, New York, 1978

Lujo Bassermann, *The Oldest Profession: A History of Prostitution*, Dorset Press, New York, 1993

K. Jack Bauer (editor), *Soldiering: The Civil War Diary of Rice C. Bull*, Berkley Books, New York, 1988 edition

Dan Baum, 'The Price of Valor', *The New Yorker*, 12 &19 July 2004, pp. 44–52

John Baynes, *Morale: A Study of Men and Courage*, Avery Publishing Group, New York, 1988 edition

Henry Baynham, *From the Lower Deck: the old navy 1780–1840*, Hutchinson, London, 1969

Charles Edwin Woodward Bean, *Official History of Australia in the War of 1914–18: Vol 1, The Story of ANZAC*, Angus & Robertson, Sydney, 1937

—— *The Official History of Australia in the War of 1914–1918: Volume IV, The AIF in France: 1917*, third edition, Angus & Robertson, Sydney, 1935

—— *The Official History of Australia in the War of 1914–1918: Volume III, The AIF in France: 1916*, fourth edition, Angus & Robertson, Sydney, 1936

Antony Beevor, *Berlin: The Downfall 1945*, Viking, New York, 2002

Antony Beevor with Luba Vinogradova, (editors and translators of) Vasily Grossman, *A Writer at War: Vasily Grossman with the Red Army 1941–1945*, Pantheon Books, New York, 2005

Johnson Beharry VC, *Barefoot Soldier: A Story of Extreme Valour*, Sphere, London, 2006

Martin Bell, *Through Gates of Fire: A Journey into World Disorder*, Phoenix, London, 2004 edition

Christopher Bellamy, *Knights in White Armour*, Pimlico edition, London, 1997

David Bellavia with John Bruning, *House to House: The Most Terrifying Battle of the Iraq War – Through the Eyes of the Man who Fought it*, Pocket Books, London, 2008 edition

Phyllis Bennis and Erik Leaver, *The Iraq Quagmire: The Mounting Costs of War and the Case for Bringing Home the Troops*, a study by the Institute for Policy Studies and Foreign Policy in Focus, 31 August 2005

Evgeni Bessonov, translated by Bair Irincheev, *Tank Rider: Into the Reich with the Red Army*, Greenhill Books, London, 2003

Gottlob Herbert Bidermann, *In Deadly Combat: A German Soldier's Memoir of the Eastern Front*, translated and edited by Derek S. Zumbro, University Press of Kansas, Lawrence, Kansas, 2000

Shelford Bidwell, *Modern Warfare: A Study of Men, Weapons and Theories*, Allen Lane, London, 1973

Linda Bilmes and Joseph E. Stiglitz, *The Economic Costs of the Iraq War: An Appraisal Three Years After the Beginning of the Conflict*, paper prepared for presentation at the ASSA meeting, Boston, January 2006

Michael Bilton and Peter Kosminsky, *Speaking Out: Untold Stories from the Falklands War*, André Deutsch, London, 1989

Jeremy M. Black, 'The Execution of Admiral Byng', *MHQ: The Quarterly Journal of Military History*, Spring 1999, Vol. 11, No. 3, pp. 98–103

Elzéar Blaze, *Captain Blaze: Life in Napoleon's Army*, Leonaur, 2007 edition

Nick Bleszynski, *Shoot Straight, You Bastards!: The Truth behind the killing of 'Breaker' Morant*, Random House, Sydney, 2003 edition

Marc Bloch, *Memoirs of War, 1914–15*, translated by Carole Fink, Cornell University Press, Ithaca, New York, 1980

Captain Walter Bloem, *The Advance from Mons*, Tandem Books, London, 1967

Daniel J. Blumlo, 'How the Common Grunt and Prostitute Changed Military Policy', thesis submitted to the Department of History, Florida State University, for the degree of Masters of Arts, 2004, http://etd.lib.fsu.edu/theses

Edmund Blunden, *Undertones of War*, Penguin Books, London, 2000 edition

Mark M. Boatner III, *The Biographical Dictionary of World War II*, Presidio, Novato, California, 1996

Adrien Jean Baptiste François Bourgogne, *Memoirs of Sergeant Bourgogne 1812–1813*, Constable, London, 1997 edition

Joanna Bourke, *An Intimate History of Killing: Face to Face Killing in Twentieth-Century Warfare*, Basic Books, London, 1999

Alex Bowlby, *The Recollections of Rifleman Bowlby*, Corgi, London, 1971

John S. Bowman (editor), *The Vietnam War: An Almanac*, Bison Books, New York, 1985

—— (editor), *The Vietnam War: Day by Day*, Bison Books, London, 1989

Russell Braddon, *The Naked Island*, Penguin, Camberwell, Victoria, 1993

Sarah Bradford, *Elizabeth: A Biography of Britain's Queen*, Farrar, Straus & Giroux, New York, 1996

Omar N. Bradley, *A Soldier's Story*, Henry Holt & Company, New York, 1951

Rick Bragg, *I Am a Soldier, Too: The Jessica Lynch Story*, Alfred A. Knopf, New York, 2003

Colin H. Brown, *Stalemate in Korea: And How we Coped 1952–1953*, Australian Military History Publications, Sydney, 1997

Malcolm Brown and Shirley Seaton, *Christmas Truce*, Pan Books, London, 2001 edition

Malcolm Brown, *The Imperial War Museum Book of the Western Front*, Pan Books, London, 2001 edition

—— *The Imperial War Museum Book of the First World War*, Pan Books, London, 2002 edition

Gary P. Bruner, 'Military–Media Relations in Recent US Operations', US Army Management Staff College, 14 November 1997

The Bryce Report: Report of the Committee on Alleged German Outrages, 12 May 1915

Alan Bullock, *Hitler: A Study in Tyranny*, Odhams Press, London, 1952

B.R. Burg (editor), *Gay Warriors: A Documentary History from the Ancient World to the Present*, New York University Press, New York, 2002

Colonel A.G. Butler, *The Official History of the Australian Army Medical Services in the War of 1914–1918: Volume 1 – Gallipoli, Palestine and New Guinea*, Australian War Memorial, Melbourne, second edition, 1938

—— *The Official History of the Australian Army Medical Services in the War of 1914–1918: Volume II – The Western Front*, Australian War Memorial, Canberra, 1940

Colby Buzzell, *My War: Killing Time in Iraq*, G.P. Putnam's Sons, New York, 2005

Colin Frederick Campbell, *Letters from Camp to his Relatives During the Siege of Sebastopol*, Richard Bentley & Son, London, 1894, pp. 100–01

Philip Caputo, *A Rumor of War*, Henry Holt & Company, New York, 1996 reprint

Baron Peter Alexander Rupert Carrington, *Reflect on Things Past: The Memoirs of Lord Carrington*, Collins, London, 1988

John O. Casler, *Four Years in the Stonewall Brigade*, University of South Carolina Press, Columbia, South Carolina, 2005 edition

Sergio Catignani, 'Motivating Soldiers: The Example of the Israeli Defense Force', *Parameters*, Autumn 2004

Captain Tania M. Chacho, 'Why Did They Fight? American Airborne Units in World War II', *Defence Studies*, Vol. 1, No. 3, Autumn 2001, pp. 59–94

John Whiteclay Chambers II (editor-in-chief), *The Oxford Companion to American Military History*, Oxford University Press, Oxford, 1999

Otto Preston Chaney Jr., *Zhukov*, David & Charles, Devon, 1972

Iris Chang, *The Rape of Nanking: The Forgotten Holocaust of World War II*, Penguin Books, New York, 1998

Guy Chapman, *A Passionate Prodigality*, Fawcett Crest edition, New York, 1967

D.E. Charlwood, *No Moon Tonight*, Pacific Books, Angus & Robertson, Sydney, 1956

Tom Clancy with General Fred Franks Jr., *Into the Storm: A Study in Command*, Berkley Books, New York, 1998 edition

Henry Clifford, *Henry Clifford VC: his letters and sketches from the Crimea*, Michael Joseph, London, 1956

Micheal Clodfelter, *Vietnam in Military Statistics: A History of the Indochina Wars, 1772–1991*, McFarland & Company, Jefferson, North Carolina, 1995

Hamlin Alexander Coe, *Mine Eyes Have Seen the Glory: Combat Diaries of Union Sergeant Hamlin Alexander Coe*, Associated University Presses, Cranbury, New Jersey, 1975

Jean-Roch Coignet, *Captain Coignet: A Soldier of Napoleon's Imperial Guard from the Italian Campaign to Waterloo*, Leonaur, 2007 edition

Tim Collins, *Rules of Engagement: A Life in Conflict*, Headline, London, 2005

George Coppard, *With a Machine Gun to Cambrai: A Story of the First World War*, Cassell, London, 1999 edition

Major General Patrick Cordingley, *In the Eye of the Storm: Commanding the Desert Rats in the Gulf War*, Coronet Books, Hodder & Stoughton, London, 1997 edition

Cathryn Corns and John Hughes-Wilson, *Blindfold and Alone: British Military Executions in the Great War*, Cassell, London, 2005 edition

General Peter Cosgrove, *My Story*, HarperCollins, Sydney, 2006

Edward Costello, *Rifleman Costello: The Adventures of a Soldier of the 95th (Rifles) in the Peninsular & Waterloo Campaigns of the Napoleonic Wars*, Leonaur, 2005 edition

John Costello, *Love, Sex and War: Changing Values 1939–45*, Collins, London, 1985

Gunnery Sergeant Jack Coughlin, USMC and Captain Casey Kuhlman, USMCR, with Donald A. Davis, *Shooter*, St Martin's Press, New York, 2006

William Craig, *Enemy at the Gates: The Battle for Stalingrad*, Penguin, London, 2000

John Crawford, *The Last True Story I'll Ever Tell: An Accidental Soldier's Account of the War in Iraq*, Riverhead Books, New York, 2005

John Crawford (editor), *No Better Death: The Great War Diaries and Letters of William G. Malone*, Reed Publishing, 2005

Major Robert Crisp, *Brazen Chariots*, Ballantine Books, New York, 1961

Louis Crompton, *Homosexuality and Civilization*, Harvard University Press, Cambridge, Massachusetts, 2003

D.G. Crotty, *Four Years Campaigning in the Army of the Potomac*, Dygert Brothers & Co., Grand Rapids, Michigan, 1874

Brigadier F.P. Crozier, *A Brass Hat in No-Man's Land*, Jonathan Cape, London, 1930

F.M. Cutlack, *Breaker Morant: A Horseman who made History*, Ure Smith, Sydney, 1962

Sameera Dalvi, 'Homosexuality and the European Court of Human Rights: Recent Judgments Against the United Kingdom and Their Impact on Other

Signatories to the European Convention of Human Rights', *University of Florida Journal of Law and Public Policy*, Vol. 15, December 2004, pp. 468–512

William C. Davis (editor), *Diary of a Confederate Soldier: John S. Jackman of the Orphan Brigade*, University of South Carolina Press, Columbia, South Carolina, 1990

—— *Rebels and Yankees: The Fighting Men of the American Civil War*, Salamander Books, London, 2001 reprint

William C. Davis, Brian C. Pohanka and Don Troiani (editors), *Civil War Journal: The Leaders*, Rutledge Hill Press, Nashville, Tennessee, 1997

Francis W. Dawson, *Reminiscences of Confederate Service 1861–1865*, Louisiana State University Press, Baton Rouge, Louisiana, 1980

Yael Dayan, *A Soldier's Diary: Sinai 1967*, Penguin, Middlesex, 1969 edition

General Sir Peter De La Billière, *Storm Command: A Personal Account of the Gulf War*, HarperCollins, London, 1992

—— *Looking for Trouble*, HarperCollins, London, 1994

Peter Dennis, Jeffrey Gery, Ewan Morris and Robin Prior, *The Oxford Companion to Australian Military History*, Oxford University Press, Melbourne, 1995

Carlo D'Este, *A Genius For War: A Life of General George S. Patton*, HarperCollins, London, 1996

Jared Diamond, *Guns, Germs, and Steel: The Fates of Human Societies*, W.W. Norton & Company, New York, 1999

Christopher Dickey, *Expats: Travels in Arabia, From Tripoli to Tehran*, Atlantic Monthly Press, New York, 1990

Paul Dickson and Thomas B. Allen, 'Marching on History', *Smithsonian*, February 2003, pp. 84–94

Norman Dixon, *On the Psychology of Military Incompetence*, Pimlico, London, 1976

Kathy Dobie, 'AWOL in America: When desertion is the only option', *Harper's*, Vol. 310, No. 1858, March 2005, pp. 33–44

James C. Donahue, *No Greater Love: A Day with the Mobile Guerrilla Force in Vietnam*, New American Library, New York, 1989

Frances Donaldson, *Edward VIII*, J.B. Lippincott Company, Philadelphia, 1975

Joseph Donaldson, *Donaldson of the 94th Scots Brigade: The Recollections of a Soldier During the Peninsula & South of France Campaigns of the Napoleonic Wars*, Leonaur, 2008 edition

David Donovan, *Once a Warrior King*, Corgi edition, London, 1988

Robert J. Donovan, *PT 109: John F. Kennedy in WW II*, McGraw-Hill, New York, 2001 edition

Walter Hubert Downing, *To the Last Ridge: The WWI Experiences of W.H. Downing*, Duffy & Snellgrove, Sydney, 1998

Frederick Downs, *The Killing Zone: My Life in the Vietnam War*, W.W. Norton & Company, New York, 2007 edition

Wayland Fuller Dunaway, *Reminiscences of a Rebel*, The Neale Publishing Company, New York, 1913

T. Duncan, M. Stout, *Official History of New Zealand in the Second World War 1939–45: Medical Services in New Zealand and the Pacific*, War History Branch, Department of Internal Affairs, Wellington, New Zealand, 1958

Hugh Dundas, *Flying Start*, Penguin, 1990

Michael J. Durant with Steven Hartov, *In the Company of Heroes*, Bantam Press, London, 2003

Gwynne Dyer, *War*, The Bodley Head, London, 1986

David Eisenhower, *Eisenhower: At War 1943–1945*, Random House, New York, 1986

John Ellis, *The Sharp End: The Fighting Man in World War II*, Charles Scribner's Sons, New York, 1980

Vasily B. Emelianenko, *Red Star Against the Swastika: The Story of a Soviet Pilot over the Eastern Front*, Greenhill Books, London, 2005

Douglas K. Evans, *Sabre Jets Over Korea: A Firsthand Account*, Tab Books, Blue Ridge Summit, Pennsylvania, 1984

Matt Eversmann and Dan Schilling (editors), *The Battle of Mogadishu: Firsthand Accounts from the Men of Task Force Ranger*, Ballantine Books, New York, 2004

Joseph H. Ewing, 'The New Sherman Letters', *American Heritage Magazine*, Vol. 38, No. 5, July–August 1987

Andrew Exum, *This Man's Army: A Soldier's Story from the Front Lines of the War on Terrorism*, Gotham Books, New York, 2005 edition

Rainer Fabian and Hans Christian Adam, translated by Fred Taylor, *Images of War: 130 Years of War Photography*, New English Library, London, 1985

Bernard B. Fall, *Hell in a Very Small Place: The Siege of Dien Bien Phu*, Pall Mall Press, Oxford, 1966

Captain Anthony Farrar-Hockley, *The Edge of the Sword*, Readers' Book Club, 1955

Niall Ferguson, *The Pity of War: Explaining World War I*, Basic Books, New York, 1999

—— *Empire: How Britain Made the Modern World*, Penguin Books, London, 2004 edition

Joachim C. Fest, *Hitler*, Weidenfeld & Nicolson, London, 1974

Nathaniel Fick, *One Bullet Away: The Making of a Marine Officer*, Weidenfeld & Nicolson, London, 2005

Laurie Field, *The Forgotten War: Australia and the Boer War*, Melbourne University Press, Melbourne, 1995 edition

William A. Fletcher, *Rebel Private: Front and Rear – Memoirs of a Confederate Soldier*, Meridian, New York, 1997 edition

John Foley, *Mailed Fist*, Mayflower, 1975 edition

George and Anne Forty, *They Also Served: A Pictorial Anthology of Camp Followers through the Ages*, Midas Books, Speldhurst, 1979

Robert Lane Fox, *Alexander the Great*, Penguin Books, London, 2004 edition

Sergeant Major Henry Franks, *Leaves from A Soldier's Note Book*, Mitre Publications, 1979 edition

General Tommy Franks with Malcolm McConnell, *American Soldier*, Regan Books, New York, 2004

Sir Lawrence Freedman, *The Official History of the Falklands Campaign, Volume II: War and Diplomacy*, Routledge, Taylor & Francis Group, London, 2005

'Front Lines and Deadlines – Perspectives on War Reporting', *Media Studies Journal*, Vol. 15, No. 1, Summer 2001

Major General John Frost, *2 Para Falklands: The Battalion at War*, Sphere Books, London, 1984

Paul Fussell, *Wartime: Understanding and Behaviour in the Second World War*, Oxford University Press, New York, 1989

—— *Doing Battle: The Making of a Skeptic*, Little Brown & Company, Boston, 1996

—— *The Great War and Modern Memory*, 25th anniversary edition, Oxford University Press, New York, 2000

Richard A. Gabriel, *No More Heroes: Madness and Psychiatry in War*, Hill & Wang, New York, 1987

Reuven Gal, 'Commitment and Obedience in the Military: An Israeli Case Study', *Armed Forces & Society*, Vol. 11, No. 4, Summer 1985, pp. 553–64

—— *A Portrait of the Israeli Soldier*, Greenwood, New York, 1986

Bill Gammage, *The Broken Years: Australian Soldiers in the Great War*, Penguin, Melbourne, 1982 edition

Atul Gawande, 'Casualties of War: Military Care for the Wounded from Iraq and Afghanistan', *New England Journal of Medicine*, Vol. 135, No. 24, pp. 2471–5

Aparisim Ghosh, 'Inside the Mind of an Iraqi Suicide Bomber', *Time*, 4 July 2005, pp. 20–5

Bobby Ghosh, 'The Enemy's New Tools', *Time*, 25 June 2007, pp. 30–4

Nancy Gibbs, 'The Lucky Ones', *Time*, 21 March 2005, pp. 24–5

Guy Gibson, *Enemy Coast Ahead – Uncensored*, Crécy Publishing, Manchester, 2005 edition

Henry G. Gole, 'Don't Kill the Messenger: Vietnam War Reporting in Context', *Parameters*, Winter 1996, pp. 144–59

Timothy Gowing, *Voice from the Ranks: A Personal Narrative of the Crimean War by a Sergeant of the Royal Fusiliers*, William Heinemann, Melbourne, 1954

Robert Graves, *Goodbye to All That*, Blue Ribbon Books, New York, 1930

J. Glenn Gray, *The Warriors: Reflections on Men in Battle*, Bison Books, Lincoln, Nebraska, 1998 edition

Paddy Griffith (editor), *British Fighting Methods in the Great War*, Frank Cass, London, 1996

Georg Grossjohann, *Five Years, Four Fronts*, translated by Ulrich Abele, The Aberjona Press, Bedford, Pennsylvania, 1999

Dave Grossman, *On Killing: The Psychological Cost of Learning to Kill in War and Society*, Little Brown & Company, Boston, 1996

—— with Loren W. Christensen, *On Combat: The Psychology and Physiology of Deadly Conflict in War and in Peace*, Warrior Science Publications, Millstadt, Illinois, second edition, 2007

General Heinz Guderian, *Panzer Leader*, Futura, London, 1974

Nicoletta F. Gullace, 'White Feathers and Wounded Men: Female Patriotism and the Memory of the Great War', *Journal of British Studies*, No. 36, April 1997, pp. 178–206

General Sir John Hackett, *I Was a Stranger*, Sphere Books, London, 1979

Colonel David H. Hackworth and Julie Sherman, *About Face*, Pan Books, Sydney, 1989

Victor Davis Hanson, *The Western Way of War: Infantry Battle in Classical Greece*, second edition, University of California Press, Berkeley, 2000

—— *The Soul of Battle: From Ancient Times to the Present Day, How Three Great Liberators Vanquished Tyranny*, Anchor Books, New York, 2001

Max Hastings and Simon Jenkins, *The Battle for the Falklands*, W.W. Norton & Company, New York, 1983

Max Hastings, *D-Day, June 6 1944*, Simon & Schuster, New York, 1984

—— *Going to the Wars*, Pan, London, 2001

Ken Hechler, *The Bridge at Remagen*, Ballantine Books, New York, 1957

Chris Hedges, *War Is a Force That Gives Us Meaning*, Anchor Books, New York, 2002

Willi Heilmann, *I Fought You From the Skies*, Award Books, New York, 1966

Herodotus, *Histories*, Book 1:105, George Rawlinson translation, Wordsworth Editions, Ware, Hertfordshire, 1996

Michael Herr, *Dispatches*, Picador, London, 1978

James Hewitt, *Love and War*, Blake, London, 1999

Christopher Hibbert, *The Destruction of Lord Raglan: A Tragedy of the Crimean War 1854–55*, Penguin, London, 1985 edition

—— *The Wheatley Diary: A Journal and Sketchbook kept during the Peninsular War and the Waterloo Campaign*, The Windrush Press, Gloucestershire, 1997 edition

—— (editor), *The Recollections of Rifleman Harris, as told to Henry Curling*, The Windrush Press, Gloucestershire, 1998 edition

Tom Hickman, *The Sexual Century*, Carlton Books Limited, London, 1999

George Hicks, *The Comfort Women: Japan's Brutal Regime of Enforced Prostitution in the Second World War*, W.W. Norton & Company, New York, 1997

Richard Hillary, *The Last Enemy: The Memoir of a Spitfire Pilot*, Burford Books, Short Hills, New Jersey, 1997

Colonel Jon T. Hoffman, *Chesty: The Story of Lieutenant General Lewis B. Puller, USMC*, Random House, New York, 2001

Richard Holmes, *Firing Line*, Pimlico, 1985

—— *Dusty Warriors*, Harper Perennial, London, 2007 edition

—— *Sahib: The British Soldier in India*, HarperCollins, London, 2005

Ole R. Holsti, 'A Widening Gap between the US Military and Civilian Society?: Some Evidence, 1976–96', *International Security*, Vol. 23, No. 3, Winter 1998–99, pp. 5–42

Homer, *The Iliad*

Commodore G.F. Hopkins (editor), *Tales from Korea: The Royal New Zealand Navy in the Korean War*, Reed Publishing, Auckland, New Zealand, 2002

Alistair Horne, *The Price of Glory: Verdun 1916*, Penguin, London, 1993 reprint

Andrew Hoskins, *Televising War: From Vietnam to Iraq*, Continuum, London, 2005 edition

Miles Hudson and John Stanier, *War and the Media*, Sutton Publishing, Gloucestershire, 1997

William Bradford Huie, *The Execution of Private Slovik*, Westholme, Yardley, Pennsylvania, 2004 edition

Human Security Report 2005, Human Security Centre, Liu Institute for Global Studies, University of British Columbia

Major Chris Hunter, *Eight Lives Down*, Corgi, London, 2008 edition

Samuel P. Huntington, *The Soldier and the State: The Theory and Politics of Civil–Military Relations*, Harvard University Press, Cambridge, Massachusetts, 1985 edition

Samuel Hynes, *Flights of Passage: Reflections of a World War II Aviator*, Bloomsbury, London, 1989 edition

—— *The Soldier's Tale: Bearing Witness to Modern War*, Pimlico, London, 1998

'Iraq Body Count: A Dossier of Civilian Casualties 2003–2005', produced by Iraq Body and Oxford Research Group, released July 2005, www.iraqbodycount.org

Lawrence James, *Warrior Race: A History of the British at War*, Abacus, London, 2002 edition

Morris Janowitz, *The Professional Soldier: A Social and Political Portrait*, The Free Press of Glencoe, Illinois, 1960

James Jones, *WWII*, Ballantine Books, New York, 1975

Ernst Jünger, *Storm of Steel*, translated by Michael Hofmann, Penguin Books, London, 2004 edition

Neil Kagan, Harris J. Andrews and Paula York-Soderlund (editors), *An Illustrated History of Courage Under Fire: Great Battles of the Civil War*, Oxmoor House, Birmingham, Alabama, 2002

Stanley Karnow, *Vietnam: A History*, Pimlico edition, London, 1994

Bill Katovsky and Timothy Carlson, *Embedded: The Media at War in Iraq*, The Lyons Press, Guilford, Connecticut, 2003

John Keegan, *The Face of Battle*, Penguin edition, London, 1978

—— *Six Armies in Normandy: From D-Day to the Liberation of Paris*, Penguin, London, 1983

—— *The Second World War*, Penguin Books, London, 1989

—— *The First World War*, Alfred A. Knopf, New York, 1999 (general edition)

—— *World War II: A Visual Encyclopedia*; PRC Publishing Ltd, London, 1999

—— *The First World War*, Vintage Books edition, New York, 2000

John Keegan and Joseph Darracott, *The Nature of War*, Jonathan Cape, London, 1981

Michael Kelly, *Martyrs' Day: Chronicle of a Small War*, Picador edition, London, 1994

Bob Kerrey, *When I Was a Young Man: A Memoir*, Harcourt, Orlando, Florida, 2003 edition

Joshua Key, *The Deserter's Tale: Why I Walked Away From the War in Iraq*, Text Publishing, Melbourne, Australia, 2007

Siegfried Knappe with Ted Brusaw, *Soldat: Reflections of a German Soldier, 1936–1949*, Dell Publishing, New York, 1993 edition

Phillip Knightley, *The First Casualty: The War Correspondent as Hero and Myth-Maker from the Crimea to Kosovo*, Johns Hopkins University Press, Baltimore, 2002 edition

Günter Koschorrek, *Blood Red Snow: The Memoirs of a German Soldier on the Eastern Front*, Greenhill Books, London, 2002

Ron Kovic, *Born on the Fourth of July*, Corgi, London, 1990

Robert Lacey, *Monarch: The Life and Reign of Elizabeth II*, The Free Press, New York, 2002

B. Ladkin and Jorn Stuphorn, 'Two Causal Analyses of the Black Hawk Shootdown during Operation Provide Comfort', Faculty of Technology, University of Bielefeld, Germany

Major General J.C. Latter, *The History of the Lancashire Fusiliers: 1914–1918*, Gale & Polden, Aldershot, 1949 (two volumes)

Brian Lavery, *Nelson's Navy: The Ships, Men and Organisation 1793–1815*, Conway Maritime Press, London, 1993 reprint

John Lawrence and Robert Lawrence MC, *When the Fighting is Over: Tumbledown – A Personal Story*, Bloomsbury, London, 1988

Robert Leckie, *Helmet for My Pillow*, ibooks edition, New York, 2001

Laurie Lee, *Moment of War: A Memoir of the Spanish Civil War*, The New Press, New York, 1991

Harold P. Leinbaugh and John D. Campbell, *The Men of Company K: The Autobiography of a World War II Rifle Company*, William Morrow & Company, New York, 1985

Lawrence LeShan, *The Psychology of War: Comprehending its Mystique and its Madness*, Helios Press, New York, 2002 edition

Joshua Levine, *Forgotten Voices of the Blitz and the Battle for Britain*, Ebury Press, London, 2006

Cecil Lewis, *Sagittarius Rising*, Penguin, London, 1977

Captain B.H. Liddell Hart (editor), *The Letters of Private Wheeler 1809–1828*, Michael Joseph, London, 1951

Gerald F. Linderman, *Embattled Courage: The Experience of Combat in the American Civil War*, The Free Press, New York, 1987

Christopher Lloyd, *The British Seaman*, London, 1968

Sean Longden, *To the Victor the Spoils – D-Day to VE Day: The Reality Behind the Heroism*, Arris Books, Gloucestershire, 2005 edition

Emilio Lussu, *Sardinian Brigade*, translated by Marion Rawson, Grove Press, New York, 1967

Douglas MacArthur, *Reminiscences*, McGraw-Hill Book Company, New York, 1964

Robert McCormick, *With the Russian Army*, The Macmillan Company, New York, 1915

George MacDonald Fraser, *Quartered Safe out Here*, HarperCollins, London, 2000 edition

James R. McDonough, *Platoon Leader: A Memoir of Command in Combat*, Ballantine Books, New York, 2003 edition

Gary McKay, *In Good Company: One Man's War in Vietnam*, Allen & Unwin, Sydney, 1987

M. McKinley McClure, *Hey! Major, Look Who's Here*, Dorrance & Company, Philadelphia, 1972

Hugh McManners, *Falklands Commando*, William Kimber, London, 1984
—— *The Scars of War*, HarperCollins, London, 1993
John C. McManus, *The Deadly Brotherhood: The American Combat Soldier in World War II*, Presidio Books, Ballantine Books, New York, 1998
Brigadier A.B. McPherson, *The Second World War 1939–1945, Army: Discipline*, War Office, London, 1950
James M. McPherson, *The Battle Cry of Freedom: The Civil War Era*, Oxford University Press, 1988
Malcolm MacPherson, *Roberts Ridge*, Corgi Books, London, 2006
Myra MacPherson, *Long Time Passing: Vietnam and the Haunted Generation*, Doubleday, New York, 1984
Josephine Maltby, 'Financial reporting and the conscription of trade and industry 1914–1918', Discussion Paper No. 2003.11, University of Sheffield Management School, July 2003
William Manchester, *American Caesar: Douglas MacArthur 1880–1964*, Little, Brown & Company, Boston, 1978
—— *Goodbye Darkness: A Memoir of the Pacific War*, Little, Brown & Company, Boston, 1980
Field Marshal Erich Von Manstein, *Lost Victories*, edited and translated by Anthony G. Powell, Greenhill Books, London, 1987 edition
David Maraniss and Ellen Nakashima, *The Prince of Tennessee: Al Gore Meets His Fate*, Touchstone, New York, 2000
Samuel Lyman Atwood Marshall, *Men Against Fire: The Problem of Battle Command*, University of Oklahoma Press, Norman, Oklahoma, 2000 edition
Robert Mason, *Chickenhawk*, Corgi Books, London, 1987 edition
John Masters, *The Road Past Mandalay*, Cassell, London, 2002 edition
Bill Mauldin, *The Brass Ring*, Berkley Medallion Books, New York, 1973 edition
—— *Up Front*, W.W. Norton & Company, New York, 2000 edition
Mark Mazzetti, 'Friendly Fire', *US News & Weekly Report*, 17 March 2003, pp. 17–20
Major General F.W. von Mellenthin, *Panzer Battles*, translated by H. Betzler, Futura, London, 1977 edition
Catherine Merridale, *Ivan's War: Life and Death in the Red Army, 1939–1945*, Metropolitan Books, New York, 2006
Kevin J. Mervin, *Weekend Warrior: A Territorial Soldier's War in Iraq*, Mainstream Publishing, Edinburgh, 2005
Henry Metelmann, *Through Hell for Hitler: A dramatic first-hand account of fighting with the Wehrmacht*, Patrick Stephens Ltd, Wellingborough, Northamptonshire, England, 1990
Martin Middlebrook, *The First Day on the Somme*, W.W. Norton & Company, New York, 1972

—— *Operation Corporate: The Story of the Falklands War, 1982*, Viking, London, 1985

—— *The Kaiser's Battle*, Penguin, London, 2000 edition

William Ian Miller, *The Mystery of Courage*, Harvard University Press, Cambridge, Massachusetts, 2002

Sergeant Dan Mills, *Sniper One: The Blistering True Story of a British Battle Group Under Siege*, Penguin, London, 2008 edition

Major T.J. Mitchell and Miss G.M. Smith, *History of the Great War Based on Official Documents: Medical Services – Casualties and Medical Statistics of the Great War*, Battery Press edition, Nashville, Tennessee, 1997

Major General Jim Molan, *Running the War in Iraq*, HarperCollins, Sydney, 2008

Darren Moore, *Duntroon, The Royal Military College of Australia: 1911–2001*, Royal Military College of Australia, Canberra, 2001

Lieutenant General Harold G. Moore and Joseph L. Galloway, *We Were Soldiers Once ... And Young*, HarperCollins, New York, 2002 edition

William Moore, *The Thin Yellow Line*, Clarke, Leo Cooper Ltd, London, 1974

Lord C.E. Moran, *The Anatomy of Courage*, Constable, London, 1966

Captain Jon Mordan, 'Press Pools, Prior Restraint and the Persian Gulf War', *Air & Space Power Chronicles – Chronicles Online Journal*, www.airpower.maxwell.af.mil, 6 June 1999

Charles C. Moskos and Thomas E. Ricks, *Reporting War When There is No War*, Cantigny Conference Series Special Report, Chicago, 1996

Farley Mowat, *And No Birds Sang*, Cassell, London, 1980 edition

Fritz Nagel, *Fritz: The World War I Memoirs of a German Lieutenant*, Der Angriff Publications, Huntington, West Virginia, 1981

Sean Naylor, *Not a Good Day to Die: The Untold Story of Operation Anaconda*, Michael Joseph, 2005

Lieutenant Commander James Newton, *Armed Action: My War in the Skies with 847 Naval Air Squadron*, Headline Review, London, 2007

Tim O'Brien, *If I Die in a Combat Zone*, HarperCollins, Flamingo Seventies Classic edition, London, 2003

Captain Scott O'Grady with Jeff Coplon, *Return with Honor*, HarperPaperbacks, New York, 1996 edition

Richard O'Neill (editor), *Patrick O'Brian's Navy: The Illustrated Companion to Jack Aubrey's World*, Salamander Books, London, 2003

Gerald Oram, *Military Executions during World War I*, Palgrave MacMillan, Basingstoke, 2003

George Orwell, *Homage to Catalonia*, Harcourt, Orlando, Florida, 1980

Richard Overy, *Russia's War*, Penguin, London, 1998

Joseph Owen, *Colder than Hell: A Marine Rifle Company at Chosin Reservoir*, Naval Institute Press, Annapolis, Maryland, 1996

John Parker, *The Gurkhas: The Inside Story of the World's Most Feared Soldiers*, Bounty Books, London, 2004 edition

General George S. Patton Jr., *War as I Knew It*, Houghton Mifflin, Boston, 1995 edition

Maurie Pears and Fred Kirkland (editors), *Korea Remembered*, Department of Defence (Australia), Sydney, 2002 edition

Geoffrey Perret, *Old Soldiers Never Die: The Life of Douglas MacArthur*, Adams Media Corporation, Holbrook, Massachusetts, 1996

Joseph E. Persico, 'Little Short of Murder', *Military History Quarterly*, Vol. 17, No. 2, Winter 2005, pp. 26–33

RAF Flight Lieutenants John Peters and John Nichol, *Tornado Down: The Horrifying True Story of their Gulf War Ordeal*, Signet, London, 1993 edition

John Pimlott and Stephen Badsey, *The Gulf War Assessed*, Arms and Armour, London, 1992

Max Plowman, *A Subaltern on the Somme*, E.P. Dutton & Company, New York, 1928

Oliver Poole, *Black Knights: On the Bloody Road to Baghdad*, HarperCollins Publishers, London, 2003

Reginald Pound, *The Lost Generation*, Constable, London, 1964

Colin Powell with Joseph E. Persico, *A Soldier's Way: An Autobiography*, Hutchinson, London, 1995

Samuel H. Preston and Emily Buzzell, 'Mortality of American Troops in Iraq', Population Studies Centre Working Paper Series, PSC 06-01, University of Pennsylvania, August 2006

Robin Prior and Trevor Wilson, 'Paul Fussell at War', *War in History*, Vol. 1, No. 1, March 1994, pp. 63–80

Tim Pritchard, *Ambush Alley: The Most Extraordinary Battle of the Iraq War*, Presidio Press, New York, 2005

Lewis B. Puller Jr., *Fortunate Son: The Autobiography of Lewis B. Puller, Jr.*, Bantam Books, New York, 1993 edition

Julian Putkowski and Julian Sykes, *Shot at Dawn: Executions in World War One by authority of the British Army Act*, Leo Cooper, London, 1992 edition

Ernie Pyle, *Brave Men*, Bison Books, Lincoln, Nebraska, 2001 edition

Peter F. Ramsberger and D. Bruce Bell, *What We Know about AWOL and Desertion: A Review of the Professional Literature for Policy Makers and Commanders*, Army Research Institute for the Behavioral and Social Sciences, Alexandria, Virginia, August 2002, p. 22 (ARI-SR-51)

Laurence Rees, *War of the Century: When Hitler Fought Stalin*, The New Press, New York, 1999

Geoffrey Regan, *Backfire: A history of friendly fire from ancient warfare to the present day*, Robson Books, London, 2002

Rekohu: A Report on Moriori and Ngati Mutunga Claims in the Chatham Islands, Waitangi Tribunal Report, WAI 64, 25 May 2001

Report of the Chairman of the Joint Chiefs of Staff Media–Military Relations Panel (known as the Sidle Panel), released 23 August 1984, provided at: http://www.ndu.edu/library/epubs/20030710a.pdf.

Anthony Rhodes, *Sword of Bone*, Buchan & Enright, London, 1986 edition

Private Frank Richards, *Old Soldiers Never Die*, The Naval and Military Press, Uckfield, East Sussex, 2001

Major General F.M. Richardson, *Fighting Spirit: A Study of the Psychological Factors in War*, Leo Cooper, London, 1978

Hans Peter Richter, *The Time of the Young Soldiers*, translated by Anthea Bell, Armada, London, 1989 edition

Thomas E. Ricks, 'The Widening Gap Between the Military and Society', *The Atlantic Monthly*, July 1997, pp. 66–78

—— *Fiasco: The American Military Adventure in Iraq*, Penguin, London, 2006

Stuart Rintoul, *Ashes of Vietnam: Australian Voices*, William Heinemann Australia, Melbourne, 1987

John Stuart Roberts, *Siegfried Sassoon*, Richard Cohen Books, London, 2000

Sidney Rogerson, *Twelve Days on the Somme*, Greenhill Books, London, 2006 edition

Kathy Roth-Douquet and Frank Schaeffer, *AWOL: The Unexcused Absence of America's Upper Classes from Military Service – and How It Hurts Our Country*, Collins, New York, 2006

Sydney Fairbairn Rowell, *Full Circle*, Melbourne University Press, Melbourne, 1974

Jules Roy, *Dienbienphu*, translated by Robert Baldick, Faber & Faber, London, 1965

Cornelius Ryan, *The Last Battle*, Simon & Schuster, New York, 1966

—— *The Longest Day*, Simon & Schuster, Touchstone edition, New York, 1994

Stephan Ryan, *Pétain the Soldier*, A.S. Barnes & Company, Cranbury, New Jersey, 1969

Saburo Saki, *Samurai*, Four Square, London, 1966

Al Santoli, *To Bear any Burden: The Vietnam War and its Aftermath in the Words of Americans and Southeast Asians*, Abacus, London, 1985

Siegfried Sassoon, *Memoirs of an Infantry Officer*, Faber & Faber, London, 1965 edition

—— *Memoirs of a Fox-hunting Man*, The Folio Society, London, 1971 edition

Guy Sajer, *The Forgotten Soldier*, translated by Lily Emmet, Weidenfeld & Nicolson, London, 1971

Jonathan Schell, *The Real War*, Corgi, London, 1989 edition

Ze'ev Schiff and Ehud Ya'ari, *Israel's Lebanon War*, edited and translated by Ina Friedman, Simon & Schuster, New York, 1984

General H. Norman Schwarzkopf and Peter Petre, *It Doesn't Take a Hero*, Linda Grey, Bantam Books, New York, 1992

Leonard Sellers, *Death for Desertion: The Story of the Court Martial and Execution of Sub Lt Edwin Dyett*, Leo Cooper, South Yorkshire, 2003

Dr Wayne S. Sellman, 'Project 100,000: Testimony and Report on the Study of Vietnam War Era Low Aptitude Military Recruits', report to the Congressional House Committee on Veteran Affairs, 28 February 1990

Geoffrey Serle, *John Monash: A Biography*, Melbourne University Press, Melbourne, 1982

Jonathan Shay, *Achilles in Vietnam: Combat Trauma and the Undoing of Character*, Touchstone, Simon & Schuster, New York, 1995

Gary D. Sheffield, *The Redcaps: A History of the Royal Military Police and its Antecedents from the Middle Ages to the Gulf War*, Brassey's, London, 1994

—— 'The Operational Role of British Military Police on the Western Front, 1914–18' in Paddy Griffith (editor), *British Fighting Methods in the Great War*, Frank Cass, London, 1996, pp. 70–86

Ben Shephard, *A War of Nerves: Soldiers and Psychiatrists 1914–1994*, Pimlico, London, 2002

Lieutenant Colonel Charles R. Shrader, *Amicicide: The Problem of Friendly Fire in Modern War*, Combat Studies Institute, Fort Leavenworth, Kansas, Research Survey No. 1, December 1982

Colin P. Sisson, *Wounded Warriors: The True Story of a soldier in the Vietnam War and the emotional wounds inflicted*, Total Press, Auckland, New Zealand, 1993

Carl G. Slater, 'The Problems of Purchase Abolition in the British Army 1856–1862', *Military History Journal*, The South African Military History Society, Vol. 4, No. 6, December 1979

Eugene B. Sledge, *With the Old Breed at Peleliu and Okinawa*, Oxford University Press, New York, 1990

Field Marshal Sir William Slim, *Unofficial History*, Corgi Books, London, 1970 edition

Lieutenant Colonel J.H.A. Sparrow, *The Second World War 1939–1945, Army: Morale*, The War Office, London, 1949

Julian Spilsbury, *The Thin Red Line: An Eyewitness History of the Crimean War*, Cassell, London, 2006 edition

Thomas G. Sticht, William B. Armstrong, Daniel T. Hickey and John S. Caylor, *Cast-off Youth: Policy and Training Methods from the Military Experience*, Praeger, New York, 1987

Samuel A. Stouffer, Arthur A. Lumsdaine, Marion Harper Lumsdaine, Robin M. Williams Jr., M. Brewster Smith, Irving L. Janis, Shirely A. Star and Leonard S. Cottrell Jr., *The American Soldier: Combat and its Aftermath*, Princeton University Press, Princeton, New Jersey, 1949

Samuel A. Stouffer, Edward A. Suchman, Leland C. DeVinney, Shirley A. Star and Robin M. Williams Jr., *The American Soldier: Adjustment During Army Life*, Princeton University Press, Princeton, New Jersey, 1949

Julie Summers, *Remembered: The History of the Commonwealth War Graves Commission*, Merrell, London, 2007

Viktor Suvorov, *My Life in the Soviet Army: The 'Liberators'*, Berkley Books, New York, 1988 edition

R.L. Swank and W.E. Marchand, 'Combat Neuroses: Development of Combat Exhaustion', *Archives of Neurology and Psychology*, Issue 55, 1946, pp. 236–47

Anthony Swofford, *Jarhead: A Marine's Chronicle of the Gulf War*, Simon & Schuster UK, London, 2003

Kazuo Tamayama and John Nunneley, *Tales by Japanese Soldiers*, Cassell, London, 2001 edition

Studs Terkel, *The Good War: An Oral History of World War Two*, Hamish Hamilton, London, 1985

John Terraine, *General Jack's Diary 1914–18: The Trench Diary of Brigadier General J.L. Jack DSO*, Cassell, London, 2000 edition

Addison Terry, *The Battle for Pusan: A Korean War Memoir*, Presido, Novato, California, 2000

Textbook of Military Medicine: War Psychiatry, Office of the Surgeon General of the United States of America, 1995

Margaret Thatcher, *The Downing Street Years: 1979–1990*, HarperCollins, New York, 1993

Julian Thompson, *No Picnic: The story of 3 Commando Brigade in the Falklands War*, Fontana/Collins, Glasgow, 1986 edition

Mark Thompson, 'The Wounded Come Home', *Time*, 10 November 2003, pp. 32–40

—— 'Broken Down', *Time*, 16 April 2007, pp. 14–21

Thucydides, *History of the Peloponnesian War*, translated by Rex Warner, 1954, Penguin Books, London, 1972

Leo Tolstoy, *War and Peace*, translated by Louise and Aylmer Maude, Wordsworth Editions, Ware, Hertfordshire, 2001 edition

Robert C. Tucker, *Stalin in Power: The Revolution from Above, 1928–1941*, W.W. Norton & Company, New York, 1990

Bill Turque, *Inventing Al Gore: A Biography*, Houghton Mifflin, Boston, 2000

Uniform Code of Military Justice of the United States

United States Congress, Office of Technology Assessment, *Who Goes There: Friend or Foe?*, US Government Printing Office, Washington, DC, June 1993

United States Naval Academy, Responsibility and Accountability Case Study No. 1, entitled 'Blackhawk Fratricide Incident 14 April 1994'

Martin Van Creveld, *Fighting Power: German Military Performance 1914–1945*, Art of War Colloquium Publication, US Army War College, Carlisle, Pennsylvania, November 1983

—— *The Sword and the Olive: A Critical History of the Israeli Defence Force*, Public Affairs, New York, 2002 edition

Major Barry E. Venable, 'The Army and the Media', *Military Review*, January–February 2002, pp. 66–71

Dmitri Volkogonov, *Trotsky: The Eternal Revolutionary*, translated and edited by Harold Shukman, The Free Press, New York, 1996

Carl Von Clausewitz, *On War*, edited and translated by Michael Howard and Peter Paret, Princeton University Press, Princeton, New Jersey, 1989 edition

Johann Voss, *Black Edelweiss: A Memoir of Combat and Conscience by a Soldier of the Waffen-SS*, The Aberjona Press, Bedford, Pennsylvania, 2002

Ed Vulliamy, *Seasons in Hell: Understanding Bosnia's War*, Simon & Schuster, London, 1994

Allan S. Walker, 'Australia in the War of 1939–1945', *Clinical Problems of War*, Series 5, Vol. 1, Australian War Memorial, Canberra, 1962 reprint

Scott Wallsten and Katrina Kosse, 'The Economic Costs of the War in Iraq', AEI-Brookings Joint Center for Regulatory Studies, Working Paper 05-19, September 2005

Jakob Walter, *The Diary of a Napoleonic Foot Soldier*, Penguin, London, 1993 edition

Graham Webster, *The Roman Imperial Army of the First and Second Centuries AD*, third edition, University of Oklahoma Press, Norman, Oklahoma, 1998

Stanley Weintraub, *Silent Night: The Story of the World War I Christmas Truce*, Plume, New York, 2001

Bing West and Major General Ray L. Smith, *The March Up: Taking Baghdad with the 1st Marine Division*, Pimlico, London, 2004

Simon Weston, *Walking Tall*, Bloomsbury, London, 1989

John W. Wheeler-Bennett, *King George VI: His Life and Reign*, St Martin's Press, New York, 1958

Craig Wilcox, *Australia's Boer War: The War in South Africa 1899–1902*, Oxford University Press, Melbourne, 2002

Kayla Williams, *Love My Rifle More Than You: Young and Female in the US Army*, Weidenfeld & Nicolson, London, 2006

Philip Williams with M.S. Power, *Summer Soldier: The True Story of the Missing Falklands Guardsman*, Bloomsbury, London, 1990

Denis Winter, *Death's Men: Soldiers of the Great War*, Penguin, London, 1979

Jay Winter and Blaine Baggett, *1914–18: The War and the Shaping of the 20th Century*, BBC Books, London, 1996

Terence Wise, *Medieval Warfare*, Hasting House, New York, 1976

Tobias Wolff, *In Pharaoh's Army: Memories of a Lost War*, Picador, London, 1994

Cecil Woodham-Smith, *The Reason Why: A behind-the-scenes account of the charge of the Light Brigade*, Penguin Books, Harmondsworth, 1981 edition

Admiral Sandy Woodward with Patrick Robinson, *One Hundred Days: The Memoirs of the Falklands Battle Group Commander*, HarperCollins, London, 1982

Evan Wright, *Generation Kill: Living dangerously on the road to Baghdad with the ultraviolent Marines of Bravo Company*, Bantam Press, London, 2004

Micah Ian Wright, *You Back the Attack! We'll Bomb Who We Want!: Remixed War Propaganda*, Seven Stories Press, New York, 2003

www.archives.gov (official website of the US National Archives and Records Administration)

www.bbc.co.uk (official website of the British Broadcasting Corporation)

www.channel4.com (official website of Channel 4 in Britain)

www.deathpenaltyinfo.org (website for the US Death Penalty Information Center)

www.icasince.org (Institute of Corean-American Studies)

www.mightyeighth.org (US 8th Air Force Museum website)

www.mlb.com (official website of Major League Baseball)

www.rafmuseum.org.uk (official website of the Royal Air Force Museum)

www.royal.gov.uk (official website of the British monarchy)

www.shotatdawn.org.uk (campaign to secure pardons for British soldiers executed during the First World War)

www.sldn.org (US Servicemembers Legal Defense Network)

www.sss.gov (official website of the US Selective Service System)

www.virtualwall.org (site dedicated to US soldiers killed during the Vietnam War)

NOTES

INTRODUCTION

1   Arthur Hadley, *The Straw Giant – America's Armed Forces: Triumphs and Failures*, Avon Books, New York, 1987 edition, p. 52.

2   Duke of Wellington, *Recollections*, Samuel Rogers, London, 1859.

3   M. Brewster Smith, Chapter 3: 'Combat Motivation Among Ground Troops', *The American Soldier: Combat and its Aftermath*, Princeton University Press, Princeton, New Jersey, 1949, p. 170.

4   James Fisher, quoted in John C. McManus, *The Deadly Brotherhood: The American Combat Soldier in World War II*, Presidio Books, Ballantine Books, New York, 1998, p. 244.

5   Guy Sajer, *The Forgotten Soldier*, translated by Lily Emmet, Weidenfeld & Nicolson, London, 1971, p. 297.

CHAPTER 1: THE SOLDIER AND THE STATE

1   Tobias Wolff, *In Pharaoh's Army: Memories of a Lost War*, Picador, London, 1994, p. 54.

2   Timothy Gowing, *Voice from the Ranks: A Personal Narrative of the Crimean War by a Sergeant of the Royal Fusiliers*, William Heinemann, Melbourne, 1954, p. 6.

3   Bernard B. Fall, *Hell in a Very Small Place: The Siege of Dien Bien Phu*, Pall Mall Press, Oxford, 1966, pp. 336–7.

4   Omar N. Bradley, *A Soldier's Story*, Henry Holt & Company, New York, 1951, p. 357.

5   Nikonor Perevalov, 22nd NVKD Rifle Regiment, quoted in Laurence Rees, *War of the Century: When Hitler Fought Stalin*, The New Press, New York, 1999, p. 197.

6   J. Glenn Gray, *The Warriors: Reflections on Men in Battle*, Bison Books, Lincoln, Nebraska, 1998 edition, p. 181.

7   D.M. Mantell, *True Americanism: Green Berets and War Resisters*, New York, 1974, quoted in Richard Holmes, *Firing Line*, Pimlico, London, 1985, p. 286.

8   Colonel David H. Hackworth and Julie Sherman, *About Face*, Pan Books, Sydney, 1989, p. 55.

9   Nathaniel Fick, *One Bullet Away: The Making of a Marine Officer*, Weidenfeld & Nicolson, London, 2005, p. 241.

10   Ibid.

11   Gary McKay, *In Good Company: One Man's War in Vietnam*, Allen & Unwin, Sydney, 1987, p. 42.

12   Lieutenant Commander James Newton, *Armed Action: My War in the Skies with 847 Naval Air Squadron*, Headline Review, London, 2007, pp. 16–17.

13   Major David Bradley, 1st Battalion, the Princess of Wales's Royal Regiment, quoted in Richard Holmes, *Dusty Warriors*, Harper Perennial, London, 2007 edition, p. 59.

14   Rick Atkinson, *In the Company of Soldiers: A Chronicle of Combat in Iraq*, Little, Brown, London, 2004, p. 72.

15   Harold P. Leinbaugh and John D. Campbell, *The Men of Company K: The Autobiography of a World War II Rifle Company*, William Morrow & Company, New York, 1985, p. 17.

16   Yael Dayan, *A Soldier's Diary: Sinai 1967*, Penguin Books, Harmondsworth, 1969 edition, p. 52.

17   Reuven Gal, *A Portrait of the Israeli Soldier*, Greenwood, New York, 1986, pp. 146–7.

18   Reuven Gal, 'Commitment and Obedience in the Military: An Israeli Case Study', *Armed Forces & Society*, Vol. 11, No. 4, Summer 1985, p. 558.

19   Ze'ev Schiff and Ehud Ya'ari, *Israel's Lebanon War*, edited and translated by Ina Friedman, Simon & Schuster, New York, 1984, pp. 215–16.

20   Lawrence James, *Warrior Race: A History of the British at War*, Abacus, London, 2003, p. 538.

21   Carl Von Clausewitz, *On War*, edited and translated by Michael Howard and Peter Paret, Princeton University Press, Princeton, New Jersey, 1989 edition, p. 95.

22   Petty Officer Sam Bishop, HMS *Antelope*, in Michael Bilton and Peter Kosminsky, *Speaking Out: Untold Stories from the Falklands War*, André Deutsch, London, 1989, p. 71.

23   RAF Flight Lieutenants John Peters and John Nichol, *Tornado Down: The Horrifying True Story of their Gulf War Ordeal*, Signet, London, 1993 edition, pp. 30–1.

24   Fick, *One Bullet Away*, p. 236.

25   Addison Terry, *The Battle for Pusan: A Korean War Memoir*, Presidio, Novato, California, 2000, p. 179.

26   Gwynne Dyer, *War*, The Bodley Head, London, 1986, p. 14.

27   Colin Powell with Joseph E. Persico, *A Soldier's Way: An Autobiography*, Hutchinson, London, 1995, p. 134.

28   Angie Cannon, 'Trying to Make a Life', *US News & World Report*, 29 November 2004, pp. 44–51.

29   Private First Class Tristan Wyatt, 43rd Combat Engineer Company, quoted in Mark Thompson, 'The Wounded Come Home', *Time*, 10 November 2003, pp. 32–40.

30   Lewis B. Puller Jr., *Fortunate Son: The Autobiography of Lewis B. Puller, Jr.*, Bantam Books, New York, 1993 edition, p. 308.

31   Ibid.

32   Robert McNamara, *In Retrospect: The Tragedy and Lessons of Vietnam*, Vintage, New York, 1996, p. 333.

33   Colonel Tom Sims, obituary for Captain Sean Patrick Sims, killed in Iraq on 13 November 2004.

34   James Coomarasamy, 'Small town America's war dead', www.bbc.co.uk, 31 March 2007.

35   Kevin J. Mervin, *Weekend Warrior: A Territorial Soldier's War in Iraq*, Mainstream Publishing, Edinburgh, 2005, p. 338.

36   James, *Warrior Race*, p. 527.

37   United States *Uniform Code of Military Justice*, Article 88.

38   *Manual for Courts-Martial United States*, 2008 edition, pp. iv–17.

39   W.A. Croffut, *Fifty Years in Camp and Field: The Diary of Major-General Ethan Allen Hitchcock*, G.P. Putnam's Sons, New York, 1909, p. 411; quoted in Morris Janowitz, *The Professional Soldier: A Social and Political Portrait*, The Free Press of Glencoe, Illinois, 1960, p. 259.

40   Jules Roy, *Dienbienphu*, translated by Robert Baldick, Faber & Faber, London, 1965, p. 295.

41   Jonathan Shay, *Achilles in Vietnam: Combat Trauma and the Undoing of Character*, Touchstone, Simon & Schuster, New York, 1995, p. 158.

42   'Surge in violence takes US toll over 1,000', *Sydney Morning Herald*, 9 September 2004.

43   Testimony of John Kerry before the Senate Committee on Foreign Relations, 22 April 1971.

44   Sergeant Tony Espera, quoted in Fick, *One Bullet Away*, p. 318.

45   Frederick Downs, *The Killing Zone: My Life in the Vietnam War*, W.W. Norton & Company, New York, 2007 edition, pp. 196–7.

46   Jean-Baptiste Barres, *Chasseur Barres*, Leonaur, 2006 edition, p. 110.

47   James Anton, *Royal Highlander: A soldier of HM 42nd (Royal) Highlanders during the Peninsular, South of France & Waterloo Campaigns of the Napoleonic Wars*, Leonaur, 2007 edition, p. 211.

48   Private P.S. Jackson, 11th Light Horse Regiment, letter dated 8 November 1916, quoted in Bill Gammage, *The Broken Years: Australian Soldiers in the Great War*, Penguin, Melbourne, 1982 edition, p. 270.

49   Gammage, *The Broken Years*, p. 270.

50   George Coppard, *With a Machine Gun to Cambrai: A Story of the First World War*, Cassell, London, 1999 edition, pp. 134–5.

51   Harry Free, quoted in Sean Longden, *To the Victor the Spoils – D-Day to VE Day: The Reality Behind the Heroism*, Arris Books, Gloucestershire, 2005 edition, p. 363.

52   Catherine Merridale, *Ivan's War: Life and Death in the Red Army, 1939–1945*, Metropolitan Books, New York, 2006, pp. 363–4.

53   John Lawrence and Robert Lawrence MC, *When the Fighting is Over: Tumbledown – A Personal Story*, Bloomsbury, London, 1988, p. 97.

54   Ibid., p. 192.

55   Kayla Williams, *Love My Rifle More Than You: Young and Female in the US Army*, Weidenfeld & Nicolson, London, 2006, pp. 286–7.

56   Robert Mason, *Chickenhawk*, Corgi Books, London, 1987 edition, p. 388.

57   Private Frank McCarthy, 1st Infantry Division, quoted in Al Santoli, *To Bear any Burden: The Vietnam War and its Aftermath in the Words of Americans and Southeast Asians*, Abacus, London, 1985, p. 111.

58   David Donovan, *Once a Warrior King*, Corgi edition, London, 1988, p. 365.

59   Ibid., p. 371.

CHAPTER 2: WHO SERVES

1   Ben Shephard, *A War of Nerves: Soldiers and Psychiatrists 1914–1994*, Pimlico, London, 2002, p. 18.

2   Cecil Woodham-Smith, *The Reason Why: A behind-the-scenes account of the charge of the Light Brigade*, Penguin Books, Harmondsworth, 1981 edition, p. 31.

3   Norman Dixon, *On the Psychology of Military Incompetence*, Pimlico, London, 1976, p. 36.

4   Major General J.F.C. Fuller, quoted in Norman Dixon, *On the Psychology of Military Incompetence*, Pimlico, London, 1976, p. 234.

5   Denis Winter, *Death's Men: Soldiers of the Great War*, Penguin, London, 1979, p. 31.

6   Richard Holmes, *In the Footsteps of Churchill*, BBC Books, London, 2005, p. 126.

7   http://www.nationalarchives.gov.uk/pathways/firstworldwar/service_records/sr_officers.htm

8   Niall Ferguson, *The Pity of War: Explaining World War I*, Basic Books, New York, 1999, p. 348.

9   John Ellis, *The Sharp End: The Fighting Man in World War II*, Charles Scribner's Sons, New York, 1980, p. 189.

10   Arthur Wellesley, 1st Duke of Wellington, dispatch, 2 July 1813, from Vitoria, Spain to Lord Bathurst, War Minister. Quoted in Philip Henry Stanhope, *Notes of Conversations with the Duke of Wellington, 1831–1851*, Oxford University Press, London, 1938.

11   Shelford Bidwell, *Modern Warfare: A Study of Men, Weapons and Theories*, Allen Lane, London, 1973, p. 38.

12   Lawrence James, *Warrior Race: A History of the British at War*, Abacus, London, 2003, p. 292.

13   Henry Clifford, *Henry Clifford VC: his letters and sketches from the Crimea*, Michael Joseph, London, 1956, p. 123.

14   John Baynes, *Morale: A Study of Men and Courage*, Avery Publishing Group, New York, 1988 edition, p. 134.

15   John Keegan, *The Face of Battle*, Penguin edition, London, 1978, p. 220.

16   Gerald Oram, *Military Executions during World War I*, Palgrave Macmillan, Basingstoke, 2003, p. 26.

17   Field Marshal Lord Wavell, quoted in John Baynes, *Morale: A Study of Men and Courage*, Avery Publishing Group, New York, 1988 edition, p. 135.

18   Fritz Nagel, *Fritz: The World War I Memoirs of a German Lieutenant*, Der Angriff Publications, Huntington, West Virginia, 1981, p. 18.

19   James, *Warrior Race*, p. 475.

20   George Coppard, *With a Machine Gun to Cambrai: A Story of the First World War*, Cassell, London, 1999 edition, p. 1.

21   Ibid., pp. 64–5.

22   Winter, *Death's Men*, pp. 31–4.

23   Siegfried Sassoon, *Memoirs of an Infantry Officer*, Faber & Faber, London, 1965 edition, p. 97.

24   Malcolm Brown, *The Imperial War Museum Book of the First World War*, Pan Books, London, 2002 edition, pp. 252, 256.

25   Baron Peter Alexander Rupert Carrington, *Reflect on Things Past: The Memoirs of Lord Carrington*, Collins, London, 1988, p. 47.

26   'Women and War', www.iwm.org.uk (Imperial War Museum); John Costello, *Love, Sex and War: Changing Values 1939–45*, Collins, London, 1985, p. 214.

27   W.E. Forster, quoted in Niall Ferguson, *Empire: How Britain Made the Modern World*, Penguin Books, London, 2004 edition, p. 171.

28   Johnson Beharry VC, *Barefoot Soldier: A Story of Extreme Valour*, Sphere, London, 2006, p. 165.

29   John Parker, *The Gurkhas: The Inside Story of the World's Most Feared Soldiers*, Bounty Books, London, 2004 edition, p. 242.

30   Richard Norton-Taylor, 'Limit on Commonwealth troops proposed to keep army "British"', *Guardian*, www.guardian.co.uk, 2 April 2007.

31   Tim Collins, *Rules of Engagement: A Life in Conflict*, Headline, London, 2005, p. 22.

32   Michael Evans, 'How British Army is fast becoming foreign legion', *The Times*, www.timesonline.co.uk, 14 November 2005.

33   James M. McPherson, *The Battle Cry of Freedom: The Civil War Era*, Oxford University Press, Oxford, 1988, p. 600.

34   D.G. Crotty, *Four Years Campaigning in the Army of the Potomac*, Dygert Brothers & Co., Grand Rapids, Michigan, 1874, pp. 104–05.

35   Peter Krass, *Carnegie*, John Wiley & Sons, Hoboken, New Jersey, 2002, p. 76.

36   McPherson, *The Battle Cry of Freedom*, pp. 601, 609–10.

37   Lawrence M. Baskir and William A. Strauss, *Change and Circumstance: The Draft, The War, and the Vietnam Generation*, Alfred A. Knopf, New York, 1978, pp. 18–19; www.sss.gov (official website of the Selective Service System).

38   Mike Mancuso, paper entitled 'Joe DiMaggio', dated 30 April 2001, www.history.acusd.edu; Jonathan Mayo, 'As good a Marine as he was a ballplayer', www.mlb.com (official website of Major League Baseball).

39   Baskir and Strauss, *Change and Circumstance*, pp. 23, 36.

40   www.sss.gov (official website of the Selective Service System); 'Vietnam Warriors: A Statistical Profile', *Veterans of Foreign Wars*, January 2003, p. 18.

41   Baskir and Strauss, *Change and Circumstance*, p. 6.

42   Paul Fussell, *Doing Battle: The Making of a Skeptic*, Little, Brown & Company, Boston, 1996, p. 257.

43   James R. McDonough, *Platoon Leader: A Memoir of Command in Combat*, Ballantine Books, New York, 2003 edition, p. 79.

44   Baskir and Strauss, *Change and Circumstance*, pp. 32, 33.

45   Ibid., pp. 34, 43–4.

46   Micheal Clodfelter, *Vietnam in Military Statistics: A History of the Indochina Wars, 1772–1991*, McFarland & Company, Jefferson, North Carolina, 1995, pp. 245–6; Baskir and Strauss, *Change and Circumstance*, p. 49.

47   Dr Wayne S. Sellman, 'Project 100,000: Testimony and Report on the Study of Vietnam War Era Low Aptitude Military Recruits', report to the Congressional House Committee on Veteran Affairs, 28 February 1990; Department of Defense report entitled 'Project One Hundred Thousand. Characteristics and Performance of "New Standards" Men', December 1969, p. 5.

48   Sellman report, 1990, op. cit.; Department of Defense report, 1969, op. cit., pp. vi, x, xiiii, 9, 14.

49   Sellman report, 1990, op. cit.; Department of Defense report, 1969, op. cit., pp. vii, xvii.

50   Lewis B. Puller Jr., *Fortunate Son: The Autobiography of Lewis B. Puller, Jr.*, Bantam Books, New York, 1993 edition, pp. 90–1.

51   Sellman report, 1990, op. cit.; Department of Defense report, 1969, op. cit., pp. vii, xvii.

52   Myra MacPherson, *Long Time Passing: Vietnam and the Haunted Generation*, Doubleday, New York, 1984, p. 30.

53   Colin Powell with Joseph E. Persico, *A Soldier's Way: An Autobiography*, Hutchinson, London, 1995, p. 148.

54   Joshua Key, *The Deserter's Tale: Why I Walked Away From the War in Iraq*, Text Publishing, Melbourne, Australia, 2007, p. 5.

55   Entry for James Robert Kalsu, www.virtualwall.org

56   Senator John McCain, eulogy for Patrick Tillman, delivered on 3 May 2004 at San José, California.

57   Kelley Beaucar Vlahos, 'Handful of Lawmakers Send Their Kids to War', Fox News, 28 March 2003.

58   Staff Sergeant Brooks Johnson, quoted in 'Senator's soldier son gets no special perks', *Houston Chronicle*, 27 March 2003, p. 6.

59   Morris Janowitz, *The Professional Soldier: A Social and Political Portrait*, The Free Press of Glencoe, Illinois, 1960, p. 40.

60   Gary Younge, 'Private Lynch's comrade in arms finds sad place in a nation's history', *Sydney Morning Herald*, 13 April 2003; David M. Halbfinger and Steven A. Holmes, 'Military Mirrors Working-Class America', *New York Times*, 30 March 2003.

61   David M. Halbfinger and Steven A. Holmes, 'Military Mirrors Working-Class America', *New York Times*, 30 March 2003.

62   Professor Charles C. Moskos, quoted in Halbfinger and Holmes, 'Military Mirrors Working-Class America'.

63   Kathy Roth-Douquet and Frank Schaeffer, *AWOL: The Unexcused Absence of America's Upper Classes from Military Service – and How It Hurts Our Country*, Collins, New York, 2006, pp. 170–1.

64   Mark Thompson, 'Broken Down', *Time*, 16 April 2007, pp. 14–21.

65   Oliver Poole, *Black Knights: On the Bloody Road to Baghdad*, HarperCollins, London, 2003, p. 110.

66   Sergeant Brent M. Williams, '82nd Airborne Div. Legal Staff Help Soldiers Gain Citizenship', www.defendamerica.mil, 29 September 2003; Maia Jachimowicz and Ramah McKay, 'Justice, Homeland Security Departments Announce Changes', www.migrationinformation.org, 1 May 2003.

67   'The Death of Lance Cpl. Gutierrez', www.cbsnews.com, 20 August 2003.

68   Thelma Gutierrez and Wayne Drash, 'Green-card Marine prepares for 3rd deployment', www.cnn.com, 25 March 2008.

69   Mark Thompson, 'Where are the New Recruits?', *Time*, 17 January 2005, pp. 26–9; memorandum from Lieutenant General James R. Helmly, Chief, Army Reserve to Chief of Staff, United States Army, dated 20 December 2004.

70   Thompson, 'Where are the New Recruits?', pp. 26–9; 'Military re-enlistment bonuses skyrocket', www.cnn.com, 11 April 2007.

71   Arthur Hadley, *The Straw Giant – America's Armed Forces: Triumphs and Failures*, Avon Books, New York, 1987 edition, p. 22.

72   Reuven Gal, *A Portrait of the Israeli Soldier*, Greenwood, New York, 1986, p. 59.

73   Israel's Defence Service Law (Consolidated Version), 1986, paragraph 40.

74   Martin Van Creveld, *The Sword and the Olive: A Critical History of the Israeli Defence Force*, Public Affairs, New York, 2002 edition, pp. 310–12.

75   General William H. Carter, 'Army as a Career', *North American*, No. 183, 1906, p. 871, quoted in Janowitz, *The Professional Soldier*, p. 244.

76   Vice President Dick Cheney told George C. Wilson of the *Washington Post* in 1989 that 'I had other priorities in the '60s than military service'.

77   George Washington, 'Sentiments on a Peace Establishment', 2 May 1783, quoted in John C. Fitzpatrick (ed.), *The Writings of George Washington from the Original Manuscript Sources, 1745–1799*, Government Printing Office, Washington, DC, 1931–44, Vol. 24, pp. 374–6 and 388–91.

CHAPTER 3: A SOLDIER'S JOURNEY

1   Tim O'Brien, *If I Die in a Combat Zone*, HarperCollins, Flamingo Seventies Classic edition, London, 2003, p. 44.

2   Poster issued by the Parliamentary Recruitment Committee, reproduced in Malcolm Brown, *The Imperial War Museum Book of the First World War*, Pan Books, London, 2002 edition, p. 192.

3   Quoted at www.spartacus.schoolnet.co.uk

4   Private S.C. Lang, quoted in Max Arthur, *Forgotten Voices of the Great War*, Ebury Press, London, 2003 edition, p. 62.

5   Charles Hamerton, quoted at The World at War Forum, http://pub65.ezboard.com/ftheworldatwar70879.

6   Account of Sergeant Thomas Painting, 1st Battalion, King's Royal Rifle Corps, referred to in Brown, *The Imperial War Museum Book of the First World War*, p. 261.

7   Remarks by President George W. Bush on Iraq, Cincinnati Museum Center, Ohio, delivered 7 October 2002, www.whitehouse.gov/news/releases/2002/10

8   Ibid.

9   Jean-Baptiste Barres, *Chasseur Barres*, Leonaur, 2006 edition, p. 16.

10   Ron Kovic, *Born on the Fourth of July*, Corgi, London, 1990, p. 61.

11   Robert Leckie, *Helmet for My Pillow*, ibooks edition, New York, 2001, pp. 6–7.

12   Samuel Hynes, *Flights of Passage: Reflections of a World War II Aviator*, Bloomsbury, London, 1989 edition, p. 41.

13   Paul Fussell, *Doing Battle: The Making of a Skeptic*, Little, Brown & Company, Boston, 1996, p. 75.

14   Guy Sajer, *The Forgotten Soldier*, translated by Lily Emmet, Weidenfeld & Nicolson, London, 1971, pp. 167–8.

15   Private Jonathan Polansky, 101st Airborne Division, quoted in Al Santoli, *To Bear any Burden: The Vietnam War and its Aftermath in the Words of Americans and Southeast Asians*, Abacus, London, 1985, p. 128.

16   Lord C.E. Moran, *The Anatomy of Courage*, Constable, London, 1966, p. x.

17   Lieutenant Colonel J.W. Appel and Captain G.W. Beebe, 'Preventive Psychiatry: An Epidemiological Approach', *Journal of the American Medical Association*, 131, 1946, p. 1470, quoted in Gwynne Dyer, *War*, The Bodley Head, London, 1986, p. 143.

18   Dyer, *War*, p. 144.

19   John Baynes, *Morale: A Study of Men and Courage*, Avery Publishing Group, New York, 1988 edition, p. 101.

20   Ernie Pyle, *Brave Men*, Bison Books, Lincoln, Nebraska, 2001 edition, p. 124.

21   Jonathan Shay, *Achilles in Vietnam: Combat Trauma and the Undoing of Character*, Touchstone, Simon & Schuster, New York, 1995, p. 34.

22   Dan Gilgoff and Elizabeth Querna, 'The Mental Toll', *US News & World Report*, 29 November 2004, pp. 48–9.

23   Phil Zabriskie, 'Wounds That Don't Bleed', *Time*, 29 November 2004, pp. 28–30.

24   Mark Thompson, 'The Iraq War Comes Home', *Time*, 19 October 2005, www.time.com.

25   Pyle, *Brave Men*, pp. 285–6.

26   Michael Herr, *Dispatches*, Picador, London, 1978, p. 101.

27   Eugene B. Sledge, *With the Old Breed at Peleliu and Okinawa*, Oxford University Press, New York, 1990, p. 5.

28   Ibid., p. 19.

29   Ibid., p. 100.

30   Ibid., p. 128.

31   Ibid., p. 154.

32   Horace L. Baker, *Argonne Days in World War I*, University of Missouri Press, Columbia, Missouri, 2007, p. 36.

33   James Jones, *WWII*, Ballantine Books, New York, 1975, p. 43.

34   Siegfried Knappe with Ted Brusaw, *Soldat: Reflections of a German Soldier, 1936–1949*, Dell Publishing, New York, 1993 edition, p. 219.

35   Vasily Grossman, *A Writer at War: Vasily Grossman with the Red Army 1941–1945*, edited and translated by Antony Beevor and Luba Vinogradova, Pantheon Books, New York, 2005, p. 96.

36   Leckie, *Helmet for My Pillow*, p. 305.

37   Philip Caputo, *A Rumor of War*, Henry Holt & Company, New York, 1996 reprint, pp. 260–1.

38   Caputo, *A Rumor of War*, p. xvi.

39   Anthony Swofford, *Jarhead: A Marine's Chronicle of the Gulf War*, Simon & Schuster UK, London, 2003, p. 83.

40   Evan Wright, *Generation Kill: Living dangerously on the road to Baghdad with the ultraviolent Marines of Bravo Company*, Bantam Press, London, 2004, p. 242.

41   Robert Graves, *Goodbye to All That*, Blue Ribbon Books, New York, 1930, p. 145.

42   Caputo, *A Rumor of War*, p. 79.

43   Baynes, *Morale*, p. 181.

44   Norman Dixon, *On the Psychology of Military Incompetence*, Pimlico, London, 1976, p. 82.

45   Hugh Dundas, *Flying Start*, Penguin, London, 1990, p. 68.

46   Joseph E. Persico, 'Little Short of Murder', *Military History Quarterly*, Vol. 17, No. 2, Winter 2005, pp. 26–33.

47   Ibid.

48   William Manchester, *Goodbye Darkness: A Memoir of the Pacific War*, Little, Brown & Company, Boston, 1980, p. 305.

49   J. Glenn Gray, *The Warriors: Reflections on Men in Battle*, Bison Books, Lincoln, Nebraska, 1998 edition, p. 183.

50   Sir Charles Symonds, quoted in Ben Shephard, *A War of Nerves: Soldiers and Psychiatrists 1914–1994*, Pimlico, London, 2002, p. 56.

51   Walter Hubert Downing, *To the Last Ridge: The WW1 Experiences of W.H. Downing*, Duffy & Snellgrove, Sydney, 1998, p. 6.

52   H.R. Williams, *Comrades of the Great Adventure*, Sydney, 1935, quoted in Bill Gammage, *The Broken Years: Australian Soldiers in the Great War*, Penguin, Melbourne, 1982 edition, p. 262.

53   Hervey Allen, *Toward the Flame: A War Diary*, University of Pittsburgh Press, Pittsburgh, 1968 edition, p. 121.

54   Colonel David H. Hackworth and Julie Sherman, *About Face*, Pan Books, Sydney, 1989, p. 63.

55   Colin P. Sisson, *Wounded Warriors: The true story of a soldier in the Vietnam War and the emotional wounds inflicted*, Total Press, Auckland, New Zealand, 1993, p. 95.

56   Guy Gibson, *Enemy Coast Ahead – Uncensored*, Crécy Publishing, Manchester, 2005 edition, p. 11.

57   Ibid., p. 109.

58   D.E. Charlwood, *No Moon Tonight*, Pacific Books, Angus & Robertson, Sydney, 1956, p. 116.

59   Ibid., p. 114.

60   Jim Bailey, *The Sky Suspended: A Fighter Pilot's Story*, Bloomsbury, London, 2005 edition, p. 79.

61   Nathaniel Fick, *One Bullet Away: The Making of a Marine Officer*, Weidenfeld & Nicolson, London, 2005, p. 168.

62   Gottlob Herbert Bidermann, *In Deadly Combat: A German Soldier's Memoir of the Eastern Front*, translated and edited by Derek S. Zumbro, University Press of Kansas, Lawrence, Kansas, 2000, p. 25.

63   Major General John Frost, *2 Para Falklands: The Battalion at War*, Sphere Books, London, 1984, p. 28.

64   Günter Koschorrek, *Blood Red Snow: The Memoirs of a German Soldier on the Eastern Front*, Greenhill Books, London, 2002, p. 93.

65   Nick Bleszynski, *Shoot Straight, You Bastards!: The Truth behind the killing of 'Breaker' Morant*, Random House, Sydney, 2003 edition, p. 143.

66   Graves, *Goodbye to All That*, p. 232.

67   Corporal W. Gallwey, 47th Battalion, letter dated 8 April 1917, quoted in Gammage, *The Broken Years*, p. 191.

68   Catherine Merridale, *Ivan's War: Life and Death in the Red Army, 1939–1945*, Metropolitan Books, New York, 2006, p. 378.

69   Allen, *Toward the Flame*, p. 121.

70 Eric Sykes, quoted in Sean Longden, *To the Victor the Spoils – D-Day to VE Day: The Reality Behind the Heroism*, Arris Books, Gloucestershire, 2005 edition, p. 37.

71 Merridale, *Ivan's War*, 2006, p. 193.

72 Caputo, *A Rumor of War*, p. 128.

73 Siegfried Knappe with Ted Brusaw, *Soldat: Reflections of a German Soldier, 1936–1949*, Dell Publishing, New York, 1993 edition, p. 169.

74 Ibid., p. 170.

75 Ibid.

76 Gray, *The Warriors*, p. 106.

77 Major Robert Crisp, *Brazen Chariots*, Ballantine Books, New York, 1961, p. 25.

78 Richard Hillary, *The Last Enemy: The Memoir of a Spitfire Pilot*, Burford Books, Short Hills, New Jersey, 1997, p. 96.

79 Michael J. Durant with Steven Hartov, *In the Company of Heroes*, Bantam Press, London, 2003, pp. 12–13.

80 Captain Hugh McManners, *Falklands Commando*, William Kimber, London, 1984, p. 223.

81 M. Brewster Smith, Chapter 3: 'Combat Motivation Among Ground Troops', *The American Soldier: Combat and its Aftermath*, Princeton University Press, Princeton, New Jersey, 1949, p. 189.

82 Ibid.

83 Fussell, *Doing Battle*, p. 122.

84 Colin Powell with Joseph E. Persico, *A Soldier's Way: An Autobiography*, Hutchinson, London, 1995, p. 134.

85 Moran, *The Anatomy of Courage*, p. 75.

86 Caputo, *A Rumor of War*, p. 85.

87 Jones, *WWII*, p. 68.

88 Jakob Walter, *The Diary of a Napoleonic Foot Soldier*, Penguin, London, 1993 edition, p. 26.

89 Adrien Jean Baptiste François Bourgogne, *Memoirs of Sergeant Bourgogne 1812–1813*, Constable, London, 1997 edition, pp. 37–8.

90 William A. Fletcher, *Rebel Private: Front and Rear – Memoirs of a Confederate Soldier*, Meridian, New York, 1997 edition, p. 88.

91 House of Commons Debates, 17 December 1969, *Hansard*, Vol. 793, c341W.

92 Max Plowman, *A Subaltern on the Somme*, E.P. Dutton & Company, New York, 1928, pp. 163–4.

93 Bernard B. Fall, *Hell in a Very Small Place: The Siege of Dien Bien Phu*, Pall Mall Press, Oxford, 1966, pp. 247, 347.

94 Fyodor Sverdlov, 19th Infantry Brigade, 49th Army, quoted in Laurence Rees, *War of the Century: When Hitler Fought Stalin*, The New Press, New York, 1999, pp. 85–6.

95 Lieutenant Colonel Maurie Pears, 'Recollections of War', reproduced in Maurie Pears and Fred Kirkland (editors), *Korea Remembered*, Department of Defence (Australia), Sydney, 2002 edition, p. 83.

96 Emilio Lussu, *Sardinian Brigade*, translated by Marion Rawson, Grove Press, New York, 1967, p. 122.

97 Ernst Jünger, *Storm of Steel*, translated by Michael Hofmann, Penguin, London, 2004 edition, p. 132.

98 Sajer, *The Forgotten Soldier*, p. 75.

99 Johann Voss, *Black Edelweiss: A Memoir of Combat and Conscience by a Soldier of the Waffen-SS*, The Aberjona Press, Bedford, Pennsylvania, 2002, p. 84.

100 Baron Peter Alexander Rupert Carrington, *Reflect on Things Past: The Memoirs of Lord Carrington*, Collins, London, 1988, pp. 57–8.

101 Joseph Owen, *Colder than Hell: A Marine Rifle Company at Chosin Reservoir*, Naval Institute Press, Annapolis, Maryland, 1996, pp. 84–5.

102 Edward Costello, *Rifleman Costello: The Adventures of a Soldier of the 95th (Rifles) in the Peninsular & Waterloo Campaigns of the Napoleonic Wars*, Leonaur, 2005 edition, p. 65.

103 Evgeni Bessonov, translated by Bair Irincheev, *Tank Rider: Into the Reich with the Red Army*, Greenhill Books, London, 2003, pp. 148–9.

104 Bill Mauldin, *Up Front*, W.W. Norton & Company, New York, 2000 edition, p. 86.

105 Jones, *WWII*, p. 122.

106 Quoted in Paul Fussell, *Wartime: Understanding and Behaviour in the Second World War*, Oxford University Press, New York, 1989, pp. 102–3.

107 Ed Laughlin, unpublished manuscript quoted in John C. McManus, *The Deadly Brotherhood: The American Combat Soldier in World War II*, Presidio Books, Ballantine Books, New York, 1998, p.78

108 Merridale, *Ivan's War*, p. 334.

109 Grossman, *A Writer at War*, pp. 74, 340.

110 Colin H. Brown, *Stalemate in Korea: And How we Coped 1952–1953*, Australian Military History Publications, Sydney, 1997, p. 50.

111 Gibson, *Enemy Coast Ahead*, p. 53.

112 Robert Mason, *Chickenhawk*, Corgi Books, London, 1987 edition, p. 127.

113 James Hewitt, *Love and War*, Blake Publishing, London, 1999, pp. 99, 113–14.

114 Colby Buzzell, *My War: Killing Time in Iraq*, G.P. Putnam's Sons, New York, 2005, pp. 220, 223.

115 Guy Chapman, *A Passionate Prodigality*, Fawcett Crest edition, New York, 1967, p. 158.

116 Paul Fussell, *The Great War and Modern Memory*, 25th Anniversary Edition, Oxford University Press, New York, 2000, p. 124.

117 Willi Heilmann, *I Fought You From the Skies*, Award Books, New York, 1966, p. 66.

118 Fred Olson, quoted in Harold P. Leinbaugh and John D. Campbell, *The Men of Company K: The Autobiography of a World War II Rifle Company*, William Morrow & Company, New York, 1985, p. 162.

119 Heilmann, *I Fought You From the Skies*, p. 158.

120 Captain Steven Hughes, in Max Arthur, *Above All, Courage – The Falklands Front Line: First-Hand Accounts*, Guild Publishing, London, 1985, p. 183.

121 Richard Holmes, *Firing Line*, Pimlico, London, 1985, p. 238.

122 Charlwood, *No Moon Tonight*, p. 172.

123 Herr, *Dispatches*, p. 52.

124 Tobias Wolff, *In Pharaoh's Army: Memories of a Lost War*, Picador, London, 1994, p. 5.

125 Raleigh Cash, 'Sue Sponte: Of Their Own Accord', *The Battle of Mogadishu: Firsthand Accounts from the Men of Task Force Ranger*, edited by Matt Eversmann and Dan Schilling, Ballantine Books, New York, 2004, p. 58.

126 Andrew Exum, *This Man's Army: A Soldier's Story from the Front Lines of the War on Terrorism*, Gotham Books, New York, 2005 edition, p. 177.

127 John Ellis, *The Sharp End: The Fighting Man in World War II*, Charles Scribner's Sons, New York, 1980, p. 100.

128 Mansur Abdulin, *Red Road from Stalingrad: Recollections of a Soviet Infantryman*, Pen and Sword Military, Barnsley, South Yorkshire, 2004, p. 62.

129 Costello, *Rifleman Costello*, p. 124.

130 Timothy Gowing, *Voice from the Ranks: A Personal Narrative of the Crimean War by a Sergeant of the Royal Fusiliers*, William Heinemann, Melbourne, 1954, p. 15.

131 K. Jack Bauer (editor), *Soldiering: The Civil War Diary of Rice C. Bull*, Berkley Books, New York, 1988 edition, p. 115.

132 Captain N.A. Nicholson, 14th Field Artillery Brigade, letter dated 29 July 1917, quoted in Gammage, *The Broken Years*, p. 262.

133 Dundas, *Flying Start*, p. 53.

134 Mel Cline, quoted in Leinbaugh and Campbell, *The Men of Company K*, p. 229.

135 Alex Bowlby, *The Recollections of Rifleman Bowlby*, Corgi, London, 1971, p. 163.

136 Herr, *Dispatches*, p. 52.

137 Hugh McManners, *The Scars of War*, HarperCollins, London, 1993, p. 149.

138 General Tommy Franks with Malcolm McConnell, *American Soldier*, Regan Books, New York, 2004, p. 79.

139 Mason, *Chickenhawk*, p. 360.

140 Irving L. Janis, Chapter 7: 'Morale Attitudes of Combat Flying Personnel in the Air Corps', *The American Soldier: Combat and its Aftermath*, Princeton University Press, Princeton, New Jersey, 1949, p. 385.

141 Pyle, *Brave Men*, p. 172.

142 Samuel Lyman Atwood Marshall, *Men Against Fire: The Problem of Battle Command*, University of Oklahoma Press, Norman, Oklahoma, 2000 edition, p. 149.

143 Lawrence Nickell, 5th Infantry Division, quoted in McManus, *The Deadly Brotherhood*, p. 275.

144 Moran, *The Anatomy of Courage*, p. 184.

145 L.H. Bartemeier et al., 'Combat Exhaustion', *Journal of Nervous and Mental Diseases*, Vol. 104, 1946, p. 370, quoted in Ellis, *The Sharp End*, p. 306.

146 Ellis, *The Sharp End*, p. 307.

147 Sledge, *With the Old Breed at Peleliu and Okinawa*, p. 98.

148 Morris Dunn, quoted in Leinbaugh and Campbell, *The Men of Company K*, p. 59.

149 Marshall, *Men Against Fire*, pp. 151, 153.

150 Marc Bloch, *Memoirs of War, 1914–15*, translated by Carole Fink, Cornell University Press, Ithaca, New York, 1980, p. 166.

151 Gary McKay, *In Good Company: One Man's War in Vietnam*, Allen and Unwin, Sydney, 1987, pp. 187–8.

152   Monica Davey, 'Eight Soldiers Plan to Sue Over Army's Stop Loss Policy', *New York Times*, 6 December 2004, p. A8.

153   Henry Clifford, *Henry Clifford VC: his letters and sketches from the Crimea*, Michael Joseph, London, 1956, p. 265.

154   Bauer, *Soldiering*, p. 124.

155   Sajer, *The Forgotten Soldier*, p. 72.

156   McKay, *In Good Company*, p. 100.

157   McManners, *Falklands Commando*, pp. 49–50.

158   Major General Patrick Cordingley, *In the Eye of the Storm: Commanding the Desert Rats in the Gulf War*, Coronet Books, Hodder and Stoughton, London, 1997 edition, p. 180.

159   Tim Collins, *Rules of Engagement: A Life in Conflict*, Headline, London, 2005, p. 119.

160   Holmes, *Firing Line*, p. 90.

161   John Fitzgerald Kennedy, letter to Inga Binga dated 26 September 1943, quoted in Nigel Hamilton, *JFK: Reckless Youth*, Random House, New York, 1992, p. 616.

162   WO2 McCullum, quoted in McManners, *The Scars of War*, p. 102.

163   Lieutenant Colonel Tim Spicer, quoted in McManners, *The Scars of War*, p. 109.

164   Bowlby, *The Recollections of Rifleman Bowlby*, p. 168.

165   Major Chris Keeble, 2nd Battalion, the Parachute Regiment, in Michael Bilton and Peter Kosminsky, *Speaking Out: Untold Stories from the Falklands War*, André Deutsch, London, 1989, p. 148.

166   George MacDonald Fraser, *Quartered Safe out Here*, HarperCollins, London, 2000 edition, p. 130.

167   General Douglas MacArthur, speaking at West Point, 12 May 1962.

CHAPTER 4: THE COST OF WAR

1   Lieutenant Sean Walsh, 'The Real Meaning of 4,000 Dead', *Time*, 26 March 2008, www.time.com.

2   James R. McDonough, *Platoon Leader: A Memoir of Command in Combat*, Ballantine Books, New York, 2003 edition, p. 18.

3   David Bellavia with John Bruning, *House to House: The Most Terrifying Battle of the Iraq War – Through the Eyes of the Man who Fought it*, Pocket Books, London, 2008 edition, p. 189.

4   RAF Flight Lieutenants John Peters and John Nichol, *Tornado Down: The Horrifying True Story of their Gulf War Ordeal*, Signet, London, 1993 edition, p. 65.

5   Hugh McManners, *The Scars of War*, HarperCollins, London, 1993, p. 109.

6   Christopher Hibbert (editor), *The Wheatley Diary: A Journal and Sketchbook kept during the Peninsular War and the Waterloo Campaign*, The Windrush Press, Gloucestershire, 1997, p. 13.

7   Timothy Gowing, *Voice from the Ranks: A Personal Narrative of the Crimean War by a Sergeant of the Royal Fusiliers*, William Heinemann, Melbourne, 1954, p. 113.

8   Letter from Colonel John Monash to his wife, dated 24 April 1915, reproduced in Bill Gammage, *The Broken Years: Australian Soldiers in the Great War*, Penguin, Melbourne, 1982 edition, p. 46.

9   John Crawford (editor), *No Better Death: The Great War Diaries and Letters of William G. Malone*, Reed Publishing, Auckland, New Zealand, 2005, pp. 297–8.

10   William Craig, *Enemy at the Gates: The Battle for Stalingrad*, Penguin, London, 2000, p. 314.

11   A translation of the last letter of Lieutenant Matsuo Keiu is displayed in the Australian War Memorial, Canberra.

12   John Lawrence and Robert Lawrence MC, *When the Fighting is Over: Tumbledown – A Personal Story*, Bloomsbury, London, 1988, p. 15.

13   Dan Box, 'Kovco told his wife to remarry', *The Australian*, 1 March 2008, www.theaustralian.news.com.au.

14   Major Phil Ashby, *Unscathed: Escape from Sierra Leone*, Pan Books, London, 2003, p. 237.

15   Wayland Fuller Dunaway, *Reminiscences of a Rebel*, The Neale Publishing Company, New York, 1913, p. 35.

16   Lord C.E. Moran, *The Anatomy of Courage*, Constable, London, 1966, p. 149.

17   Anthony Swofford, *Jarhead: A Marine's Chronicle of the Gulf War*, Simon & Schuster UK, London, 2003, p. 212.

18   Robert Graves, *Goodbye to All That*, Blue Ribbon Books, New York, 1930, p. 145.

19   Christopher Hibbert (editor), *The Recollections of Rifleman Harris, as told to Henry Curling*, The Windrush Press, Gloucestershire, 1998 edition, pp. 37–8.

20   D.G. Crotty, *Four Years Campaigning in the Army of the Potomac*, Dygert Brothers & Co., Grand Rapids, Michigan, 1874, p. 48.

21   Captain B.H. Liddell Hart (editor), *The Letters of Private Wheeler 1809–1828*, Michael Joseph, London, 1951, p. 88.

22   *Observer* (London), No. 353, 18 November 1822.

23   Max Plowman, *A Subaltern on the Somme*, E.P. Dutton & Company, New York, 1928, p. 93.

24   Sidney Rogerson, *Twelve Days on the Somme*, Greenhill Books, London, 2006 edition, p. 92.

25   Edmund Blunden, *Undertones of War*, Penguin Books, London, 2000 edition, p. 11.

26   Walter Hubert Downing, *To the Last Ridge: The WW1 Experiences of W.H. Downing*, Duffy & Snellgrove, Sydney, 1998, p. 17.

27   Alistair Horne, *The Price of Glory: Verdun 1916*, Penguin, London, 1993 edition, p. 176.

28   Rogerson, *Twelve Days on the Somme*, p. 5.

29   Private Frank Richards, *Old Soldiers Never Die*, The Naval and Military Press, Uckfield, East Sussex, 2001, p. 198.

30   Mansur Abdulin, *Red Road from Stalingrad: Recollections of a Soviet Infantryman*, Pen and Sword Military, Barnsley, South Yorkshire, 2004, p. 32.

31   Joseph Owen, *Colder than Hell: A Marine Rifle Company at Chosin Reservoir*, Naval Institute Press, Annapolis, Maryland, 1996, p. 203.

32   Jules Roy, *Dienbienphu*, translated by Robert Baldick, Faber and Faber, London, 1965, p. 269.

33   Guy Sajer, *The Forgotten Soldier*, translated by Lily Emmet, Weidenfeld & Nicolson, London, 1971, p. 21.

34 Catherine Merridale, *Ivan's War: Life and Death in the Red Army, 1939–1945*, Metropolitan Books, New York, 2006, p. 140.

35 Plowman, *A Subaltern on the Somme*, p. 42.

36 Martin Middlebrook, *The Kaiser's Battle*, Penguin, London, 2000 edition, pp. 319–20.

37 Johann Voss, *Black Edelweiss: A Memoir of Combat and Conscience by a Soldier of the Waffen-SS*, The Aberjona Press, Bedford, Pennsylvania, 2002, p. 144.

38 Hamlin Alexander Coe, *Mine Eyes Have Seen the Glory: Combat Diaries of Union Sergeant Hamlin Alexander Coe*, Associated University Presses, Cranbury, New Jersey, 1975, p. 120.

39 Addison Terry, *The Battle for Pusan: A Korean War Memoir*, Presidio, Novato, California, 2000, p. 106.

40 Colby Buzzell, *My War: Killing Time in Iraq*, G.P. Putnam's Sons, New York, 2005, p. 71. See also Rick Reilly, 'Where Have All the Young Men Gone?', *Time*, 17 February 2003, p. 104.

41 Darren Moore, *Duntroon – The Royal Military College of Australia: 1911–2001*, Royal Military College of Australia, Canberra, 2001, pp. 185–6.

42 Carlo D'Este, *A Genius For War: A Life of General George S. Patton*, HarperCollins, London, 1996, pp. 797–8, 802.

43 Major General Sir Fabian Ware, quoted in Philip Longworth, *The Unending Vigil: A History of the Commonwealth War Graves Commission*, Leo Cooper, London, 1985, p. 14, reproduced in Julie Summers, *Remembered: The History of the Commonwealth War Graves Commission*, Merrell, London, 2007, p. 15.

44 Major General Sir Fabian Ware, press statement for the Imperial War Graves Commission dated 20 November 1918, quoted in Summers, *Remembered*, p. 25.

45 William Manchester, *Goodbye Darkness: A Memoir of the Pacific War*, Little, Brown & Company, Boston, 1980, p. 276.

46 Letter from General George S. Patton to Frederick Ayer, dated 9 June 1943, quoted in D'Este, *A Genius For War*, p. 798.

47 Peter Arnett, *Live from the Battlefield: From Vietnam to Baghdad, 35 Years in the World's War Zones*, Touchstone edition, Simon & Schuster, New York, 1995, pp. 151–2.

48 Major Bob Leitch, quoted in McManners, *The Scars of War*, p. 345.

49 US Department of Defense, Memorandum for Correspondents, dated 26 March 1997 (www.dod.gov/news/Mar1997); US Department of Defense, News Release No. 417-03, 'Top US POW/MIA Official Visits Vietnam', 13 June 2003.

50 Summers, *Remembered*, pp. 29–30.

51 Vince Crawley and Karen Jowers, 'Pentagon rejects mass-cremation option', *Army Times*, 3 March 2003, p. 10.

52 Moran, *The Anatomy of Courage*, p. 128.

53 Paul Dubrulle, *Mon Régiment*, Paris, 1917, quoted in Horne, *The Price of Glory*, p. 177.

54 Brigadier F.P. Crozier, *A Brass Hat in No-Man's Land*, Jonathan Cape, London, 1930, pp. 94–5.

55 Craig, *Enemy at the Gates*, p. 241.

56 Manchester, *Goodbye Darkness*, p. 340.

57   Günter Koschorrek, *Blood Red Snow: The Memoirs of a German Soldier on the Eastern Front*, Greenhill Books, London, 2002, p. 97.

58   Gary McKay, *In Good Company: One Man's War in Vietnam*, Allen and Unwin, Sydney, 1987, p. 54.

59   Tim O'Brien, *If I Die in a Combat Zone*, HarperCollins, Flamingo Seventies Classic edition, London, 2003, p. 126.

60   Ibid., p. 127.

61   Philip Caputo, *A Rumor of War*, Henry Holt & Company, New York, 1996 reprint, p. 288.

62   Ibid., p. 167.

63   Robert Mason, *Chickenhawk*, Corgi Books, London, 1987 edition, p. 155.

64   Congressional Research Service Report for Congress, 'Improvised Explosive Devices (IEDs) in Iraq and Afghanistan: Effects and Countermeasures', Order Code RS22330, updated 28 August 2007.

65   Major Chris Hunter, *Eight Lives Down*, Corgi, London, 2008 edition, pp. 309–10.

66   Hibbert, *The Recollections of Rifleman Harris*, p. 41.

67   Colin Frederick Campbell, *Letters from Camp to his Relatives During the Siege of Sebastopol*, Richard Bentley & Son, London, 1894, pp. 100–01.

68   John O. Casler, *Four Years in the Stonewall Brigade*, University of South Carolina Press, Columbia, South Carolina, 2005 edition, p. 89.

69   Lieutenant J.H. Sandoe, 45th Battalion, letter dated 19 September 1915, quoted in Gammage, *The Broken Years*, p. 106.

70   Gottlob Herbert Bidermann, *In Deadly Combat: A German Soldier's Memoir of the Eastern Front*, translated and edited by Derek S. Zumbro, University Press of Kansas, Lawrence, Kansas, 2000, p. 30.

71   Ibid., p. 52.

72   Koschorrek, *Blood Red Snow*, p. 84.

73   J. Glenn Gray, *The Warriors: Reflections on Men in Battle*, Bison Books, Lincoln, Nebraska, 1998 edition, p. 105.

74   Philip Williams with M.S. Power, *Summer Soldier: The True Story of the Missing Falklands Guardsman*, Bloomsbury, London, 1990, pp. 33–4.

75   Georg Grossjohann, *Five Years, Four Fronts*, translated by Ulrich Abele, The Aberjona Press, Bedford, Pennsylvania, 1999, p. 54.

76   Evgeni Bessonov, *Tank Rider: Into the Reich with the Red Army*, translated by Bair Irincheev, Greenhill Books, London, 2003, p. 115.

77   Edwin A. Weinstein, 'Chapter 14 – Disabling and Disfiguring Injuries', *Textbook of Military Medicine: War Psychiatry*, Office of the Surgeon General, United States Army, 1995, p. 372.

78   Colonel David H. Hackworth and Julie Sherman, *About Face*, Pan Books, Sydney, 1989, p. 496.

79   Dan Schilling, 'On Friendship and Firefights', *The Battle of Mogadishu: Firsthand Accounts from the Men of Task Force Ranger*, edited by Matt Eversmann and Dan Schilling, Ballantine Books, New York, 2004, pp. 195–6.

80 Stephen E. Ambrose, *Band of Brothers: E Company, 506th Regiment, 101st Airborne from Normandy to Hitler's Eagle's Nest*, Simon & Schuster, New York, 2001 edition, p. 97.

81 Frank Hunt, quoted in Stuart Rintoul, *Ashes of Vietnam: Australian Voices*, William Heinemann Australia, Melbourne, 1987, p. 75.

82 Simon Weston, *Walking Tall*, Bloomsbury, London, 1989, p. 154.

83 Bellavia, *House to House*, p. 175.

84 Stephen E. Ambrose, *Citizen Soldiers: The US Army from the Normandy Beaches to the Bulge to the Surrender of Germany*, Touchstone edition, New York, 1998, pp. 143–4.

85 Lawrence, *When the Fighting is Over*, p. 137.

86 Francis Davis, 'Storming The Home Front', *The Atlantic Monthly*, Vol. 291, No. 2, March 2003, p. 129.

87 Ron Kovic, *Born on the Fourth of July*, Corgi, London, 1990, p. 86.

88 Julius Bonello, 'Civil War Medicine', *The Surgical Technologist*, Vol. 32, No. 1, January 2000, pp. 14–16.

89 K. Jack Bauer (editor), *Soldiering: The Civil War Diary of Rice C. Bull*, Berkley Books, New York, 1988 edition, p. 74.

90 Major T.J. Mitchell and Miss G.M. Smith, *History of the Great War Based on Official Documents: Medical Services – Casualties and Medical Statistics of the Great War*, Battery Press edition, Nashville, 1997, pp. 315.

91 Niall Ferguson, *The Pity of War: Explaining World War I*, Basic Books, New York, 1999, p. 437.

92 Anonymous, 'Vietnam Warriors: A Statistical Profile', *Veterans of Foreign Wars*, January 2003, p. 18.

93 Lewis B. Puller Jr., *Fortunate Son: The Autobiography of Lewis B. Puller, Jr.*, Bantam Books, New York, 1993 edition, pp. 187–8.

94 Atul Gawande, 'Casualties of War: Military Care for the Wounded from Iraq and Afghanistan', *New England Journal of Medicine*, Vol. 135, No. 24, pp. 2471–5; Nancy Gibbs, 'The Lucky Ones', *Time*, 21 March 2005, pp. 24–5.

95 R.W. Zimmermann, 'Don't forget number of injured in big picture', *Army Times*, April 2003, p. 54.

96 Sergeant Erick Castro, 43rd Combat Engineer Company, quoted in Mark Thompson, 'The Wounded Come Home', *Time*, 10 November 2003, pp. 32–40.

97 Cathy Booth Thomas, 'The Scars of War', *Time*, 20 February 2006, p. 55.

98 Nancy Shute, 'Cheating Grim Death', *US News & World Report*, 29 November 2004, pp. 40–3; Angie Cannon, 'Trying to Make a Life', *US News & World Report*, 29 November 2004, pp. 44–51; Raja Mishra, 'Amputation rate for US troops twice that of past wars', *Boston Globe*, 9 December 2004, www.boston.com; Michael Weisskopf, 'A Grim Milestone: 500 Amputees', *Time*, 18 January 2007, www.time.com.

99 Lawrence F. Kaplan, 'Survivor: Iraq – America's near-invisible wounded', *The New Republic*, 13 & 20 October 2003, p. 20; Shute, 'Cheating Grim Death', pp. 40–3.

100 Gene Bolles, quoted by Lisa Marshall in 'Area Surgeon Aids Troops', *Boulder Daily Camera*, 5 April 2003, www.dailycamera.com.

101 Rod Nordland and T. Trent Gegax, ' Stressed Out at the Front', *Newsweek*, 12 January 2004, pp. 34–7.

102 Merridale, *Ivan's War*, p. 215.

103 Omar N. Bradley, *A Soldier's Story*, Henry Holt & Company, New York, 1951, p. 321.

104 Captain Desmond Allhusen, 3rd Battalion, Kings Royal Rifle Corps, quoted in Malcolm Brown, *The Imperial War Museum Book of the First World War*, Pan Books, London, 2002 edition, p. 272.

105 Graves, *Goodbye to All That*, p. 140.

106 Ibid., p. 163.

107 Casler, *Four Years in the Stonewall Brigade*, p. 213.

108 Diary entry dated 30 July 1916 for Lieutenant T.J. Richards, 1st Field Ambulance, quoted in Gammage, *The Broken Years*, p. 170.

109 Vyacheslav Kondrat'ev, quoted in Merridale, *Ivan's War*, p. 270.

110 Koschorrek, *Blood Red Snow*, pp. 113–14.

111 Manchester, *Goodbye Darkness*, p. 383.

112 Harry Arnold, quoted in John C. McManus, *The Deadly Brotherhood: The American Combat Soldier in World War II*, Presidio Books, Ballantine Books, New York, 1998, p. 147.

113 Private David Webster, quoted in Ambrose, *Band of Brothers*, p. 169.

114 Michael Herr, *Dispatches*, Picador, London, 1978, p. 70.

115 Richards, *Old Soldiers Never Die*, p. 60.

116 Captain N.A. Nicholson, 14th Field Artillery Brigade, letter dated 5 February 1917, quoted in Gammage, *The Broken Years*, p. 191.

117 Captain Maberly Esler, Royal Army Medical Corps, quoted in Max Arthur, *Forgotten Voices of the Great War*, Ebury Press, London, 2003 edition, p. 88.

118 Joshua Key, *The Deserter's Tale: Why I Walked Away From the War in Iraq*, Text Publishing, Melbourne, Australia, 2007, p. 128.

119 Casler, *Four Years in the Stonewall Brigade*, p. 24.

120 General George S. Patton Jr., *War as I Knew It*, Houghton Mifflin, Boston, 1995 edition, pp. 236–7.

121 Hackworth, *About Face*, p. 484.

122 Vasily Grossman, *A Writer at War: Vasily Grossman with the Red Army 1941–1945*, edited and translated by Antony Beevor and Luba Vinogradova, Pantheon Books, New York, 2005, pp. 19–20.

123 Henry Metelmann, *Through Hell for Hitler: A dramatic first-hand account of fighting with the Wehrmacht*, Patrick Stephens Ltd, Wellingborough, Northamptonshire, 1990, p. 152.

124 Figure quoted in Gammage, *The Broken Years*, p. 219.

125 Harold P. Leinbaugh and John D. Campbell, *The Men of Company K: The Autobiography of a World War II Rifle Company*, William Morrow & Company, New York, 1985, p. 281.

126 Eugene Sledge, quoted in Studs Terkel, *The Good War: An Oral History of World War Two*, Hamish Hamilton, London, 1985, p. 60.

CHAPTER 5: LOVE, SEX AND WAR

1   J. Glenn Gray, *The Warriors: Reflections on Men in Battle*, Bison Books, Lincoln, Nebraska, 1998 edition, pp. 61–2.

2   Robin M. Williams Jr. and M. Brewster Smith, 'Chapter 2 – General Characteristics of Ground Combat', *The American Soldier: Combat and its Aftermath*, Princeton University Press, Princeton, New Jersey, 1949, p. 80.

3   George Hicks, *The Comfort Women: Japan's Brutal Regime of Enforced Prostitution in the Second World War*, W.W. Norton & Company, New York, 1997, p. 29.

4   Susan Brownmiller, *Against Our Will: Men, Women and Rape*, Simon & Schuster, New York, 1975, p. 38.

5   'Jack Nastyface', quoted in Henry Baynham, *From the Lower Deck: the old navy 1780–1840*, Hutchinson, London, 1969, pp. 68–9; Christopher Lloyd, *The British Seaman*, Collins, London, 1968, p. 246.

6   Admiral Hawkins, quoted in Lloyd, *The British Seaman*, p. 247.

7   Herodotus, *Histories*, Book 1:105, George Rawlinson translation, Wordsworth Editions, Ware, Hertfordshire, 1996, p. 50.

8   Memorandum from Lord Kitchener to the Army in India on Venereal Disease, dated 1905, quoted in John Baynes, *Morale: A Study of Men and Courage*, Avery Publishing Group, New York, 1988 edition, p. 269.

9   Major T.J. Mitchell and Miss G.M. Smith, *History of the Great War Based on Official Documents: Medical Services – Casualties and Medical Statistics of the Great War*, Battery Press edition, Nashville, 1997, p. 73.

10   Colonel A.G. Butler, *The Official History of the Australian Army Medical Services in the War of 1914–1918: Vol. 1 – Gallipoli, Palestine and New Guinea*, Australian War Memorial, Melbourne, second edition, 1938, pp. 25, 76–7.

11   Butler, *The Official History of the Australian Army Medical Services in the War of 1914–1918: Vol. 1*, pp. 415–16; Patsy Adam-Smith, *The Anzacs*, Nelson, Melbourne, 1985 reissue, p. 70.

12   Allan S. Walker, *Australia in the War of 1939–1945, Series 5, Vol. 1: Clinical Problems of War*, Australian War Memorial, Canberra, 1962 reprint, pp. 264, 266; Mitchell and Smith, *History of the Great War Based on Official Documents*, pp. 74, 75; Colonel A.G. Butler, *The Official History of the Australian Army Medical Services in the War of 1914–1918: Vol. 2 – The Western Front*, Australian War Memorial, Canberra, 1940, p. 410.

13   Brigadier F.P. Crozier, *A Brass Hat in No-Man's Land*, Jonathan Cape, London, 1930, p. 127.

14   Private Frank Richards, *Old Soldiers Never Die*, The Naval and Military Press, Uckfield, East Sussex, 2001, pp. 118–19.

15   Richard Holmes, *Firing Line*, Pimlico, London, 1985, p. 95.

16   Philip Caputo, *A Rumor of War*, Henry Holt & Company, New York, 1996 reprint, p. 143.

17   Franklin D. Jones, 'Chapter 3 – Disorders of Frustration and Loneliness', *Textbook of Military Medicine: War Psychiatry*, Office of the Surgeon General, United States Army, 1995, p. 74.

18   John Ellis, *The Sharp End: The Fighting Man in World War II*, Charles Scribner's Sons, New York, 1980, pp. 273–4.

19   Ibid., p. 270.

20 N. McCallum, *Journey with a Pistol*, Gollancz, London, 1959, p. 75, quoted in Ellis, *The Sharp End*, p. 270.

21 Ellis, *The Sharp End*, p. 271.

22 Tom Hickman, *The Sexual Century*, Carlton Books, London, 1999, p. 75.

23 Paul Fussell, *Doing Battle: The Making of a Skeptic*, Little Brown & Company, Boston, 1996, p. 79.

24 Propaganda card A10-046-8-44, National Archives, reproduced in John Costello, *Love, Sex and War: Changing Values 1939–45*, Collins, London, 1985, illustration no. 32.

25 Herbert A. Friedman, 'Sex and Psychological Operations', http://www.psywarrior.com/sexandprop.html.

26 Private Robert Waterfield, papers located in the Oriental and India Office Collection of the British Library, quoted in Richard Holmes, *Sahib: The British Soldier in India*, HarperCollins, London, 2005, p. 494.

27 John Masters, *The Road Past Mandalay*, Cassell, London, 2002 edition, p. 85.

28 Paul Fussell, *Wartime: Understanding and Behaviour in the Second World War*, Oxford University Press, New York, 1989, p. 253.

29 Lieutenant Colonel J.H.A. Sparrow, *The Second World War 1939–1945, Army: Morale*, The War Office, London, 1949, p. 9.

30 Colonel David H. Hackworth and Julie Sherman, *About Face*, Pan Books, Sydney, 1989, p. 213.

31 Anthony Swofford, *Jarhead: A Marine's Chronicle of the Gulf War*, Simon & Schuster UK, London, 2003, pp. 91–2.

32 Ibid., p. 110.

33 Joshua Key, *The Deserter's Tale: Why I Walked Away From the War in Iraq*, Text Publishing, Melbourne, Australia, 2007, pp. 72–3.

34 Ibid., p. 124.

35 Chaplain James Pritchard, quoted in David Smith, 'Divorces inflict home front damage on US troops as Iraq war drags on', *Observer*, 1 June 2008, www.guardian.co.uk/world/2008/jun/01/usa.usforeignpolicy.

36 Sparrow, *The Second World War 1939–1945, Army: Morale*, p. 10.

37 N. Frankel and L. Smith, *Patton's Best*, Hawthorn Books, New York, 1978, p. 75, quoted in Ellis, *The Sharp End*, p. 275.

38 Russell Braddon, *The Naked Island*, Penguin, Camberwell, Victoria, 1993, pp. 29, 31.

39 Sean Longden, *To the Victor the Spoils – D-Day to VE Day: The Reality Behind the Heroism*, Arris Books, Gloucestershire, 2005 edition, p. 81.

40 Hans Peter Richter, *The Time of the Young Soldiers*, translated by Anthea Bell, Armada, London, 1989 edition, pp. 82–3.

41 George Hicks, *The Comfort Women*, pp. 31–2.

42 Lujo Bassermann, *The Oldest Profession: A History of Prostitution*, Dorset Press, New York, 1993 edition, p. 249.

43 Iris Chang, *The Rape of Nanking: The Forgotten Holocaust of World War II*, Penguin Books, New York, 1998, pp. 4, 6, 49, 89–91, 95.

44 Ibid., p. 53; Hicks, *The Comfort Women*, pp. 86, 94–5.

45   United Nations Press Release HR/CN/94/26, dated 16 February 1994; Chunhgee
     Sarah Soh, lectures delivered to the Institute for Corean-American Studies entitled
     'Human Rights and Humanity: The Case of the "Comfort Woman"', No. 98-1204-
     CSSb, and 'Japan's Responsibility Towards Comfort Women Survivors', No. 2001-
     0501-CSS, www.icasinc.org (Institute of Corean-American Studies); Hicks, *The
     Comfort Women*, pp. 48, 94.

46   United Nations Press Release HR/CN/94/26; Soh, 'Human Rights and Humanity: The
     Case of the "Comfort Woman"', and 'Japan's Responsibility Towards Comfort Women
     Survivors'; Hicks, *The Comfort Women*, p. 20.

47   Hicks, *The Comfort Women*, pp. 16, 74.

48   Ibid., p. 160.

49   'Documents: US troops used "comfort women" after WWII', www.cnn.com, 25 April
     2007.

50   Hicks, *The Comfort Women*, pp. 161–2; 'Documents: US troops used "comfort
     women" after WWII', www.cnn.com.

51   Engine Room Artificer W.S. Watson, HMNZS *Tutira*, quoted in Commodore G.F.
     Hopkins (editor), *Tales from Korea: The Royal New Zealand Navy in the Korean War*,
     Reed Publishing, Auckland, New Zealand, 2002, p. 30.

52   Bernard B. Fall, *Hell in a Very Small Place: The Siege of Dien Bien Phu*, Pall Mall Press,
     Oxford, 1966, pp. 403–04; James Pringle, 'Meanwhile: Au revoir, Dien Bien Phu',
     International Herald Tribune, 31 March 2004, www.iht.com.

53   Robert Mason, *Chickenhawk*, Corgi Books, London, 1987 edition, pp. 195–6.

54   Ibid., p. 196.

55   Ibid., p. 287.

56   Jonathan Shay, *Achilles in Vietnam: Combat Trauma and the Undoing of Character*,
     Touchstone, Simon & Schuster, New York, 1995, p. 160.

57   Caputo, *A Rumor of War*, pp. 38–9.

58   Jonathan Schell, *The Real War*, Corgi, London, 1989 edition, p. 119.

59   Gary McKay, *In Good Company: One Man's War in Vietnam*, Allen & Unwin, Sydney,
     1987, p. 103.

60   Brownmiller, *Against Our Will*, pp. 94–5.

61   Frederick Downs, *The Killing Zone: My Life in the Vietnam War*, W.W. Norton &
     Company, New York, 2007 edition, pp. 18–19.

62   Ibid., p. 19.

63   Barry Kelly, quoted in Stuart Rintoul, *Ashes of Vietnam: Australian Voices*, William
     Heinemann Australia, Melbourne, 1987, p. 88.

64   Jones, 'Chapter 3 – Disorders of Frustration and Loneliness', *Textbook of Military
     Medicine*, pp. 66–7, 74–6.

65   Donald Macintyre, 'Special Report: Asia's Slave Trade', *Time* (Asia), 12 August 2002,
     www.time.com.

66   *Manual of Military Law*, War Office, 1914, para 42, p. 97.

67   Ibid., p. 118.

68   Niall Ferguson, *The Pity of War: Explaining World War I*, Basic Books, New York,
     1999, p. 349.

69   Costello, *Love, Sex and War*, pp. 160–1.

70   William Manchester, *Goodbye Darkness: A Memoir of the Pacific War*, Little, Brown & Company, Boston, 1980, p. 100.

71   Stephen E. Ambrose, *Citizen Soldiers: The US Army from the Normandy Beaches to the Bulge to the Surrender of Germany*, Touchstone edition, New York, 1998, pp. 266–7.

72   www.hmsrichmond.org/rnarticles.htm.

73   United States *Uniform Code of Military Justice*, Article 125.

74   www.sldn.org (Servicemembers Legal Defense Network); Bill Clinton, *My Life*, Hutchinson, London, 2004, p. 485.

75   'Fewer Dismissals of Gay Soldiers', *Associated Press*, 10 March 2003

76   Clinton, *My Life*, p. 450.

77   Plutarch, *Life of Pelopidas*, 18.1–18.4, translated by John Dryden, The Internet Classics Archive, http://classics.mit.edu.

78   Louis Crompton, *Homosexuality and Civilization*, Harvard University Press, Cambridge, Massachusetts, 2003, pp. 69–74.

79   Homer, *The Iliad*, Book 18, 22–4.

80   Crompton, *Homosexuality and Civilization*, p. 6.

81   Quintus Curtius, *History of Alexander*, translated by J.C. Rolfe, Harvard University Press, Cambridge, Massachusetts, 1946, Vol. 2, p. 47, quoted in Crompton, *Homosexuality and Civilization*, p. 76.

82   Plutarch, *Lives*, translated by Bernadotte Perrin, Harvard University Press, Cambridge, Massachusetts, 1919, Vol. 7, p. 413, quoted in Crompton, *Homosexuality and Civilization*, p. 77.

83   Robert Lane Fox, *Alexander the Great*, Penguin Books, London, 2004 edition, p. 56.

84   Crompton, *Homosexuality and Civilization*, p. 78.

85   Suetonius, 'Julius Caesar', *Lives of the Twelve Caesars*, translated by H.M. Bird, Wordsworth Editions, Ware, Hertfordshire, 1997, p. 50.

86   Crompton, *Homosexuality and Civilization*, pp. 358, 403, 420, 422, 508.

87   Tom Hickman, *The Sexual Century*, Carlton Books Limited, London, 1999, pp. 59–60.

88   Manchester, *Goodbye Darkness*, p. 148.

89   Anthony Rhodes, *Sword of Bone*, Buchan & Enright, London, 1986 edition, p. 7.

90   Marian Orley, Women's Auxiliary Air Force, quoted in Joshua Levine, *Forgotten Voices of the Blitz and the Battle for Britain*, Ebury Press, London, 2006, p. 426.

91   Dellie Hahne, quoted in Studs Terkel, *The Good War: An Oral History of World War Two*, Hamish Hamilton, London, 1985, p. 117.

92   Rick Atkinson, *In the Company of Soldiers: A Chronicle of Combat in Iraq*, Little, Brown, London, 2004, p. 26.

93   'Do You? Do You?', *Newsweek*, 27 January 2003, p. 11.

94   J. Glenn Gray, *The Warriors: Reflections on Men in Battle*, Bison Books, Lincoln, Nebraska, 1998 edition, p. 72.

95   *History of the Directorate of Repatriation*, Department of National Defense, February 1947, pp. 30–1; Melynda Jarratt, thesis entitled 'The War Brides of New Brunswick', Masters of Arts, University of New Brunswick, 1995.

96   Costello, *Love, Sex and War*, pp. 322–6.

97   Ibid., pp. 323, 327.

CHAPTER 6: KILL OR BE KILLED: LIVE AND LET LIVE

1   Guardsman F.E. Noakes, 3rd Battalion, Coldstream Guards, quoted in Malcolm Brown, *The Imperial War Museum Book of the Western Front*, Pan Books, London, 2001 edition, p. 333.

2   Günter Koschorrek, *Blood Red Snow: The Memoirs of a German Soldier on the Eastern Front*, Greenhill Books, London, 2002, p. 84.

3   Ibid., p. 128.

4   Samuel Lyman Atwood Marshall, *Men Against Fire: The Problem of Battle Command*, University of Oklahoma Press, Norman, Oklahoma, 2000 edition, p. 78.

5   Anatoly Chekhov, 13th Guards Rifle Division, quoted in Vasily Grossman, *A Writer at War: Vasily Grossman with the Red Army 1941–1945*, edited and translated by Antony Beevor and Luba Vinogradova, Pantheon Books, New York, 2005, p. 157.

6   Chaplain Montague Bere, quoted in Brown, *The Imperial War Museum Book of the Western Front*, p. 334.

7   Lieutenant Colonel Fred Swan, quoted in Julian Borger and Stuart Millar, '2pm: Saddam is spotted. 2.48pm: pilots get their orders. 3pm: 60ft crater at target', *Guardian*, 9 April 2003, www.guardian.co.uk.

8   Gwynne Dyer, *War*, The Bodley Head, London, 1986, p. 119.

9   Samuel Hynes, *Flights of Passage: Reflections of a World War II Aviator*, Bloomsbury, London, 1989 edition, p. 216.

10 Dave Grossman, *On Killing: The Psychological Cost of Learning to Kill in War and Society*, Little Brown & Company, Boston, 1996, pp. 97–8.

11 John Ciardi, quoted in Studs Terkel, *The Good War: An Oral History of World War Two*, Hamish Hamilton, London, 1985, p. 200.

12 Alexander Aitken, *Gallipoli to the Somme: Recollections of a New Zealand Infantryman*, Oxford University Press, London, 1963, pp. 33–4.

13 Ken Hechler, *The Bridge at Remagen*, Ballantine Books, New York, 1957, p. 102.

14 James R. McDonough, *Platoon Leader: A Memoir of Command in Combat*, Ballantine Books, New York, 2003 edition, p. 199.

15 Sergeant Stefan Westmann, 29th Division, German Army, quoted in Max Arthur, *Forgotten Voices of the Great War*, Ebury Press, London, 2003 edition, pp. 70–1.

16 Private A. M. Simpson, 13th Battalion, letter dated May 1915, quoted in Bill Gammage, *The Broken Years: Australian Soldiers in the Great War*, Penguin, Melbourne, 1982 edition, pp. 104–05.

17 Guy Sajer, *The Forgotten Soldier*, translated by Lily Emmet, Weidenfeld & Nicolson, London, 1971, p. 300.

18 William Manchester, *Goodbye Darkness: A Memoir of the Pacific War*, Little, Brown & Company, Boston, 1980, p. 7.

19 Private James Sims, 2nd Battalion, the Parachute Regiment, quoted in Sean Longden, *To the Victor the Spoils – D-Day to VE Day: The Reality Behind the Heroism*, Arris Books, Gloucestershire, 2005 edition, p. 31.

20 Private James Sims, 2nd Battalion, the Parachute Regiment, quoted in Max Arthur, *Forgotten Voices of the Second World War*, Ebury Press, London, 2004, p. 357.

21 Gunnery Sergeant Jack Coughlin, USMC and Captain Casey Kuhlman, USMCR, with Donald A. Davis, *Shooter*, St Martin's Press, New York, 2006, p. 250.

22 Sergeant Dan Mills, *Sniper One: The Blistering True Story of a British Battle Group Under Siege*, Penguin, London, 2008 edition, pp. 5–6.

23 Evan Wright, *Generation Kill: Living dangerously on the road to Baghdad with the ultraviolent Marines of Bravo Company*, Bantam Press, London, 2004, p. 109.

24 Sergeant Antonio Espera, quoted in Wright, *Generation Kill*, pp. 217–18.

25 Chekhov, quoted in Grossman, *A Writer at War*, p. 158.

26 Colin P. Sisson, *Wounded Warriors: The True Story of a soldier in the Vietnam War and the emotional wounds inflicted*, Total Press, Auckland, New Zealand, 1993, p. 101.

27 Warrant Officer Class Two Dave Falconer, 1st Battalion, the Princess of Wales's Royal Regiment, quoted in Richard Holmes, *Dusty Warriors*, Harper Perennial, London, 2007 edition, p. 317.

28 Sergeant Chris Broome, 1st Battalion, the Princess of Wales's Royal Regiment, quoted in Holmes, *Dusty Warriors*, p. 239.

29 Dave Pelkey, quoted in Gerald Wright, 'Picking up the pieces', *Sydney Morning Herald*, 24 March 2007, www.smh.com.au.

30 First Lieutenant John Yaros, 3rd Infantry Division, quoted in Oliver Poole, *Black Knights: On the Bloody Road to Baghdad*, HarperCollins, London, 2003, p. 154.

31 Captain Robert Ross, 3rd Infantry Division, quoted in Poole, *Black Knights*, p. 255.

32 Poole, *Black Knights*, p. 145.

33 Ibid., p. 153.

34 Grossman, *On Killing*, p. 119.

35 Al Slater, quoted in Wright, 'Picking up the pieces', www.smh.com.au.

36 Hervey Allen, *Toward the Flame: A War Diary*, University of Pittsburgh Press, Pittsburgh, 1968 edition, p. 115.

37 Paul Fussell, *Doing Battle: The Making of a Skeptic*, Little Brown & Company, Boston, 1996, p. 108.

38 Frederick Downs, *The Killing Zone: My Life in the Vietnam War*, W.W. Norton & Company, New York, 2007 edition, p. 263.

39 Bernard Szapiel, quoted in Stuart Rintoul, *Ashes of Vietnam: Australian Voices*, William Heinemann Australia, Melbourne, 1987, p. 48.

40 Joshua Key, *The Deserter's Tale: Why I Walked Away From the War in Iraq*, Text Publishing, Melbourne, Australia, 2007, pp. 46–7.

41 Ibid., p. 49.

42 Colonel David H. Hackworth and Julie Sherman, *About Face*, Pan Books, Sydney, 1989, p. 49.

43 Samuel A. Stouffer, Arthur A. Lumsdaine and Marion Harper Lumsdaine, 'Chapter 1 – Attitudes Before Combat and Behavior in Combat', *The American Soldier: Combat and its Aftermath*, Princeton University Press, Princeton, New Jersey, 1949, p. 34.

44 Stephen E. Ambrose, *Citizen Soldiers: The US Army from the Normandy Beaches to the Bulge to the Surrender of Germany*, Touchstone edition, New York, 1998, p. 228.

45 Paul Fussell, *Wartime: Understanding and Behaviour in the Second World War*, Oxford University Press, New York, 1989, pp. 116–17.

46 Grossman, *On Killing*, p. 164.

47 William A. Fletcher, *Rebel Private: Front and Rear – Memoirs of a Confederate Soldier*, Meridian, New York, 1997 edition, p. 154.

48 Hamlin Alexander Coe, *Mine Eyes Have Seen the Glory: Combat Diaries of Union Sergeant Hamlin Alexander Coe*, Associated University Presses, Cranbury, New Jersey, 1975, p. 92.

49 K. Jack Bauer (editor), *Soldiering: The Civil War Diary of Rice C. Bull*, Berkley Books, New York, 1988 edition, p. 207.

50 *The Bryce Report: Report of the Committee on Alleged German Outrages*, 12 May 1915, copy of report provided at http://www.firstworldwar.com/source/brycereport.htm.

51 M. McKinley McClure, *Hey! Major, Look Who's Here*, Dorrance & Company, Philadelphia, 1972, p. 79.

52 Fritz Nagel, *Fritz: The World War I Memoirs of a German Lieutenant*, Der Angriff Publications, Huntington, West Virginia, 1981, p. 21.

53 Phillip Knightley, *The First Casualty: The War Correspondent as Hero and Myth-Maker from the Crimea to Kosovo*, Johns Hopkins University Press, Baltimore, 2002 edition, p. 88.

54 Ibid., p. 487.

55 Chris Hedges, *War is a Force That Gives Us Meaning*, Anchor Books, New York, 2002, p. 145.

56 Grossman, *On Killing*, p. 169.

57 James Hewitt, *Love and War*, Blake Publishing, London, 1999, p. 128.

58 Flight Lieutenant Frank Carey, 43rd Squadron, Royal Air Force, quoted in Arthur, *Forgotten Voices of the Second World War*, p. 88.

59 Sergeant Ray Holmes, 504th Squadron, Royal Air Force, quoted in Arthur, *Forgotten Voices of the Second World War*, p. 92.

60 Lieutenant Commander James Newton, *Armed Action: My War in the Skies with 847 Naval Air Squadron*, Headline Review, London, 2007, p. 8.

61 Lieutenant Alastair Mitchell, 2nd Battalion, the Scots Guards, in Michael Bilton and Peter Kosminsky, *Speaking Out: Untold Stories from the Falklands War*, André Deutsch, London, 1989, p. 204.

62 Referred to in Grossman, *On Killing*, p. 143.

63 Grossman, *On Killing*, pp. 143, 152.

64 Marshall, *Men Against Fire*, p. 56.

65 Max Plowman, *A Subaltern on the Somme*, E.P. Dutton & Company, New York, 1928, p. 179.

66 Grossman, *On Killing*, pp. 253–4.

67 Nagel, *Fritz*, p. 27.

68 Philip Caputo, *A Rumor of War*, Henry Holt & Company, New York, 1996 reprint, p. 93.

69 Emilio Lussu, *Sardinian Brigade*, translated by Marion Rawson, Grove Press, New York, 1967, p. 169.

70 Robert Graves, *Goodbye to All That*, Blue Ribbon Books, New York, 1930, p. 164.

71 Eugene B. Sledge, *With the Old Breed at Peleliu and Okinawa*, Oxford University Press, New York, 1990, p. 116.

72 Robert Rasmus, quoted in Terkel, *The Good War*, pp. 44–5.

73 James C. Donahue, *No Greater Love: A Day with the Mobile Guerrilla Force in Vietnam*, New American Library, New York, 1989, p. 158.

74 Ibid., p. 159.

75 Joseph Donaldson, *Donaldson of the 94th Scots Brigade: The Recollections of a Soldier During the Peninsula & South of France Campaigns of the Napoleonic Wars*, Leonaur, 2008 edition, p. 122.

76 Sergeant Major Henry Franks, *Leaves from A Soldier's Note Book*, Mitre Publications, Brightlingsea, Essex, 1979 edition, p. 84.

77 Gefreiter Michael Pitsch, 16th Reserve Jäger Battalion, quoted in Martin Middlebrook, *The Kaiser's Battle*, Penguin, London, 2000 edition, pp. 175–6.

78 Johann Voss, *Black Edelweiss: A Memoir of Combat and Conscience by a Soldier of the Waffen-SS*, The Aberjona Press, Bedford, Pennsylvania, 2002, p. 188.

79 Caputo, *A Rumor of War*, p. 120.

80 Ibid., p. 124.

81 William Bradford Huie, *The Execution of Private Slovik*, Westholme, Yardley, Pennsylvania, 2004 edition, p. 86.

82 Jean-Baptiste Barres, *Chasseur Barres*, Leonaur, 2006 edition, p. 41.

83 Ernst Jünger, *Storm of Steel*, translated by Michael Hofmann, Penguin, London, 2004 edition, p. 5.

84 Laurie Lee, *Moment of War: A Memoir of the Spanish Civil War*, The New Press, New York, 1991, pp. 12–13.

85 Samuel Hynes, *The Soldiers' Tale: Bearing Witness to Modern War*, Pimlico, London, 1998, p. 111.

86 Christopher Hibbert (editor), *The Wheatley Diary: A Journal and Sketchbook kept during the Peninsular War and the Waterloo Campaign*, The Windrush Press, Gloucestershire, 1997, p. 60.

87 Fletcher, *Rebel Private*, p. 26.

88 Plowman, *A Subaltern on the Somme*, p. 10.

89 Ibid., p. 11.

90 Jack Short, quoted in Terkel, *The Good War*, p. 144.

91 John Foley, *Mailed Fist*, Mayflower, London, 1975 edition, p. 44.

92 Fay Lewis, quoted in Rintoul, *Ashes of Vietnam*, p. 39.

93 Jünger, *Storm of Steel*, p. 284.

94 Stephen E. Ambrose, *Band of Brothers: E Company, 506th Regiment, 101st Airborne from Normandy to Hitler's Eagle's Nest*, Simon & Schuster, New York, 2001 edition, pp. 84–5.

95 Ibid., p. 117.

96 Ibid., pp. 117–18.

97 Ernie Pyle, *Brave Men*, Bison Books, Lincoln, Nebraska, 2001 edition, p. 364.

98 Koschorrek, *Blood Red Snow*, pp. 98–9.

99 Chaplain David Cooper, quoted in Hugh McManners, *The Scars of War*, HarperCollins, London, 1993, p. 124.

100 Stouffer, Lumsdaine and Lumsdaine, 'Chapter 1 – Attitudes Before Combat and Behavior in Combat', *The American Soldier*, p. 4.

101 Major Philip Neame, quoted in Max Arthur, *Above All, Courage – The Falklands Front Line: First-Hand Accounts*, Guild Publishing, London, 1985, p. 196.

102 Cooper, quoted in McManners, *The Scars of War*, pp. 182–3.

103 Graves, *Goodbye to All That*, p. 229.

104 George Coppard, *With a Machine Gun to Cambrai: A Story of the First World War*, Cassell, London, 1999 edition, p. 109.

105 M. Brewster Smith, 'Chapter 3 – Combat Motivation Among Ground Troops', *The American Soldier: Combat and its Aftermath*, Princeton University Press, Princeton, New Jersey, 1949, pp. 109–10.

106 Ibid., p. 169.

107 Hedges, *War is a Force That Gives Us Meaning*, p. 38.

108 Max Hastings, *Going to the Wars*, Pan Books, London, 2001, p. 104.

109 *The Air Officer's Guide*, Department of Defense, 1952, p. 104, cited in Morris Janowitz, *The Professional Soldier: A Social and Political Portrait*, The Free Press of Glencoe, Illinois, 1960, p. 221.

110 Robin M. Williams Jr. and M. Brewster Smith, 'Chapter 2 – General Characteristics of Ground Combat', *The American Soldier*, p. 98.

111 Hewitt, *Love and War*, pp. 102–03.

112 Brigadier Julian Thompson, 3rd Commando Brigade, in Bilton and Kosminsky, *Speaking Out*, p. 230.

113 Captain L.C. Roth, 2nd Pioneer Battalion, letter dated 1 January 1915, quoted in Gammage, *The Broken Years*, p. 102.

114 Conrad Wilson, 87th Infantry Division, quoted in John C. McManus, *The Deadly Brotherhood: The American Combat Soldier in World War II*, Presidio Books, Ballantine Books, New York, 1998, p. 147.

115 Mansur Abdulin, *Red Road from Stalingrad: Recollections of a Soviet Infantryman*, Pen and Sword Military, Barnsley, South Yorkshire, 2004, p. 4.

116 Adrien Jean Baptiste François Bourgogne, *Memoirs of Sergeant Bourgogne 1812–1813*, Constable, London, 1997 edition, p. 71.

117 Captain Hugh McManners, *Falklands Commando*, William Kimber, London, 1984, p. 50.

118 A. Muir, *The First of Foot*, private publication, Edinburgh, 1961, pp. 143–4, quoted in John Ellis, *The Sharp End: The Fighting Man in World War II*, Charles Scribner's Sons, New York, 1980, pp. 314.

119 Michael Herr, *Dispatches*, Picador, London, 1978, p. 112.

120 Hastings, *Going to the Wars*, p. 346.

121 Lieutenant B.W. Champion, 1st Battalion, diary entry dated 27 July 1916, quoted in Gammage, *The Broken Years*, p. 162.

122 Grossman, *A Writer at War*, p. 165.

123 Bernard B. Fall, *Hell in a Very Small Place: The Siege of Dien Bien Phu*, Pall Mall Press, Oxford, 1966, p. 145.

124 Manchester, *Goodbye Darkness*, p. 391.

125 Hackworth and Sherman, *About Face*, p. 69.

126 Phil Zabriskie, 'Wounds That Don't Bleed', *Time*, 29 November 2004, pp. 28–30.

127 Caputo, *A Rumor of War*, p. xvii.

128 Captain Ken Moorefield, 1st Infantry Division, quoted in Al Santoli, *To Bear any Burden: The Vietnam War and its Aftermath in the Words of Americans and Southeast Asians*, Abacus, London, 1985, p. 191.

129 Michael J. Durant with Steven Hartov, *In the Company of Heroes*, Bantam Press, London, 2003, p. 313.

130 Donaldson, *Donaldson of the 94th Scots Brigade*, p. 76.

131 Siegfried Sassoon, *Memoirs of a Fox-hunting Man*, The Folio Society, London, 1971 edition, p. 284.

132 Siegfried Sassoon, *Memoirs of an Infantry Officer*, Faber and Faber, London, 1965 edition, pp. 58, 60.

133 Jünger, *Storm of Steel*, p. 55.

134 Koschorrek, *Blood Red Snow*, p. 212.

135 George MacDonald Fraser, *Quartered Safe out Here*, HarperCollins, London, 2000 edition, p. 127.

136 Caputo, *A Rumor of War*, p. 124.

137 Ibid., p. 231.

138 Gary McKay, *In Good Company: One Man's War in Vietnam*, Allen & Unwin, Sydney, 1987, p. 87.

139 Ted Cowell, quoted in Rintoul, *Ashes of Vietnam*, p. 119.

140 Private Patricio Perez, 3rd Infantry Regiment (Argentina), in Bilton and Kosminsky, *Speaking Out*, p. 192.

141 Koschorrek, *Blood Red Snow*, pp. 69–70.

142 Gottlob Herbert Bidermann, *In Deadly Combat: A German Soldier's Memoir of the Eastern Front*, translated and edited by Derek S. Zumbro, University Press of Kansas, Lawrence, Kansas, 2000, p. 52.

143 Tim Collins, *Rules of Engagement: A Life in Conflict*, Headline, London, 2005, p. 356.

144 Jonathan Shay, *Achilles in Vietnam: Combat Trauma and the Undoing of Character*, Touchstone, Simon & Schuster, New York, 1995, p. 81.

145 Unnamed Vietnam veteran, quoted in Shay, *Achilles in Vietnam*, pp. 78–9.

146 Tim O'Brien, *If I Die in a Combat Zone*, HarperCollins, Flamingo Seventies Classic edition, London, 2003, p. 123.

147 Sisson, *Wounded Warriors*, p. 71.

148 David Bellavia with John Bruning, *House to House: The Most Terrifying Battle of the Iraq War – Through the Eyes of the Man who Fought it*, Pocket Books, London, 2008 edition, p. 60.

149 Edward Costello, *Rifleman Costello: The Adventures of a Soldier of the 95th (Rifles) in the Peninsular & Waterloo Campaigns of the Napoleonic Wars*, Leonaur, 2005 edition, p. 44.

150 Ibid., p. 63.

151 Coe, *Mine Eyes Have Seen the Glory*, p. 173.

152 Ibid., p. 174.

153 Maurie Pears, 'Recollections of War', reproduced in Maurie Pears and Fred Kirkland (editors), *Korea Remembered*, Department of Defence (Australia), Sydney, 2002 edition, pp. 94–5.

154 Charles Sorley, *The Letters of Charles Sorley*, Cambridge University Press, 1919, p. 283, quoted in Tony Ashworth, *Trench Warfare 1914–1918: The Live and Let Live System*, Pan Books, London, 2000, p. 130.

155 Horace L. Baker, *Argonne Days in World War I*, University of Missouri Press, Columbia, Missouri, 2007, p. 110.

156 Fussell, *Doing Battle*, p. 110.

157 Anthony Swofford, *Jarhead: A Marine's Chronicle of the Gulf War*, Simon & Schuster UK, London, 2003, p. 199.

158 Stanley Weintraub, *Silent Night: The Story of the World War I Christmas Truce*, Plume, New York, 2001, pp. 4–7.

159 Ibid., p. 16.

160 Ibid., pp. 25–7, 57, 62, 84, 101, 107, 155.

161 Ibid., p. 149.

162 Jünger, *Storm of Steel*, pp. 58–9.

163 Second Lieutenant Edward Beddington-Behrens, Royal Field Artillery, quoted in Brown, *The Imperial War Museum Book of the First World War*, p. 83.

164 Sidney Rogerson, *Twelve Days on the Somme*, Greenhill Books, London, 2006 edition, p. 62.

165 Jünger, *Storm of Steel*, pp. 56–7.

166 Winston Churchill, letter to Clementine Churchill dated 30 December 1915, Spencer-Churchill papers, quoted in Martin Gilbert, *The Challenge of War: Winston S. Churchill 1914–1916*, Minerva, London, 1990 edition, p. 624.

167 Coppard, *With a Machine Gun to Cambrai*, p. 25.

168 Ashworth, *Trench Warfare 1914–1918*, p. 195.

169 Plowman, *A Subaltern on the Somme*, p. 205.

170 Carl Von Clausewitz, *On War*, edited and translated by Michael Howard and Peter Paret, Princeton University Press, Princeton, New Jersey, 1989 edition, p. 138.

171 Weintraub, *Silent Night*, p. 8.

172 P. Gosse, *Memoirs of a Camp Follower*, Longmans, London, 1934, pp. 25–6, quoted in Ashworth, *Trench Warfare 1914–1918*, p. 225.

173 Faris R. Kirkland, 'Chapter 12 – Postcombat Reentry', *Textbook of Military Medicine: War Psychiatry*, Office of the Surgeon General, United States Army, 1995, p. 293.

174 Ibid., p. 293.

175 Manchester, *Goodbye Darkness*, p. 258.

176 J. Glenn Gray, *The Warriors: Reflections on Men in Battle*, Bison Books, Lincoln, Nebraska, 1998 edition, p. 17.

177 Private Norman Demuth, 1/5th Battalion, London Regiment, quoted in Arthur, *Forgotten Voices of the Great War*, p. 169.

178 George Orwell, *Homage to Catalonia*, Harcourt, Orlando, Florida, 1980, p. 65.

179 Gray, *The Warriors*, p. 135.

180 McManners, *Falklands Commando*, p. 81.

181 Donaldson, *Donaldson of the 94th Scots Brigade*, pp. 162–3.

182 Hibbert, *The Wheatley Diary*, p. 64.

183 Bauer, *Soldiering*, p. 82.

184 Sergeant A.L. de Vine, 4th Battalion, diary entry dated 24 May 1915, quoted in Gammage, *The Broken Years*, p. 92.

185 Gammage, *The Broken Years*, p. 92.

186 Tribute to the Anzac soldiers killed at Gallipoli by Atatürk (Mustafa Kemal) in 1934 and reproduced on the Kemal Atatürk Memorial on ANZAC Parade, Canberra, Australia.

187 Warrant Officer First Class Joe Vezgoff, 'The Great Adventure 1950–1951', reproduced in Pears and Kirkland, *Korea Remembered*, p. 9.

188 Lawrence Nickell, 5th Infantry Division, quoted in McManus, *The Deadly Brotherhood*, pp. 185–6.

189 Henry Metelmann, *Through Hell for Hitler: A dramatic first-hand account of fighting with the Wehrmacht*, Patrick Stephens Ltd, Wellingborough, Northamptonshire, 1990, p. 92.

190 Sapper Stanley Fennell, 23rd Field Company, Royal Engineers, quoted in Arthur, *Forgotten Voices of the Second World War*, p. 267.

191 Jünger, *Storm of Steel*, p. 19.

192 Niall Ferguson, *The Pity of War: Explaining World War I*, Basic Books, New York, 1999, p. 39.

193 Coppard, *With a Machine Gun to Cambrai*, p. 59.

194 Sassoon, *Memoirs of an Infantry Officer*, p. 207.

195 Kevin J. Mervin, *Weekend Warrior: A Territorial Soldier's War in Iraq*, Mainstream Publishing, Edinburgh, 2005, pp. 191–2.

196 Lieutenant Commander James Newton, *Armed Action: My War in the Skies with 847 Naval Air Squadron*, Headline Review, London, 2007, pp. 101–02.

197 Manchester, *Goodbye Darkness*, p. 242.

198 Hal Bernton, 'Woman loses her job over coffins photo', *Seattle Times*, 22 April 2004; 'Amanda Ripley, 'An Image of Grief Returns', *Time*, 3 May 2004, pp. 18–19.

199 Jonathan Evans, 'Commentary', *US News & World Report*, 2 February 2004, p. 35.

200 Paul McGeough, 'Spinning in their graves', *Sydney Morning Herald*, 15 November 2003, www.smh.com.au.

201 John Lawrence and Robert Lawrence MC, *When the Fighting is Over: Tumbledown – A Personal Story*, Bloomsbury, London, 1988, p. 52.

202 Ibid., p. 96.

203 John Terraine, *General Jack's Diary 1914–18: The Trench Diary of Brigadier General J.L. Jack DSO*, Cassell, London, 2000 edition, p. 105.

204 Private David Webster, quoted in Ambrose, *Band of Brothers*, pp. 232–3.

205 Farley Mowat, *And No Birds Sang*, Cassell, London, 1980 edition, p. 161.

206 Ambrose, *Citizen Soldiers*, p. 205.

207 Dwight D. Eisenhower, Farewell Address to the Nation, 17 January 1961.

208 Herr, *Dispatches*, p. 20.

209 www.vhpa.org/heliloss.pdf (Vietnam Helicopter Pilots Association).

210 These figures are drawn from Professor Josephine Maltby, 'Financial reporting and the conscription of trade and industry 1914–1918', Discussion Paper No. 2003.11, University of Sheffield Management School, July 2003.

211 www.hmrc.gov.uk/history/taxhis (official website of HM Revenue & Customs).

212 Craig Tolliver, 'Mutual Funds for War: S&P Provides List of Top Funds if the Shooting Starts', www.CBSMarketwatch.com, 9 January 2003.

213 Captain Charles Delvert, quoted in Alistair Horne, *The Price of Glory: Verdun 1916*, Penguin, London, 1993 edition, p. 195.

214 Metelmann, *Through Hell for Hitler*, pp. 180–1.

215 Bob Gibson, quoted in Rintoul, *Ashes of Vietnam*, p. 197.

216 Lewis B. Puller Jr., *Fortunate Son: The Autobiography of Lewis B. Puller, Jr.*, Bantam Books, New York, 1993 edition, p. 208.

217 Lieutenant Commander Patrick Kettle, HMS *Sheffield*, in Bilton and Kosminsky, *Speaking Out*, p. 54.

218 Private Horacio Benitez, 5th Infantry Regiment (Argentina), in Bilton and Kosminsky, *Speaking Out*, pp. 189–90.

219 Private Troy Samuels, 1st Battalion, the Princess of Wales's Royal Regiment, quoted in Johnson Beharry VC, *Barefoot Soldier: A Story of Extreme Valour*, Sphere, London, 2006, p. 349.

220 Lieutenant Colonel Matt Maer, 1st Battalion, the Princess of Wales's Royal Regiment, quoted in Holmes, *Dusty Warriors*, p. 326.

221 Andrew Exum, *This Man's Army: A Soldier's Story from the Front Lines of the War on Terrorism*, Gotham Books, New York, 2005 edition, pp. 229–30.

222 Kayla Williams, *Love My Rifle More Than You: Young and Female in the US Army*, Weidenfeld & Nicolson, London, 2006, p. 274.

223 Colonel Teddy R. Spain, commander, 18th Military Police Brigade, US Army, quoted in Thomas E. Ricks, *Fiasco: The American Military Adventure in Iraq*, Penguin, London, 2006, p. 253.

224 Key, *The Deserter's Tale*, p. 179.

225 Lieutenant General Harold G. Moore and Joseph L. Galloway, *We Were Soldiers Once … And Young*, HarperCollins, New York, 2002 edition, p. 2.

226 Swofford, *Jarhead*, p. 233.

CHAPTER 7: KILLING YOUR OWN – THE DEATH PENALTY

1 Christopher Hibbert (editor), *The Recollections of Rifleman Harris, as told to Henry Curling*, The Windrush Press, Gloucestershire, 1998 edition, p. 9.

2 Edward Mead Earle (editor), *Makers of Modern Strategy*, Atheneum, New York, 1966, p. 56, quoted in Gwynne Dyer, *War*, The Bodley Head, London, 1986, p. 63.

3 Lieutenant Colonel James Wolfe, pamphlet entitled 'Instructions for the guidance of the 20th Foot should the French effect a landing', 1755, quoted in Richard Holmes, *Firing Line*, Pimlico, London, 1985, p. 336.

4 Mutiny Act (1689), reproduced in part at: http://www.constitution.org/sech/sech_120.htm.

5 Royal Navy, The Articles of War (1757), http://www.hmsrichmond.org/rnarticles.htm.

6 Voltaire, *Candide*, 1759.

7 Captain B.H. Liddell Hart (editor), *The Letters of Private Wheeler 1809–1828*, Michael Joseph, London, 1951, pp. 68, 133.

8 Hibbert, *The Recollections of Rifleman Harris*, p. 2.

9 Ibid., p. 3.

10   Jean-Roch Coignet, *Captain Coignet: A Soldier of Napoleon's Imperial Guard from the Italian Campaign to Waterloo*, Leonaur, 2007 edition, pp. 208, 209.

11   D.G. Crotty, *Four Years Campaigning in the Army of the Potomac*, Dygert Brothers & Co., Grand Rapids, Michigan, 1874, p. 123.

12   Frank Wilkeson, *Recollections of a Private Soldier in the Army of the Potomac*, G.P. Putnam's Sons, New York, 1886, quoted in Gerald F. Linderman, *Embattled Courage: The Experience of Combat in the American Civil War*, The Free Press, New York, 1987, p. 172.

13   Linderman, *Embattled Courage*, p. 173.

14   Ibid., pp. 174, 176.

15   Ibid., p. 175.

16   John O. Casler, *Four Years in the Stonewall Brigade*, University of South Carolina Press, Columbia, South Carolina, 2005 edition, p. 190.

17   Nick Bleszynski, *Shoot Straight, You Bastards!: The Truth behind the killing of 'Breaker' Morant*, Random House, Sydney, 2003 edition, pp. 318–24, 335–6, 340–2, 347.

18   Craig Wilcox, *Australia's Boer War: The War in South Africa 1899–1902*, Oxford University Press, Melbourne, 2002, p. 288.

19   Ibid., p. 276.

20   Ibid.

21   Statement to the House of Commons by Dr John Reid, Minister of State for the Armed Forces, 24 July 1998.

22   Gerald Oram, *Military Executions during World War I*, Palgrave Macmillan, Hampshire, 2003, p. 26.

23   Ibid., p. 26.

24   *Manual of Military Law*, War Office, 1914, p. 721.

25   George Coppard, *With a Machine Gun to Cambrai: A Story of the First World War*, Cassell, London, 1999 edition, pp. 75–6.

26   Cathryn Corns and John Hughes-Wilson, *Blindfold and Alone: British Military Executions in the Great War*, Cassell, London, 2005 edition, p. 450.

27   Public Record Office, War Office 95/25, quoted in Corns and Hughes-Wilson, *Blindfold and Alone*, p. 118.

28   Public Record Office, Papers of Judge Advocate General's Office (Capital trials), War Office 71/485.

29   Ibid., War Office 71/492.

30   Ibid., War Office 71/520.

31   Ibid., War Office 71/556.

32   Brigadier A.B. McPherson, *The Second World War 1939–1945, Army: Discipline*, War Office, London, 1950, p. 29.

33   Brigadier F.P. Crozier, *A Brass Hat in No-Man's Land*, Jonathan Cape, London, 1930, pp. 81–2.

34   Julian Putkowski, address given at the 'Unquiet Graves' International Conference on Executions, held in Flanders in May 2000, and 'To Encourage Others', www.shotatdawn.org.uk.

35   Army Act (1914), Part 1, Section 52.

36 John Terraine, *General Jack's Diary 1914–18: The Trench Diary of Brigadier General J.L. Jack DSO*, Cassell, London, 2000 edition, p. 56.

37 Coppard, *With a Machine Gun to Cambrai*, p. 49.

38 Captain M.S. Esler, medical officer, 2nd Middlesex, quoted in Malcolm Brown, *The Imperial War Museum Book of the Western Front*, Pan Books, London, 2001 edition, pp. 171–2.

39 Private John McCauley, 'A Manxman's Diary', Imperial War Museum, pp. 29, 32, quoted in Oram, *Military Executions during World War I*, p. 92.

40 'Part XXIII – Discipline', *The Statistics of the Military Effort of the British Empire During the Great War*, War Office, London, 1922.

41 Ben Shephard, *A War of Nerves: Soldiers and Psychiatrists 1914–1994*, Pimlico, London 2002, pp. 30–1.

42 Charles Myers, quoted in Corns and Hughes-Wilson, *Blindfold and Alone*, p. 452.

43 Lord C.E. Moran, *The Anatomy of Courage*, Constable, London, 1966, p. 186.

44 Shephard, *A War of Nerves*, p. 29.

45 Siegfried Sassoon, *Memoirs of an Infantry Officer*, Faber & Faber, London, 1965 edition, p. 26.

46 Public Record Office, Papers of Judge Advocate General's Office (Capital trials), War Office 71/1027.

47 Captain Geoffrey Donaldson, Warwickshire Regiment, quoted in Brown, *The Imperial War Museum Book of the Western Front*, p. 157.

48 R. Blake (editor), *The Private Papers of Douglas Haig, 1914–19*, Eyre & Spottiswode, London, 1952, p. 79, quoted in G.D. Sheffield, 'The Operational Role of British Military Police on the Western Front, 1914–18', in Paddy Griffith (editor), *British Fighting Methods in the Great War*, Frank Cass, London, 1996, p. 74.

49 Sheffield, 'The Operational Role of British Military Police on the Western Front, 1914–18', p. 77.

50 Crozier, *A Brass Hat in No-Man's Land*, pp. 110, 181.

51 Robert Graves, *Goodbye to All That*, Blue Ribbon Books, New York, 1930, p. 227.

52 Private J. Kirham, 5th Manchester Pals, quoted in Martin Middlebrook, *The First Day on the Somme*, W.W. Norton and Company, New York, 1972, p. 223.

53 Lieutenant Colonel Seton Hutchinson, *History and Memoir of the 33rd Battalion, Machine Gun Corps*, quoted in William Moore, *The Thin Yellow Line*, Clarke, Leo Cooper, London, 1974, p. 143.

54 Terraine, *General Jack's Diary 1914–18*, p. 55.

55 Emilio Lussu, *Sardinian Brigade*, translated by Marion Rawson, Grove Press, New York, 1967, p. 194.

56 John Major, letter to Andrew MacKinlay, MP, quoted in *Independent*, 16 August 1993, quoted in Shephard, *A War of Nerves*, p. 67.

57 Shephard, *A War of Nerves*, p. 70.

58 Public Record Office, Papers of Judge Advocate General's Office (Capital trials), War Office 71/509.

59 Ibid.

60 Ibid.

61 Ibid.

62   Ibid.

63   Ibid.

64   'Intention to Pardon Executed World War I Soldiers', Government News Network, 137013P/227 of 2006, www.gnn.gov.uk.

65   Alistair Horne, *The Price of Glory: Verdun 1916*, Penguin, London, 1993 edition, p. 271.

66   Lussu, *Sardinian Brigade*, p. 257.

67   Oram, *Military Executions during World War I*, pp. 32–3.

68   Charles Edwin Woodward Bean, *The Official History of Australia in the War of 1914–1918: Volume V, The AIF in France: 1918*, first edition, Angus & Robertson, Sydney, 1937, p. 29.

69   Ibid., pp. 27, 31–2; Geoffrey Serle, *John Monash: A Biography*, Melbourne University Press, Melbourne, 1982, p. 395.

70   Bill Gammage, *The Broken Years: Australian Soldiers in the Great War*, Penguin, Melbourne, 1982 edition, p. 236.

71   Bean, *The Official History of Australia in the War of 1914–1918: Volume V*, pp. 30–1.

72   Lieutenant Colonel Lambert Ward, Army Act Debate, March 1927, quoted in Moore, *The Thin Yellow Line*, p. 62.

73   Alex Bowlby, *The Recollections of Rifleman Bowlby*, Corgi, London, 1971, p. 93.

74   General Sir Claude Auchinleck, quoted in Moore, *The Thin Yellow Line*, p. 225.

75   McPherson, *The Second World War 1939–1945, Army: Discipline*, pp. 48–9, 54, 57–8.

76   Dmitri Volkogonov, *Trotsky: The Eternal Revolutionary*, translated and edited by Harold Shukman, The Free Press, New York, 1996, p. 175.

77   Mark M. Boatner III, *The Biographical Dictionary of World War II*, Presidio, Novato, California, 1996, p. 418.

78   Catherine Merridale, *Ivan's War: Life and Death in the Red Army, 1939–1945*, Metropolitan Books, New York, 2006, pp. 80–1, 88.

79   Fyodor Sverdlov, 19th Infantry Brigade, 49th Army, quoted in Laurence Rees, *War of the Century: When Hitler Fought Stalin*, The New Press, New York, 1999, p. 84.

80   Recollections of A.Z. Akimenko, 1953, unpublished, quoted in A.A. Maslov, 'How Were Soviet Blocking Detachments Employed?', translated by Colonel David M. Glantz, Foreign Military Studies Office, Fort Leavenworth, Kansas, leav-www.army. mil/fmso/documents/blockdet.htm.

81   William Craig, *Enemy at the Gates: The Battle for Stalingrad*, Penguin, London, 2000, p. 72.

82   Richard Overy, *Russia's War*, Penguin, London, 1998, p. 158.

83   Vladimir Kantovski, 54th Penal Company, quoted in Rees, *War of the Century*, p. 138.

84   Craig, *Enemy at the Gates*, p. 43.

85   Merridale, *Ivan's War*, p. 158.

86   Overy, *Russia's War*, p. 160.

87   Martin Van Creveld, *Fighting Power: German Military Performance 1914–1945*, Art of War Colloquium Publication, US Army War College, Carlisle, Pennsylvania, November 1983, pp. 130–1.

88   Ken Hechler, *The Bridge at Remagen*, Ballantine Books, New York, 1957, p. 195.

89 Willi Heilmann, *I Fought You From the Skies*, Award Books, New York, 1966, p. 178.

90 Guy Sajer, *The Forgotten Soldier*, translated by Lily Emmet, Weidenfeld & Nicolson, London, 1971, p. 390.

91 Van Creveld, *Fighting Power*, p. 134.

92 William Bradford Huie, *The Execution of Private Slovik*, Westholme, Yardley, Pennsylvania, 2004 edition, pp. 32, 51, 53.

93 Ibid., pp. 68, 70.

94 Ibid., pp. 131–3.

95 Ibid., pp. 140–1.

96 Ibid., pp. 142–3.

97 Ibid., pp. 162, 164, 167–8.

98 Colonel Guy Williams, quoted in Huie, *The Execution of Private Slovik*, p. 169.

99 Ibid., p. 170.

100 Huie, *The Execution of Private Slovik*, pp. 152, 177–8.

101 Major General Normal D. Cota, quoted in Huie, *The Execution of Private Slovik*, p. 177.

102 Huie, *The Execution of Private Slovik*, p. 188.

103 Uzal W. Ent, 'The Sad Story of Private Eddie Slovik', *World War II*, November 1999, reproduced at: http://www.28-110-k.org/sad_story_of_private_eddie_slovi.html; Huie, *The Execution of Private Slovik*, pp. 18–19, 201, 232–4.

104 Lieutenant Colonel James E. Rudder, Commander, 109th Infantry Regiment, 29th Infantry Division, message to regiment dated 31 January 1945, reproduced in Huie, *The Execution of Private Slovik*, p. 117.

105 Huie, *The Execution of Private Slovik*, p. 11.

106 Ibid., pp. 174, 228.

107 Lieutenant Colonel Henry J. Sommer, quoted in Huie, *The Execution of Private Slovik*, p. 174.

108 David Eisenhower, *Eisenhower: At War 1943–1945*, Random House, New York, 1986, p. 643.

109 Bernard B. Fall, *Hell in a Very Small Place: The Siege of Dien Bien Phu*, Pall Mall Press, Oxford, 1966, pp. 286–8.

110 Robert Carter, *Four Brothers in Blue*, University of Texas Press, Austin, Texas, 1978, p. 346, quoted in Linderman, *Embattled Courage*, p. 58.

111 Huie, *The Execution of Private Slovik*, p. 218.

112 Field Marshal Erich Von Manstein, *Lost Victories*, edited and translated by Anthony G. Powell, Greenhill Books, London, 1987 edition, pp. 221–2.

113 General George S. Patton Jr., *War as I Knew It*, Houghton Mifflin, Boston, 1995 edition, p. 362.

CHAPTER 8: KILLING YOUR OWN – FRIENDLY FIRE

1 Tom Newton Dunn, 'The tape they wanted to hide', *Sun*, 6 February 2007, www.thesun.co.uk.

2 Adrien Jean Baptiste François Bourgogne, *Memoirs of Sergeant Bourgogne 1812–1813*, Constable, London, 1997 edition, p. 60.

3   Mansur Abdulin, *Red Road from Stalingrad: Recollections of a Soviet Infantryman*, Pen and Sword Military, Barnsley, South Yorkshire, 2004, p. 31.

4   Major Robert Crisp, *Brazen Chariots*, Ballantine Books, New York, 1961, p. 148.

5   Ron Kovic, *Born on the Fourth of July*, Corgi, London, 1990, p. 144.

6   Ibid., p. 149.

7   Guy Chapman, *A Passionate Prodigality*, Fawcett Crest edition, New York, 1967, p. 68.

8   Captain Walter Bloem, *The Advance from Mons*, Tandem Books, London, 1967, p. 102.

9   Eugene B. Sledge, *With the Old Breed at Peleliu and Okinawa*, Oxford University Press, New York, 1990, p. 69.

10 Evgeni Bessonov, translated by Bair Irincheev, *Tank Rider: Into the Reich with the Red Army*, Greenhill Books, London, 2003, p. 59.

11 Joseph Owen, *Colder than Hell: A Marine Rifle Company at Chosin Reservoir*, Naval Institute Press, Annapolis, Maryland, 1996, p. 160.

12 Specialist 4 Galen Bungum, quoted in Lieutenant General Harold G. Moore and Joseph L. Galloway, *We Were Soldiers Once ... And Young*, HarperCollins, New York, 2002 edition, p. 399.

13 Major Philip Neame, quoted in Max Arthur, *Above All, Courage – The Falklands Front Line: First-Hand Accounts*, Guild Publishing, London, 1985, p. 198.

14 Anthony Swofford, *Jarhead: A Marine's Chronicle of the Gulf War*, Simon & Schuster UK, London, 2003, p. 219.

15 General H. Norman Schwarzkopf and Peter Petre, *It Doesn't Take a Hero*, Linda Grey, Bantam Books, New York, 1992, p. 184; Lieutenant Colonel Charles R. Shrader, *Amicicide: The Problem of Friendly Fire in Modern War*, Combat Studies Institute, Fort Leavenworth, Kansas, Research Survey No. 1, December 1982, p. x.

16 Schwarzkopf and Petre, *It Doesn't Take a Hero*, pp. 183–4.

17 Sledge, *With the Old Breed at Peleliu and Okinawa*, p. 108.

18 Rick Atkinson, *Crusade: The Untold Story of the Gulf War*, HarperCollins, London, 1993, p. 316.

19 Timothy Gowing, *Voice from the Ranks: A Personal Narrative of the Crimean War by a Sergeant of the Royal Fusiliers*, William Heinemann, Melbourne, 1954, pp. 111, 112.

20 John Stuart Roberts, *Siegfried Sassoon*, Richard Cohen Books, London, 2000 edition, p. 128.

21 Edmund Blunden, *Undertones of War*, Penguin Books, London, 2000 edition, p. 39.

22 Sergeant Fred J. Kluge, quoted in Moore and Galloway, *We Were Soldiers Once*, p. 381.

23 Moore and Galloway, *We Were Soldiers Once*, p. 381.

24 Shrader, *Amicicide*, p. 91.

25 Hugh McManners, *The Scars of War*, HarperCollins, London, 1993, p. 231.

26 Ibid., pp. 232–4.

27 Hugh Dundas, *Flying Start*, Penguin, 1990, p. 120.

28 J. Douglas Harvey, *Boys, Bombs and Brussels Sprouts*, Goodread Biography, Halifax, 1983, p. 185, quoted in Paul Fussell, *Wartime: Understanding and Behaviour in the Second World War*, Oxford University Press, New York, 1989, p. 21.

29   Sergeant Leslie Batt, 238 Squadron, Royal Air Force, quoted in Joshua Levine, *Forgotten Voices of the Blitz and the Battle for Britain*, Ebury Press, London, 2006, p. 160.

30   Guy Gibson, *Enemy Coast Ahead – Uncensored*, Crécy Publishing, Manchester, 2005 edition, p. 233.

31   Arthur Hadley, *The Straw Giant – America's Armed Forces: Triumphs and Failures*, Avon Books, New York, 1987 edition, p. 60.

32   Jim Bailey, *The Sky Suspended: A Fighter Pilot's Story*, Bloomsbury, London, 2005 edition, p. 166.

33   www.commandosupremo.com, profile of Italo Balbo; Mark M. Boatner III, *The Biographical Dictionary of World War II*, Presidio, Novato, California, 1996, p. 25; Geoffrey Regan, *Backfire: A history of friendly fire from ancient warfare to the present day*, Robson Books, London, 2002, p. 181.

34   Regan, *Backfire*, pp. 181–2; www.cv6.org (website for the USS *Enterprise* CV-6).

35   Regan, *Backfire*, pp. 184–5; Shrader, *Amicicide*, p. 68.

36   Regan, *Backfire*, pp. 186–8; Shrader, *Amicicide*, p. 67.

37   Regan, *Backfire*, pp. 190–1; Shrader, *Amicicide*, p. 67.

38   Regan, *Backfire*, p. 191; Shrader, *Amicicide*, p. 67.

39   Shrader, *Amicicide*, p. 68.

40   Regan, *Backfire*, p. 196.

41   Ibid., pp. 190–5.

42   Regan, *Backfire*, p. 216.

43   Shrader, *Amicicide*, p. 84.

44   Yael Dayan, *A Soldier's Diary: Sinai 1967*, Penguin, Harmondsworth, 1969 edition, p. 76.

45   Shrader, *Amicicide*, p. 21; Regan, *Backfire*, p. 156.

46   Barry Broadfoot (editor), *Six Years War, 1939–1945: Memories of Canadians at Home and Abroad*, Doubleday, Toronto, 1974, p. 289, quoted in Fussell, *Wartime*, p. 23.

47   Omar N. Bradley, *A Soldier's Story*, Henry Holt & Company, New York, 1951, p. 151.

48   Lieutenant Charles Stockwell, 2nd Division, quoted in Stephen E. Ambrose, *Citizen Soldiers: The US Army from the Normandy Beaches to the Bulge to the Surrender of Germany*, Touchstone edition, New York, 1998, p. 247.

49   General Heinz Guderian, *Panzer Leader*, Futura, London, 1974, p. 113.

50   Richard Holmes, *Firing Line*, Pimlico, London, 1985, p. 173.

51   John Keegan, *The Face of Battle*, Penguin edition, London, 1978, p. 195.

52   Edward Costello, *Rifleman Costello: The Adventures of a Soldier of the 95th (Rifles) in the Peninsular & Waterloo Campaigns of the Napoleonic Wars* , Leonaur, 2005 edition, p. 275.

53   Christopher Hibbert (editor), *The Wheatley Diary: A Journal and Sketchbook kept during the Peninsular War and the Waterloo Campaign*, The Windrush Press, Gloucestershire, 1997, p. 4.

54   William A. Fletcher, *Rebel Private: Front and Rear – Memoirs of a Confederate Soldier*, Meridian, New York, 1997 edition, p. 82.

55   Shrader, *Amicicide*, p. 2.

56   Regan, *Backfire*, p. 87.

57 Alistair Horne, *The Price of Glory: Verdun 1916*, Penguin, London, 1993 edition, pp. 98–9.

58 Regan, *Backfire*, pp. 114–17; Charles Edwin Woodward Bean, *The Official History of Australia in the War of 1914–1918: Volume IV, The AIF in France: 1917*, third edition, Angus & Robertson, Sydney, 1935, pp. 886–7.

59 Shrader, *Amicicide*, p. 11.

60 Abdulin, *Red Road from Stalingrad*, p. 101.

61 John Keegan, *The Second World War*, Penguin Books, London, 1989, p. 358; Shrader, *Amicicide*, p. 35.

62 Shrader, *Amicicide*, p. 40.

63 Regan, *Backfire*, pp. 202–03.

64 Shrader, *Amicicide*, p. 41; Regan, *Backfire*, pp. 203–04.

65 Shrader, *Amicicide*, p. 42; Regan, *Backfire*, pp. 205–06.

66 US Congress, Office of Technology Assessment, *Who Goes There: Friend or Foe?*, US Government Printing Office, Washington, DC, June 1993, p. 14.

67 Ernie Pyle, *Brave Men*, Bison Books, Lincoln, Nebraska, 2001 edition, p. 462.

68 Shrader, *Amicicide*, p. 42.

69 Regan, *Backfire*, p. 208.

70 Bradley, *A Soldier's Story*, p. 349.

71 Terry Copp, 'Reassessing Operation Totalize', *Legion Magazine*, September/October 1999, www.legionmagazine.com.

72 Fussell, *Wartime*, pp. 18–19.

73 'Friendly Fire on Hill 282', www.britains-smallwars.com; Pat Quinn, 'A Rude Awakening', Land Forces of Britain, the Empire and the Commonwealth – The Argyll and Sutherland Highlanders, www.regiments.org.

74 Moore and Galloway, *We Were Soldiers Once*, p. 211.

75 Colonel David H. Hackworth and Julie Sherman, *About Face*, Pan Books, Sydney, 1989, p. 528.

76 Ibid., p. 542.

77 Jonathan S. Wiarda, 'The US Coast Guard in Vietnam: Achieving Success in a Difficult War', *Naval War College Review*, Volume LI, No. 2, Spring 1998, pp. 43–4.

78 Rick Atkinson, *Crusade: The Untold Story of the Gulf War*, HarperCollins, London, 1993, p. 315; www.gruntonline.com, entry on Battle of Hill 875.

79 Final Report to Congress, 'Conduct of the Persian Gulf War', US Government, April 1992, p. 591.

80 Atkinson, *Crusade*, p. 200.

81 Ibid., pp. 206–7.

82 Ibid., p. 464.

83 Written statement by the Minister for Armed Forces, Archie Hamilton, 24 July 1991, *Hansard (House of Commons Debates: Written Answers to Questions)*, Column 705; Atkinson, *Crusade*, p. 464.

84 General Sir Peter De La Billière, *Storm Command: A Personal Account of the Gulf War*, HarperCollins, London, 1992, pp. 299–300.

85 Hamilton, *Hansard*, Column 706.

86 Stewart M. Powell, 'Friendly Fire', *Airforce*, Vol. 74, No. 12, December 1991, pp. 58, 60.

87 Final Report to Congress, 'Conduct of the Persian Gulf War', p. 589; Major Bradford G. Washabaugh, 'Friendly Fire: Time For Action', paper reproduced at www.globalse-curity.org/military/library/report/1992.

88 Colonel David H. Hackworth, 'Friendly Fire Casualties', *Marine Corps Gazettes*, Vol. 76, No. 3, March 1992, p. 46.

89 Atkinson, *Crusade*, p. 316.

90 Final Report to Congress, 'Conduct of the Persian Gulf War', p. 590.

91 US Congress, *Who Goes There: Friend or Foe?*, p. 38.

92 General Accounting Office, testimony before the US House of Representatives on 18 July 1998, pertaining to the GAO review of the Air Force Investigation of Black Hawk Fratricide Incident on Operation Provide Comfort.

93 'American error kills four Canadian servicemen', *New York Times*, 19 April 2002, Section A, p. 15; transcript of radio traffic for the 18 April fratricide incident, repro-duced at: www.cbc.ca/new/background/friendlyfire (CBC News).

94 Dave Moniz, 'Pentagon probes bombing that killed 4 Canadians', *USA Today*, 19 April 2002, www.usatoday.com.

95 'The "Friendly Fire" Reports', CBC News, 22 October 2003, www.cbc.ca/news.

96 Sergeant 1st Class Kathleen T. Rhem, 'Two US Troops Die in 2nd "Friendly Fire" Accident', DefenseLINK News, 5 December 2001, www.defenselink.mil/news.

97 'US investigates friendly fire deaths of 3 soldiers', www.cnn.com, 5 December 2001.

98 Ibid.

99 George Cross citation for Trooper Christopher Finney; Patrick Barkham, 'Troops' anger over US "friendly fire"', BBC News, 31 March 2003, www.bbc.co.uk.

100 'Wounded British soldiers condemn US "cowboy" pilot', *Guardian*, 31 March 2003.

101 Audrey Gillan and Richard Norton-Taylor, 'Trooper who saved friend in face of friendly fire awarded George Cross', *Guardian*, 31 October 2003; 'Wounded British soldiers condemn US "cowboy" pilot', *Guardian*, 31 March 2003; Patrick Barkham, 'Troops' anger over US "friendly fire"', BBC News, 31 March 2003.

102 'Wounded British soldiers condemn US "cowboy' pilot", *Guardian*, 31 March 2003.

103 'US can release cockpit footage', BBC News, 6 February 2006, www.bbc.co.uk.

104 Newton Dunn, 'The tape they wanted to hide', *Sun*, 6 February 2007.

105 United States Central Command, 'Investigation of Suspected Friendly Fire Incident Near An Nasiriyah, Iraq, 23 March 2003', dated 6 March 2004; 'Marine captain faulted in "friendly fire" incident', www.cnn.com, 29 March 2004.

106 1st Lieutenant Michael Seeley, quoted in 'Marine captain faulted in "friendly fire" incident', www.cnn.com, 29 March 2004.

107 'Friendly fire hits Kurdish convoy', BBC News, 6 April 2003, www.bbc.co.uk; 'Apparent "friendly fire" kills 18, Kurdish officials say', www.cnn.com, 6 April 2003.

108 Luke Harding and Michael Howard, '18 die as US plane bombs Kurdish convoy in worst "friendly fire" incident', *Guardian*, 7 April 2003.

109 Shrader, *Amicicide*, p. x.

110 Hackworth and Sherman, *About Face*, p. 594.

111 US Congress, *Who Goes There: Friend or Foe?*, p. 1.

112 Evan Wright, *Generation Kill: Living dangerously on the road to Baghdad with the ultraviolent Marines of Bravo Company*, Bantam Press, London, 2004, pp. 100, 102, 163, 299, 305.

CHAPTER 9: THE MILITARY VERSUS THE MEDIA

1 Phillip Knightley, *The First Casualty: The War Correspondent as Hero and Myth-Maker from the Crimea to Kosovo*, Johns Hopkins University Press, Baltimore, 2002 edition, p. 492.

2 H.D. Laswell, *Propaganda Technique in the World War*, London, 1927, p. 192, quoted in Niall Ferguson, *The Pity of War: Explaining World War I*, Basic Books, New York, 1999, p. 11.

3 Trevor Royle, *War Report: The War Correspondent's View of Battle from the Crimea to the Falklands*, Mainstream Publishing, London, 1987, pp. 224–5.

4 Max Hastings, *Going to the Wars*, Pan, London, 2001, p. 320.

5 Sergeant Dan Mills, *Sniper One: The Blistering True Story of a British Battle Group Under Siege*, Penguin, London, 2008 edition, p. 173.

6 Letter to Peter Arnett from the men of A Battery, 3rd Battalion, 6th Artillery, quoted in Peter Arnett, *Live from the Battlefield: From Vietnam to Baghdad, 35 Years in the World's War Zones*, Touchstone edition, Simon & Schuster, New York, 1995, p. 262.

7 Figures are drawn from Knightley, *The First Casualty*.

8 Martin Bell, *Through Gates of Fire: A Journey into World Disorder*, Phoenix, London, 2004 edition, p. 170.

9 Knightley, *The First Casualty*, pp. 296–7.

10 Arnett, *Live from the Battlefield*, p. 140.

11 Joseph L. Galloway, quoted in Greg Barber, 'Reporting the Story', *PBS Online NewsHour*, 20 April 2000, www.pbs.org.

12 Henry G. Gole, 'Don't Kill the Messenger: Vietnam War Reporting in Context', *Parameters*, Winter 1996, pp. 144–59.

13 Robert Graves, *Goodbye to All That*, Blue Ribbon Books, New York, 1930, p. 230.

14 Alistair Horne, *The Price of Glory: Verdun 1916*, Penguin, London, 1993 edition, p. 196.

15 Siegfried Sassoon, 'Fight to a Finish', *War Poems*, Faber & Faber, London, 1983, p. 85.

16 Bell, *Through Gates of Fire*, p. 54.

17 Admiral Sir Jeremy Black, letter to Miles Hudson, quoted in Miles Hudson and John Stanier, *War and the Media*, Sutton Publishing, Gloucestershire, 1997, pp. 170–1.

18 Nathaniel Fick, *One Bullet Away: The Making of a Marine Officer*, Weidenfeld & Nicolson, London, 2005, p. 184.

19 Major Chris Hunter, *Eight Lives Down*, Corgi, London, 2008 edition, pp. 101–02.

20 James Kitfield, 'Lessons from Kosovo', *Media Studies Journal: Front Lines and Deadlines – Perspectives on War Reporting*, Vol. 15, No. 1, Summer 2001, p. 34.

21 Letter from Sir Arthur Wellesley to Lord Liverpool, quoted in Royle, *War Report*, p. 17.

22 Elzéar Blaze, *Captain Blaze: Life in Napoleon's Army*, Leonaur, 2007 edition, p. 176.

23 'The War: The British Expedition', *The Times*, 23 October 1854, p. 7.

24   Letter from Lord Raglan to the Duke of Newcastle, dated 13 November 1854, quoted in Christopher Hibbert, *The Destruction of Lord Raglan: A Tragedy of the Crimean War 1854–55*, Penguin, London, 1985 edition, p. 158.

25   Letter from Lord Raglan to the Duke of Newcastle, dated 1 May 1855, quoted in Hibbert, *The Destruction of Lord Raglan*, p. 160.

26   Hibbert, *The Destruction of Lord Raglan*, p. 160.

27   Knightley, *The First Casualty*, p. 15.

28   Joseph H. Ewing, 'The New Sherman Letters', *American Heritage Magazine*, Vol. 38, No. 5, July–August 1987, www.americanheritage.com.

29   *New York Times*, 17 July 1861, quoted in Ewing, 'The New Sherman Letters', www.americanheritage.com.

30   General William Tecumseh Sherman, letter to Senator Thomas Ewing dated 6 February 1863, quoted in Ewing, 'The New Sherman Letters', www.americanheritage.com.

31   Captain Hugh McManners, *Falklands Commando*, William Kimber, London, 1984, p. 119.

32   Sir Lawrence Freedman, *The Official History of the Falklands Campaign, Volume II: War and Diplomacy*, Routledge, Taylor & Francis Group, London, 2005, pp. 560, 563–4.

33   Max Hastings and Simon Jenkins, *The Battle for the Falklands*, W.W. Norton & Company, New York, 1983, pp. 255–6.

34   Hastings, *Going to the Wars*, p. 354.

35   Ibid., p. 353.

36   Kenneth Bacon, 'The Pentagon and the Press', *PBS Online NewsHour*, 6 April 1999, www.pbs.org.

37   Ibid.

38   Chris Plante, 'Fox News, military reach deal on Rivera', www.cnn.com, 1 April 2003.

39   Kevin Maguire and Andy Lines, 'Bush Plot to Bomb his Arab Ally', www.mirror.co.uk, 22 November 2005; John Plunkett, 'Bush claim revives al-Jazeera bombing fears', www.guardian.co.uk, 23 November 2005; 'Al-Jazeera Kabul offices hit in US raid', www.bbc.co.uk, 13 November 2001.

40   Robert Fisk, 'Closing Down the Press: Did the US Murder Journalists?', *Counterpunch*, 29 April 2003, www.counterpunch.org.

41   Fick, *One Bullet Away*, p. 184.

42   Ibid., p. 182.

43   Lieutenant Colonel Dave Grossman with Loren W. Christensen, *On Combat: The Psychology and Physiology of Deadly Conflict in War and in Peace*, Warrior Science Publications, Millstadt, Illinois, 2007, second edition, p. 361.

44   US Marine Corps Press Release No. 0505-05-0611, 'Marine involved in mosque shooting will not face court martial', 4 May 2005; 'Open Letter to the Devil Dogs of the 3.1', www.kevinsites.net, 21 November 2004.

45   'Open Letter to the Devil Dogs of the 3.1', www.kevinsites.net.

46   Hague Convention on Laws and Customs of War on Land, signed 18 October 1907, Annex to the Convention: Regulations Respecting the Laws and Customs of War on Land, Section II: Hostilities, Article 23.

47 Guy Chapman, *A Passionate Prodigality*, Fawcett Crest edition, New York, 1967, pp. 80–1.

48 Ibid., p. 81.

49 Captain D.V. Mulholland, 1st Machine Gun Battalion, letter dated 17 March 1917, quoted in Bill Gammage, *The Broken Years: Australian Soldiers in the Great War*, Penguin, Melbourne, 1982 edition, p. 258.

50 Graves, *Goodbye to All That*, p. 224.

51 Stephen E. Ambrose, *Citizen Soldiers: The US Army from the Normandy Beaches to the Bulge to the Surrender of Germany*, Touchstone edition, New York, 1998, pp. 352–3.

52 Harold P. Leinbaugh and John D. Campbell, *The Men of Company K: The Autobiography of a World War II Rifle Company*, William Morrow & Company, New York, 1985, p. 59.

53 General George S. Patton Jr., *War as I Knew It*, Houghton Mifflin, Boston, 1995 edition, p. 351.

54 Timothy Gowing, *Voice from the Ranks: A Personal Narrative of the Crimean War by a Sergeant of the Royal Fusiliers*, William Heinemann, Melbourne, 1954, p. 21.

55 *The Bryce Report: Report of the Committee on Alleged German Outrages*, 12 May 1915, copy of report provided at www.firstworldwar.com/source/brycereport.htm.

56 Denis Winter, *Death's Men: Soldiers of the Great War*, Penguin, London, 1979, p. 214.

57 Günter Koschorrek, *Blood Red Snow: The Memoirs of a German Soldier on the Eastern Front*, Greenhill Books, London, 2002, p. 152.

58 Colonel David H. Hackworth and Julie Sherman, *About Face*, Pan Books, Sydney, 1989, p. 106.

59 'Military investigates shooting of wounded insurgent', www.cnn.com, 16 November 2004.

60 'Marine taped shooting man in Iraq mosque won't face court martial', www.cnn.com, 4 May 2005.

61 Marine Corps Press Release No. 0505-05-0611, 'Marine involved in mosque shooting will not face court martial'.

62 Ibid.

63 Ibid.

64 Lieutenant Tom Gibson, quoted in Stephen E. Ambrose, *Band of Brothers: E Company, 506th Regiment, 101st Airborne from Normandy to Hitler's Eagle's Nest*, Simon & Schuster, New York, 2001 edition, p. 206.

65 James R. McDonough, *Platoon Leader: A Memoir of Command in Combat*, Ballantine Books, New York, 2003 edition, p. 174.

66 Martin Van Creveld, *The Sword and the Olive: A Critical History of the Israeli Defence Force*, Public Affairs, New York, 2002 edition, pp. 347–8.

67 'Open Letter to the Devil Dogs of the 3.1', www.kevinsites.net, 21 November 2004.

68 Andrew Exum, *This Man's Army: A Soldier's Story from the Front Lines of the War on Terrorism*, Gotham Books, New York, 2005 edition, p. 198.

69 Ibid.

70 Ibid.

71 Major General Jim Molan, *Running the War in Iraq*, HarperCollins, Sydney, 2008, p. 226.

72 Ibid., p. 251.

73 Associate Justice Hugo L. Black, opinion in *New York Times Co. v. United States, 403 US 713* (1971), decided 30 June 1971.

CHAPTER 10: IS THERE A NEED FOR WAR?

1   George Santayana, 'Soliloquy No. 25 – Tipperary', *Soliloquies*, Scribners, New York, 1924, p. 102.

2   Vegetius, *Epitome Rei Militaris*, 373 CE?, Prologue, Part 3.

3   Thucydides, *History of the Peloponnesian War*, translated by Rex Warner, 1954, Penguin Books, London, 1972, pp. 400–02.

4   Ibid., pp. 402–08.

5   *Rekohu: A Report on Moriori and Ngati Mutunga Claims in the Chatham Islands*, Waitangi Tribunal Report, WAI 64, 25 May 2001, p. 1.

6   Ron Brunton, 'Ignoring Sins of the Father', *Courier Mail*, 23 June 2001, www.news.com.au/couriermail; *Rekohu*, Waitangi Tribunal Report, pp. 35–7.

7   *Rekohu*, Waitangi Tribunal Report, pp. 38–46.

8   Evidence of Rakatau Katihe in Chatham Island minute book, cited in Michael King, *Moriori: A People Rediscovered*, Viking, Auckland, New Zealand, 1989, p. 66.

9   Evidence of Hirawana Tapu, quoted in Jared Diamond, *Guns, Germs, and Steel: The Fates of Human Societies*, W.W. Norton & Company, New York, 1999, p. 53.

10 *Rekohu*, Waitangi Tribunal Report, p. 63.

11 Thucydides, *History of the Peloponnesian War*, p. 405.

12 Jonathan Schell, *The Real War*, Corgi, London, 1989 edition, pp. 9–10.

13 Eugene B. Sledge, *With the Old Breed at Peleliu and Okinawa*, Oxford University Press, New York, 1990, p. 156.

14 Private Paul Curtis, quoted in Jonathan Darman, 'The Wages of War', *The Bulletin*, 15 November 2005, p. 49.

15 Dennis Marshall-Hasdell, quoted in Hugh McManners, *The Scars of War*, HarperCollins, London, 1993, p. 387.

16 Senator Chuck Hagel, quoted in Michael Hirsh, 'Hawks, Doves and Dubya', *Newsweek*, 2 September 2002, p. 27.

17 General Heinz Guderian, *Panzer Leader*, Futura, London, 1974, p. 463.

18 Carl Von Clausewitz, *On War*, edited and translated by Michael Howard and Peter Paret, Princeton University Press, Princeton, New Jersey, 1989 edition, p. 69.

19 Ibid., p. 579.

20 Ibid., p. 87.

21 John Keegan, *The Face of Battle*, Penguin edition, London, 1978, p. 59.

22 Rick Atkinson, *In the Company of Soldiers: A Chronicle of Combat in Iraq*, Little, Brown, London, 2004, p. 176.

23 Clausewitz, *On War*, p. 606.

24 Ibid., p. 584.

25 Ibid., p. 92.

26 Ibid.

27 Ibid., p. 94.

28 Hague Convention on Laws and Customs of War on Land, signed 18 October 1907, Annex to the Convention: Regulations Respecting the Laws and Customs of War on Land, Section III: Military Authority over the Territory of the Hostile State, Article 43.

29 Christopher Bellamy, *Knights in White Armour*, Pimlico edition, London, 1997, pp. 3, 5.

30 Shelford Bidwell, *Modern Warfare: A Study of Men, Weapons and Theories*, Allen Lane, London, 1973, p. 216.

31 Nathaniel Fick, *One Bullet Away: The Making of a Marine Officer*, Weidenfeld & Nicolson, London, 2005, p. 106.

32 Ibid., p. 203.

33 Chris Hedges, *War Is a Force That Gives Us Meaning*, Anchor Books, New York, 2002, p. 3.

34 Bill Mauldin, quoted in Studs Terkel, *The Good War: An Oral History of World War Two*, Hamish Hamilton, London, 1985, p. 362.

35 General Dwight D. Eisenhower, speech to the Canadian Club in Ottawa, Canada on 10 January 1946.

INDEX